JAMES DYSON, CBE, has spent 30 years inventii
more efficient. His achievements so far include
Sea Truck boat and the Evolutionary Ballbari
chairman of the Design Museum.

~ppointed

ROBERT UHLIG is the technology correspondent for the *Daily Telegraph*.

Praise for this book

'A fascinating reference book you'll want to read.'

'Intriguing little gems are contained in this lavishly illustrated book.'

'Today James Dyson ranks alongside the likes of Laszlo Biro and John Logie Baird.'

Contributors: Jim Bennett, Keeper, Museum of the History of Science, Oxford; Jonathan Betts, Curator of Horology, National Maritime Museum; Neil Brown, Curator of Classical Physics, Time and Microscopes, the National Museum of Science and Technology; R. Angus Buchanan, Professor of the History of Technology and Director, Centre for the History of Technology, Science and Society, University of Bath; Martin Cambell-Kelly, Reader in Computer Science, University of Warwick; Colin Chant, Lecturer in History of Science and Technology, The Open University; David Derbyshire, Science Correspondent, *The Daily Telegraph*; Glyn Davies, Professor Emeritus and Honorary Fellow, University of Wales; Andrew English, Motoring Correspondent, *The Daily Telegraph*; Nick Flowers, Space Scientist, Mullard Space Laboratory; Ian Gibb, Former Head of Agricultural Engineering, University of Reading; Colin Harding, Curator of Photographic Technology, National Museum of Photography, Film and Television; Mike Harding, Assistant Curator of Mechanical Engineering, The National Museum of Science and Technology; Michael Harvey, Curator of Cinematography and John Trenouth, Senior Curator of Television, National Museum of Photograph, Film and Television; Roger Highfield, Science Editor, *The Daily Telegraph*; Robert Matthews, Science Editor, *The Sunday Telegraph*; Robert Maybury, Technology writer; Keith Parker, Assistant Curator of Communications, The National Museum of Science and Technology; Malcolm Pein, Chess Correspondent, *The Daily Telegraph*; David Penney, Editor, *Antiquarian Horology*; Jon Pratty, Technology writer; Mike Rentell, Secretary, The Airship Association; Colin Russell, Professor Emeritus, History of Science, The Open University; Andrew Scott, Head of Museum, National Railway Museum; Fred Wilkinson, Consultant to the Royal Amouries

Mammoth Book of

GREAT
INVENTIONS

Edited by James Dyson and Robert Uhlig

ROBINSON
London

Constable & Robinson Ltd
3 The Lanchesters
162 Fulham Palace Road
London W6 9ER
www.constablerobinson.com

First published in hardback as *James Dyson's History of Great Inventions*
by Constable, an imprint of Constable & Robinson Ltd, 2001

This Mammoth edition first published in the UK by Robinson,
an imprint of Constable & Robinson Ltd, 2004

ISBN 1–84119–903–6

Printed and bound in India

1 3 5 7 9 10 8 6 4 2

Contents

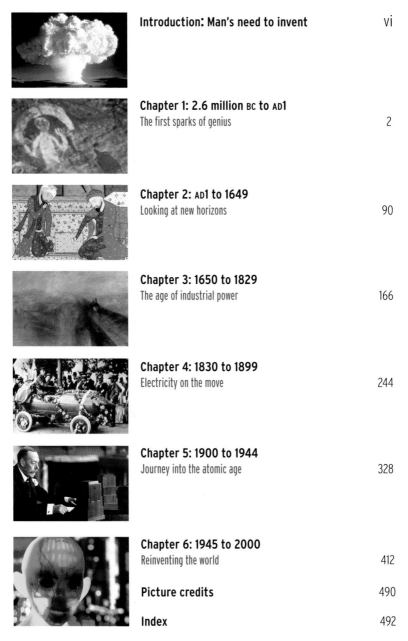

Introduction
Man's need to invent

We humans are astonishingly inventive. The proof is everywhere: the paper you hold in your hands now, the ink, the presses that printed it, even the notion of using words to put these thoughts into your mind. They are all inventions, acts of creation by someone, somewhere.

In this book, the stories behind the most important of these inventions will be told. And what amazing stories they are, spanning a range of human emotions and motivations: fear, lust, greed and altruism.

It has certainly prompted me, as an inventor, to ponder why I do what I do. It can be enormous fun to think up new ways of doing things and then to prove they are no mere pipe dreams by turning them into real, live products sold around the world. But another big motivation for inventors has been frustration: frustration about things that have been around for years, doing an okay job, but not anything like as good a job as they could. The breakthrough was realizing that things don't have to be the way they are.

Frustration over the countless ships lost at sea through navigation errors led to John Harrison's invention of extraordinarily accurate clocks. Frustration at the appalling death rate in maternity wards led the Hungarian doctor Ignaz Semmelweiss to invent antiseptic practice. And frustration at the many blunders in mathematical tables led the Victorian mathematician Charles Babbage to take the first steps towards the computer.

But inventions have come about for many other reasons. Pure serendipity did the trick in some cases. The way to make glass probably came from someone just happening to find tiny globules of the stuff in the embers of a huge beach fire thousands of years ago. The anti-impotence drug Viagra was originally intended to treat angina: then male patients sheepishly reported to their doctors that the drug was having a far more impressive effect on their private lives. And when cocklebur seeds stuck to George de Mestral's jacket, he was prompted to invent Velcro.

It's not all romantic nonsense. If an invention is to get a patent, there's actually a legal requirement that it be based on a leap of imagination or a happy accident. Indeed, the very word "invention" comes from the Latin verb meaning "to come upon". So, obvious little tweaks to existing ideas will get pretty short shrift from the Patent Office.

Necessity has indeed been mother to countless inventions: the Sumerians invented written language to keep their businesses in order. In 1856, Henry Bessemer invented steel to make stronger cannons. And today's computers, for example, are direct descendants of the electronic machine invented by Post Office engineers during the Second World War to break Adolf Hitler's most secret ciphers.

What all inventions have in common is that they start from a knowledge of what's already out there. It may be cockleburs or calculus, but every invention draws in some way on the huge accumulation of know-how built up over millennia. Any inventor must understand what I see as two of the key intellectual breakthroughs of the past 2,000 years. The first is the recognition (due mainly to Newton) that the laws of science don't just apply here and there. They are literally universal. Spotting the underlying unity in apparently totally unconnected things can lead to world-beating inventions.

The other key idea is that of the "scientific method", that is, studying systematically things that you don't understand. You have your idea, you design an experiment to test it, and then draw your conclusions. We owe that to the 17th-century thinker Francis Bacon, and it sounds trite until you try the alternative: random guessing and wishful thinking.

But for most of recorded history, having a bright idea was no protection against being ripped off by the unscrupulous. Thomas Edison, the doyen of inventors, said it first: "No sooner does a fellow succeed in making a good thing, than some other fellows pop up and tell you they did it years ago." It is not a whole lot better now, but there is something that, in theory at least, makes sure that the credit and the money for the invention go where they are due: patents.

The idea that ingenious people should benefit from their creativity emerged first in 15th-century Italy, and spread to the rest of Europe via émigré Venetian glass blowers. The earliest known English patent was granted to John of Utynam, a Flemish stained-glass maker, in 1449 by the high-minded Henry VI.

John received a 20-year monopoly to exploit the fruits of his ingenuity. In return, he was required to teach his process to native others. That, too, is still part of the philosophy behind the modern patent: that it doesn't just encourage innovation, but also the spread of that innovation. Inventors have to disclose how they did it.

And what trouble that has caused ever since. For by revealing exactly what you have done and how, you are putting your intellectual crown jewels right where other people can get them. Not only that, but by stating what is new about your invention, you are revealing your likely marketing strategy.

Patent violations are a constant nightmare. It is a common misconception that when you have paid for your patent, the shining knights of the Patent Office will ride to your defence. If only. In fact, they just take your money for a non-binding examination, and then walk away. You have to fight your own fights and even then not always to any purpose. Edison chased his rival, George Westinghouse, through the courts for years over the invention of the light bulb, while Westinghouse went on making a mint. Alfred Nobel, inventor of dynamite, had to defend his patents against violation by the entire British government.

I want to see a change in the patent law, so that if someone can show they have created something, they own that idea. I see this as a fundamental human right. As it stands at present, copyright law gives pop songwriters far more protection for a three-minute ditty than a patent gives an engineer for a fundamental breakthrough in propulsion. And that just can't be right.

To my mind, the key test for any invention worthy of the name is whether it can be made practicable and put into production. Making an idea actually work and perform well is what inventing is about. Indeed, overcoming manufacturing challenges has led to some of the most important of all inventions.

Take the ship's pulley block. In the early 1800s, the Royal Navy needed 100,000 wooden pulley blocks for its ships every year. That's about 10 an hour, every hour, month in, month out. Producing so many while maintaining quality was not easy, but in 1801 a patent was granted to an engineer who claimed to have the answer: Marc Brunel, the French-born father of Isambard. He built the first mass-production facility at Portsmouth, its machines turning out more blocks of higher quality with 10 men than had previously been possible using 100.

Henry Ford famously made a science of mass production, cutting the time needed to make a car chassis from more than 12 hours to just 24 seconds. People like to point out that he got his ideas from watching a meat-rack system, but this vastly underplays the skill Ford and his colleagues put into optimizing the whole process.

Some of the most important inventions of the last hundred years have been not products at all, but the methods used to make them. Methods such as job-scheduling and linear programming for optimizing production in the face of all kinds of constraints, and Toyota's famous Just In Time method for operating minimal inventory.

Knowing when to ignore experts is also important. Edison's proposal for electric lighting circuitry was greeted with total scepticism from eminent scientists, until he lit up whole streets with his lights. Similarly, in the 1830s, Brunel faced scepticism as to whether his steamship the Great Western would reach America. It couldn't carry enough coal for the trip, the experts declared. Brunel went ahead anyway, and when the Great Western docked in New York after a 15-day trip (half the usual time), it still held 200 tons of coal.

The Wright brothers faced no less "incontrovertible" proof of the futility of their task. Leading scientists derided the idea of heavier-than-air flying vehicles as impossible. Years later, they were saying the same about supersonic travel.

Being able to focus on problems, and stay focused for days, even years, is critical. Edison never stopped until his ideas were on the market. If anything, he stayed too focused: he discovered radio waves during his lab experiments, but failed to pursue the discovery because of his determination to get products out of the door.

I certainly hope that this book kills off the myth of the madcap inventor in his shed having a Eureka moment and making a fortune. Having the bright idea is crucial, of course, but it is just the start of the process of making successful inventions. One per cent inspiration, 99 per cent perspiration, as Edison so accurately put it. Whatever obstacles are thrown in their way, people will continue to invent: it is part of the human drive to create, to improve our lot.

It is my sincere hope that this book will fascinate and inspire in equal measure. For invention – the chance to make something work a whole lot better – is something within the reach of everyone. You may not make a fortune with your invention, or become famous. But you may one day hear someone say, "Hey, I've bought one of those. It's brilliant."

And that's priceless.

James Dyson

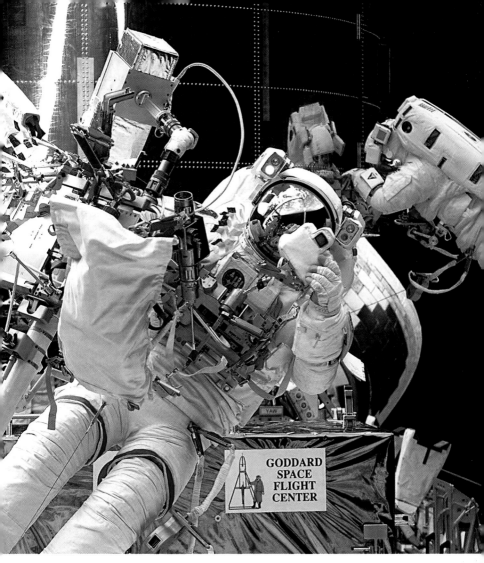

Over the past century the world of science and
invention has progressed at unimaginable speed. In
this photograph, from 1999, members of
the Discovery shuttle crew walk in space to
complete their mission to repair America's
12-ton, four-storey Hubble space observatory.
It was less than a hundred years since the Wright
brothers achieved the world's first
powered flight at Kitty Hawk, North Carolina.

Chapter 1
The first sparks of genius
2.6 million BC to AD 1

From man's first steps to the birth of Christ was a period like no other in the history of invention. Never again would man's survival be so dependent on his ability to invent ways to solve fundamental problems. Never again would man's technological creativity be the most significant factor in his evolution and the establishing of civilization.

By the time modern man (homo sapiens or man the wise) appeared, probably in Africa between 100,000 and 250,000 years ago, his forefathers, the early hominids, had already invented stone tools. It is possible that they had also manufactured crude canoes and shelters. However, it would take many more years and a succession of vital inventions for man to evolve from a primitive hunter-gatherer to the highly technologically literate citizen of the time of Christ.

We like to think that we are currently living through a period when technology has an unparalleled hold on society, but it is nothing compared with that of the ancient world, when invention and technology were the most powerful forces shaping civilization. Technology was the one factor that made all the other changes – social, political and cultural – possible. Without the inventions of ink and papyrus, many of man's ideas would not have spread as fast nor as widely. Without weapons and, later, the wheel, armies would not have won territories as quickly.

The single largest step in early man's social evolution came around 10,000 years ago with the invention of animal husbandry and agriculture. This enabled him to progress from living in nomadic communities to settling in villages and small towns. The progress was brought about by a combination of climatic change and man's invention of more efficient hunting tools, of a means of controlling and utilizing fire to clear undergrowth and of ways of building lasting shelters. It led to a massive growth in population, which in turn triggered a further rapid increase in technological innovation. Most of this change took place in the eastern Mediterranean, where the climate and the annual flooding of fertile soils favoured the development of agriculture and later of cities such as Babylon. By around 6500 BC, Jericho is believed to have been the largest city in the world, with a population of 2,500.

Four thousand years later, the urban revolution had brought about a momentous cultural transition that in turn generated new needs. These were met by a quantum leap in technological innovation and the establishment of craftsmen and scientists. For the first time, manufacturing became established as man invented ways of making textiles, firing ceramics, producing metalwork and processing foodstuffs. This prompted barter methods to evolve into more sophisticated trading arrangements, culminating in the invention of tokens or early money.

With these technological changes came a corresponding increase in the complexity of the social and political organization of human groups, which in turn necessitated the invention of written language, first to keep track of trading arrangements, then to communicate and record events, processes, philosophies and, of course, inventions.

The history of invention is littered with devices that had little or no purpose and never caught on, but this was still a period of invention for necessity's sake. It would be some time before an invention would be greeted with questions as to its role – and even longer until Michael Faraday would retort, "What use is a baby?" when asked what use his dynamo had.

It was also a period when science and technology's symbiotic relationship was reversed. Technology, now often the application of scientific discovery and observation, predated science and in this period was empirical and handed down through the generations. By the time the city states were flowering in the early centuries BC, scientist-inventors began to emerge. Figures such as Hero, Strato, Ctesibius and Philon used observations and measurements of the physical and natural world to devise inventions. However, they were all minnows when compared with Archimedes. Here was a man of the calibre that the world would not see again until Sir Isaac Newton in the 17th century. The inventor had truly arrived.

Tools

2.6 m^BC   Africa

E vidence that hominids invented, refined and used tools over the past two and a half million years is widely scattered, and the fact that they roamed the world in small numbers means that such evidence is usually fragmentary and difficult to interpret. The first hominids emerged in central Africa, where the vegetation and climate were conducive to the survival of a weak and vulnerable biped dependent on his intelligence and his ability to extend his power by the use of tools. The earliest tool fragments come from the Olduvai Gorge on the Serengeti Plain in East Africa and date to 2.6 million BC.

This flint axe-head, found in southern England, dates back to 350,000 BC.

From this, hominids appear to have moved gradually into all accessible parts of the Earth, unless prevented from doing so by excessively inhospitable conditions. These migratory movements were themselves made possible only by the increasing hominid proficiency with tools.

Human beings are thus, virtually by definition, tool-making creatures. They discovered the facility of extending the powers provided by their unique combination of hand and eye: the hand, with the remarkable dexterity and precision derived from the prehensile thumb, working with the opposite fingers, and the eye, guiding the observational and deductive powers of a larger brain. With this natural equipment they were able to respond to their environment by devising and making appropriate tools.

It can be safely assumed that the first tools were fashioned in wood, reed, bone or some other comparatively easily worked material. Such material rarely

survives millennia of decay and does not leave any evidence for archaeological examination. Some bone digging tools have been found in early mines, such as Grime's Graves in Norfolk, where flints were mined on a large scale, but for the most part we depend upon the survival of stone fragments to indicate the existence of early tools.

Stone tools gave the aeons of prehistory – from the first evidence of hominid existence down to the second or third millennium BC – its name, the Stone Age. This long period is divided into Old and New Stone Ages, the Palaeolithic and Neolithic respectively, according to the tool-making capacity of the hominids concerned.

Up to around 10,000 BC, in the Palaeolithic period, we can be sure most hominids existed in small tribal groups that survived by hunting, fishing and gathering fruit and natural sources of edible matter. They were impelled to lead a nomadic life as these sources of food changed. Their tools consisted of stone instruments for cutting and killing, such as axes, adzes (with a blade shaped to scoop wood out of a trunk, and so fashion a trough or a boat), spear-points and arrow-heads.

Man discovered quite early that some hard and fine-grained stones such as flint could be worked in order to give them a sharp cutting edge. This was done by striking the flint at an

By 3000 BC flints such as this from Egypt were being finely worked.

2.6 m BC Earliest tool fragments, believed to have been made by Homo habilis (Handy Man).

A god with a sickle, from the Tisza culture, Hungary, shows that by 4000 BC Neolithic man had begun to cultivate the land.

angle with a hard pebble, causing it to flake. Repeated striking in this way could form an efficient blade or point. The shaped stone could then be bound to a piece of bone or wood to make cutting tools, arrows, saws and spears. Equipped with tools and weapons, our ancestors were able to colonize all parts of the Earth.

Around the end of the last Ice Age, about 15,000 years ago, the transition from Old to New Stone Age began. This transition was associated with a shift to a more settled form of life with the discovery of techniques of animal husbandry and the invention of agriculture. The acquisition of domesticated animals meant that communities were able to live off the products of their flocks such as milk and other dairy produce, skins, furs and leather.

Agriculture began with the systematic cultivation of certain crops of cereals, which involved many new skills in preparing the ground by regular ploughing, harvesting the crop with sickles or other sharp cutting tools, and threshing the crop to separate the valuable grain from the chaff. Thus began the New Stone Age, encouraging the development of a range of important tools. It brought a remarkably rapid increase in population, which was both a cause and a consequence of the more settled form of life. This introduced other skills of a more domestic nature in using fire – long since tamed as a means of self-defence against predators, and for simple cooking – in order to bake grain into bread, to bake clay into brick, to shape clay into pots and to harden it in the fire, and to brew and distil strong beverages.

The principal cutting tools of the Neolithic period were still

1.6 m BC Homo erectus evolves, probably from Homo habilis, and is first to develop the tool of fire and live in caves.

The evolution of tools, from Paleolithic to Neolithic, from left: arrowhead, Italy; streaked iron stone, South Africa; spear head, Africa; flint arrowhead, Wiltshire; flint axe-head, Denmark; dagger, Denmark; polished axe-head, Wiltshire.

made of stone, but they acquired more sophistication in their finish than the comparatively clumsy tools of the Old Stone Age, being frequently polished. The combination of an increasing population and new skills of a settled lifestyle amounted to a profound transformation of human society.

These changes occurred first in those parts of the world which were most highly favoured for a settled agricultural economy. These were the great river flood plains of sub-tropical latitudes: the Nile, Tigris, Euphrates, Indus, Yangtze and Yellow River. It was here that the first urban communities arose, and thus the first civilizations, with skills of literacy and numeracy making possible large-scale administration and the growth of empires. These developments brought a wealth of inventions as the ingenuity and creativity of the human mind responded to new challenges and new opportunities.

The invention of the wheel found multiple applications, in wagons, chariots and in simple mechanisms such as the potter's wheel, whereby the facility for creating ceramics was greatly expanded.

The bow and arrow had already emerged as a powerful weapon. Now it was adapted for woodworking in the form of the bow-lathe, in which the twine from the upper end of a bow is turned round the object to be cut. This spun first one way then the other way by the pressure of the foot pulling against the resilience of the bow, while the operator applied hand-held cutting tools to the spinning material.

It is important to recognize that all these early tools were probably invented many times, in different communities around the world. However, without somebody's having the idea, and the idea then being selected from the ferment of human creativity as worthy of development to meet a particular set of circumstances, it would simply not have occurred. It is intriguing, for example, that such a basic tool as the wheel was not invented in the indigenous civilizations of the New World. This shows clearly the importance of adequate communication between human societies in order to ensure the distribution of new ideas, and especially of good tools.

Fire

1.4m BC Africa

T he single most significant invention in the history of mankind is the tool of fire. Without it, we would be like early man, dependent on lightning or some other fortuitous spark in order to cook our food, ward off predators, or heat our caves. Fire has been essential at every stage in civilization.

Like us, early hominids needed fire for warmth and to cook food. Neolithic man used it to clear forests into open grassland for agriculture and to produce fertilizer, enabling him to establish farms and villages. Later, it was needed to fire pottery and to extract metal from ore, and more recently to power engines, to aid chemical reactions and generate electricity.

Evidence of the earliest use of fire comes from Kenya and South Africa, where traces of cooking can be found in the charred bones of animals around hearths that have been dated to around 1,420,000 years ago. From these beginnings, when man was dependent on keeping fires permanently alight, it took many hundreds of thousands of years before man could create fire himself.

By about 7000 BC, Neolithic man had discovered that friction between pieces of wood, and possibly sparks from striking flints, could kindle a fire.

An ingenious fire piston was invented in South East Asia. By compressing air with a plunger in a small bamboo tube, the holder generated enough heat to encourage fire.

Then came the tinderbox, which made the tools of fire portable, but a reliable and portable source of fire was not to come until the invention of the friction match in 1827.

Buildings

Japan

With tools made only of stone or bone there were limits to what Stone-Age man could achieve. The oldest artificial structure yet found was discovered in 1999 on a hillside site at Chichibu, Japan. The Chichibu shelter is thought to be half a million years old and much older than the previous oldest remains at Terra Amata in France, a mere 200,000 to 400,000 years old.

All that was left were 10 post holes in the ground but the depth within the soil at which the depressions were found and the accompanying volcanic ash enabled the archaeologists, a team from Meiku University, Tokyo, to date the site quite precisely.

500,000 BC

400,000 BC
Homo sapiens evolves, probably from Homo erectus, and is first to develop cutting tools made by striking flakes off flint stones.

The Great Pyramid, the apogee of stone block building.

By the Middle Ages brick and stone were both common building materials in Europe.

Tents, tee-pees or yurt-like constructions would commonly have housed the travelling tribes that roamed across Europe and Asia, but we know little of these, perhaps because of their temporary nature. The oldest remains of a tented village date only to 40,000 BC, found in Moldova in Russia.

The materials used in Moldova tell an intriguing story. The shelter-makers were clearly mammoth hunters, for the entire structure was constructed of mammoth bones, intricately interlocking, stacked up high and originally covered with mammoth skin. A curious feature of these bone structures was a custom of painting them with red pigment.

The pharaoh Djoser built the first pyramids in ancient Egypt, using stone blocks, at Memphis.

With the coming of modern homo sapiens, around 40,000 years ago, a greater sophistication and permanence is evident in the building of settlements. From around 10,000 BC in Europe, once these peoples had learnt to cultivate crops reliably, they needed to stay in one place to tend them, so more lasting habitations developed.

In the valley of the River Jordan, around 9000 BC, round houses called *tholoi* were built, and by 7000 BC, they had evolved into the square-shaped dwellings seen all over the Middle East and of the type mentioned in the Old Testament. They were single-storey mud-brick constructions, grouped together in clusters, built on stone foundations. These Middle Eastern settlements were vigorous agrarian communities with evidence of animal domestication and a collective way of living.

A prehistoric mud brick from Jericho.

40,000 BC Rapid increase in uses of technology sees primitive lamps, bone needles and thread, mortars and pestles, chisels and spears in use.

In parts of South America, and reaching as far north as New Mexico, a similar form of mud brick, known as adobe, was common. Adobe was a mixture of clay and quartzite sediments which gave the brick extreme hardness when dried out. Unfortunately it wasn't waterproof and so adobe buildings typically had wide overhangs to keep the infrequent rains off the precious walls.

Improving the durability of brick was vital. By 3000 BC, the Egyptians were mixing chopped straw into the clay of their building bricks. However, only low-grade domestic structures were built with bricks because they had little or no structural strength. The first pyramids were built at Memphis with monumental stone blocks within the ceremonial complex of Saqqara designed by the architect Imhotep under the pharaoh Djoser. The key to building stone structures such as these was the development of jointing methods, involving metal keys to link blocks, and earthen ramps up which the enormous stones were pushed and pulled with ropes, rollers and the block and tackle.

After this, the Great Pyramid of Cheops at Giza was begun around 2900 BC, followed by the Great Sphinx. These structures were an extraordinary achievement: Cheops was 147 metres high. It took until the 19th-century for taller buildings to be built anywhere. But they were, in a sense, a dead end in building evolution. Giant solid structures like the pyramids were the apogee of this kind of stone block building.

Later, Greek, Etruscan and then Roman architects all tried to make substantial structures using stone columns with timber or iron beams across them.

Strong frames of masonry provided strength at the core of the great temples and monuments of the Middle East, such as the complex at Karnak. The invention in Mesopotamia of the kiln-fired brick in around 1500 BC allowed ever more ambitious construction.

As civic buildings increased in size, a pressing need to cover large spaces developed. The Egyptians had the manpower to haul giant slabs of stone on to the roofs of columned structures, but later builders explored other ways of spanning space. The arch, the corbelled vault of brick or stone and, finally, the invention by the Romans of a form of concrete led to modern building technology as we know it.

Angling
Africa

I n the beginning there was hunger and the pressing need to feed a family. Homo sapiens looked around and by about 38,000 BC was using crudely fashioned bone to hook food.

By 25,000 BC Palaeolithic man was using a short stick or bone called a gorge. Needle-sharp at each end, it was baited and tied in the middle to a short line. But the gorge didn't provide enough fish for the larger settlements in the Neolithic period. By about 8000 BC traps woven from reeds and branches were strung across rivers.

Fishing lines were made of animal gut or vegetable matter, but they would not reach far enough from the river bank of water harbouring larger fish. For this, man needed a rod.

By around 4000 BC, the Chinese were using bamboo rods and silk line, and the Greeks in about 2000 BC used branches. But rods in their modern form did not catch fish suppers until around the 4th century AD, when the Romans started using jointed rods up to two metres long. The earliest evidence of fly fishing comes from the Roman rhetorician Claudius Aelian (*c*.AD 170–235).

38,000 BC

Egyptian fish hooks, 3000 BC.

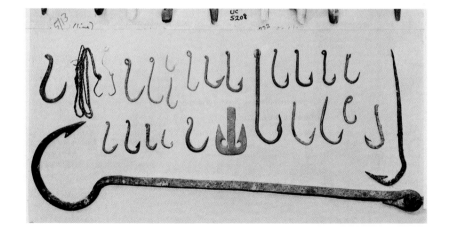

Hunting implements

Africa

The bow and arrow was invented in Africa around 30,000 BC

For prehistoric man, for whom a dead animal provided food, skins for clothing, and bones and horns for tools, inventing new devices to ensnare, injure or to swiftly kill prey was paramount.

Early hunting implements were no more sophisticated than a sharpened stick or a heavy rock. From these evolved the spear, later with a barbed tip and a separate shaft, and variations on the stick, such as the boomerang, the knobkerrie of southern Africa and the trombash of the Upper Nile.

Initially, these were weighted sticks but in time they evolved so that the boomerang returned to the thrower and the trombash acquired metal prongs to deliver a deadlier blow.

By 30,000 BC, the bow and arrow had been invented in Africa and it enabled man to kill outside his throwing range, as did the blowpipe.

Equally important were the implements of entrapment, such as nooses, snares and poison, all of which are believed to have been in use by 20,000 BC.

Art
Lower Austria

28,000 BC

A round 30,000 years ago Cro-Magnon cave dwellers in southern Europe are thought to have used flint tools to chip, scrape and gouge early pictorial communications. Since these people were cave dwellers, it was the invention of the lamp that then may have fuelled a growing sophistication in these works. The oldest lamp found so far, from France, is around 17,000 years old and made from a lump of animal fat.

Early art comes in two forms: two-dimensional wall paintings, the most famous of which are at Lascaux in France, and three-dimensional carved objects, such as the Venus Of Willendorf, a stumpy looking figure of a female discovered in Lower Austria.

Generally accepted to be the earliest known piece of prehistoric art, the Venus was a limestone sculpture about 11cm high. It may have been carried by the male hunter-gatherer as a reminder of his mate back home.

Carved between 28,000 and 25,000 BC, this relic predates the earliest known wall paintings, and neatly illustrates the different functions early artistic efforts fulfilled. Religious or shaman significance may have been important to early artists.

Extraction of fuels
Europe

17,000 BC

T he earliest fuel lamps, from at least 17,000 BC, were inefficient. The animal fat man had extracted produced sooty flames and generated little heat. Bronze-Age man in Wales, between three and four thousand years ago, must therefore have been pleased to discover deposits of coal.

This was a more useful fuel, burning well and keeping alight all night. Archaeologists found cinder remains of fires around Roman border garrisons dating back to AD 400.

Terracotta oil lamp, 800 BC.

12,000 BC The dog becomes the first animal to be domesticated, followed by the cow, sheep, goat and pig over the next 6000 years.

It wasn't just coal that was valued by the ancients. Naturally occurring deposits of bitumen and crude oil were found in the Middle East as long ago as the 3rd and 4th-centuries BC.

Thriving petroleum-based economies sprang up in the Arabian Gulf and the first war over oil may have been fought in 88 BC when Antiochus XII of Greece attacked the oil-rich Middle-Eastern kingdom of Nabatea to capture its bitumen and oil reserves.

Boats

8000 BC Europe

Population migration across the Pacific Islands and Australia indicates that it is possible that as far back as 50,000 years ago people may have been able to cross quite large tracts of ocean. However, there is no evidence of how they did it. The most likely explanation is that they used the dugout.

The earliest water-borne craft yet found is the Pese canoe, from Holland. This four-metre dugout boat is thought to be around 10,000 years old, and would have been propelled by the paddle.

Most people may think of dugout canoes as being primitive craft – vessels honed from a single log of wood – but they paved the way for thousands of years of wooden boat building.

The invention of metal tools, such as the adze, used for crude shaping and gouging, allowed more complex construction methods and the development of more ambitious vessels.

Egyptian, Scandinavian and Celtic boat builders in the Bronze and Iron Ages adapted and extended dugouts with extra planking to form the foundations of much larger boats. The Chinese used the principles of the dugout boat to make immensely rigid and strong sailing vessels such as the Junk that is still familiar today.

Around 5000 BC, Mesopotamian traders built sophisticated trading ships from bundles of reeds, and wooden ships were known to ply the Nile around 5500 BC.

As well as keeping the economy going, the ships of ancient Egypt had immense symbolic value. Rich noblemen or merchants were sometimes buried complete with their ships, which were thought to take them into the next world.

Pottery

China

7900 BC

The oldest known piece of pottery was found in China and dates back to 7900 BC. A Neolithic settlement site at Catalhüyük in modern Turkey revealed crumbling, unfired pots and terracotta figures made around 7000 bc. Like the Chinese finds, the early pots were made by pressing clay, piece by piece, into a mould. They were then left to bake slowly out in the sun.

By 6500 BC, Neolithic man had discovered that firing clay at temperatures of 500°C would change the chemical consistency of the pot. Fired pots have been found that date back to 6500 BC, but it was another 5,000 years before the kiln was invented in Mesopotamia in order to bake bricks for building construction.

Weaving

Judaea

6500 BC

In a cave at Nahal Hemer in the Judaean desert, archaeologists found evidence of textile manufacturing from at least 6500 BC. We do not know exactly how the linen collection was produced, but it is too sophisticated to have been man's first attempt to weave textiles and it indicates that the first textiles may have been woven up to 10,000 years ago.

In Neolithic communities textile weaving is likely to have evolved from basket weaving. All it required was the development of spun yarns to enable the replacement of reeds with cotton, silk, wool and flax. Early looms, such as that used to weave the Nahal Hemer linen, were made by driving four pegs into the ground in a square. The looms were stretched across pegs, which were dismantled once the cloth had been woven. Consequently, there are no remaining weaving looms.

6500 BC Brewing beer becomes popular in areas of the Middle East and Africa too hot to support grape cultivation.

6000 BC Many grains, such as wheat and barley are grown widely.

Weaving in Ancient
Greece

We know that by 4400 BC, the process had been improved with the invention of lease rods, used to separate the odd and even warp threads so that the weft yarn could be pulled through the weave more speedily. Vertical looms were invented by 3000 BC, with weights made of chalk, clay or ceramics to keep the warp threads hanging from a crossbar taut.

Although the earliest evidence of weaving comes from the eastern Mediterranean, the Chinese developed advanced techniques, using a foot-powered treadle, at an earlier date. By 2500 BC the Chinese were weaving intricate damask twills which would not have been possible on a simple loom.

By the early centuries AD, very sophisticated weaving, using silks and wools, was practised throughout the Ancient World in Persian carpets and tapestries.

Intoxicants

5400 BC Mesopotamia

Alcohol is the oldest and probably the most widely used drug in the world. It is thought that man stumbled on the first intoxicant by sampling fermented fruit, or the contents of a jar of honey that had been left unattended longer than usual. Contamination by airborne yeasts acted on natural sugars in fruit or honey to convert them to alcohol by fermentation.

There are many ancient references to alcohol's role in society. Drinking houses were regulated by the oldest known code of laws, that of Hammurabi of Babylonia in around 1770 BC. Sumerian pharmacists of about 2100 BC describe beer in pharmacopoeias found on clay tablets. The later Egyptian doctors mentioned beer or wine in many prescriptions.

Alcohol was frequently used in religious ceremonies by early civilizations. Its effect was attributed to gods and the intoxicating effect helped the shaman or priest achieve the state of frenzy necessary for some religious rites. But it did not take long for liquor to pass from religious to social use. Egyptian and Mesopotamian records show ample evidence of widespread public drunkenness.

It has even been suggested that the first generation of relatively weak alcoholic beverages may have aided Western civilization by offering a safer alternative to water. With the rise of agriculture, people were encouraged to gather in villages and towns, and the water supply became polluted with their waste.

Residues of liquid in a pottery jar from Hajji Firuz Tepe, a Neolithic village in the Zagros mountains of Iran, show a resinated wine similar to Greek retsina, revealing that wine dates back to 5400–5000 BC.

This yellowish residue was analysed by a team led by Patrick McGovern, of the University of Pennsylvania Museum. Using a spectroscope they measured light absorbed by the residue at various infrared and ultraviolet frequencies. They found the residue contained resin from the terebinth tree. This is known to have been used in antiquity as a medicinal agent and additive to inhibit the growth of the bacteria that converted wine to vinegar.

Beer also has been available for many millennia. Perhaps even before the grape was cultivated, ancient people had found that the starchy grains of barley, wheat, millet and corn could be fermented. The trick was to allow the grain to sprout, at which point enzymes were released to chew up the starch molecules and release its simpler component chemical units, notably glucose. In nature this provided the sprouting seedling with an energy supply, but in brewing it supplied the glucose needed to support the growth of yeasts.

King Hammurabi of Babylonia regulated drinking when he established the oldest known code of laws.

Alcohol escaped the clutches of the priests to become the source of social ills and of humour.

·THE ·LAST ·POST·

La dernière étape.

Analysis of a 4th millennium BC jar from the site of Godin Tepe in the Middle Zagros mountains of Iran, suggests beer was brewed in Mesopotamia. We know for sure that barley and wheat beers were being brewed in Egypt as long ago as the 3rd millennium BC.

The Egyptians used a sophisticated, two-step process, resembling some African brewing methods today. The grain was divided into two batches. One was made into malt by sprouting and gentle drying, then coarsely ground. The other batch may have been malted or not, but was coarsely ground and heated in water. The two were then mixed together. The active enzymes from the uncooked malt broke down the starch in the cooked grain.

At this point, the chaff was still included in the mixture. It was removed by sieving and the sweet cloudy liquid was inoculated with yeast and fermented.

After thousands of years of drinking relatively low alcohol beer and wine, the West was faced with alcohol in highly concentrated forms thanks to the development of distillation in about AD 700 by Arab chemists.

The still was invented to extract essential oils for making perfumes, but it was soon being used to overcome the drawback that the yeasts that produce alcohol can tolerate concentrations only of about 16 per cent.

Distillation increases the percentage of alcohol beyond this natural limit by exploiting the difference in the boiling points of alcohol and water. If a liquid containing alcohol was heated and the vapour condensed, the condensate had a higher alcohol strength and resulted in spirits.

The balance

5000 BC Egypt

The first device that could measure things and thereby turn man's subjective impression of objects into precise mathematical terms was the equal-arm balance, thought to have been invented by the Egyptians in around 5000 BC in order to weigh gold dust.

Initially, there was no empirical definition of weights and merchants would assess weights of goods by comparing them to one another on a beam scale supported at its centre and with two pans hanging down.

It took until around 3000 BC for the Egyptians to invent the *kite*, the unit of their first weight system, but this was possibly predated by Babylonian *mina*, which spread throughout the ancient world to become the standard unit of weight. The standard measurement of length was the Egyptian *cubit*, which began life as the distance measured from the elbow to the tip of extended fingers. This was soon standardized on a royal

A painted wooden box, 2000 BC, shows Anubis, the Egyptian god of the dead, using a balance to weigh up the souls of the dead.

master cubit fashioned of black granite, against which all Egyptian cubit sticks were regularly checked. The next innovation in weighing scales – and it seems surprising that it took 5,000 years, until the time of the birth of Christ, to occur – was the addition of a pin through the centre of the beam balance to make it a reliable measuring device. Before this invention, two different weights could be made to appear equal by moving the fulcrum of the equal-arm balance towards the end supporting the heavier weight.

Weighing scales became reliable, precise and accurate only in the 18th century with the invention of the knife-edge. Until then the beam was balanced with a margin of error that invariably favoured its owner.

The balance is one of the first uses of the lever, one of the simplest yet one of the most remarkable machines invented by man. If you want super-human strength, you don't have to spend years of your life training in a gym. All you need to do is to find yourself a lever.

I know of no faster way of making oneself seem amazingly strong: I still recall how easy it was to flip a Mini on its side to get some welding done in my garage, using just my weight, boosted by the lever effect of a railway sleeper.

However, it is not always easy to spot when a device is using the lever effect. Bottle openers, crowbars and seesaws are all obvious levers. Each of them has the load at one end, the effort at the other and the fulcrum in between. An oar is another example. This is simply a crowbar which, through the rowing action, is rotating about the vertical, instead of the horizontal, axis. But there are many levers in disguise. For example, a spade is a lever whose fulcrum is the edge of the soil being dug, while scissors and pliers are just two levers working back to back.

Both nutcrackers and wheelbarrows, with their pivot at one end rather than between the load and the effort are another kind of lever.

But in all these cases, the lever effect routinely multiplies your strength by a factor of three or more and, with a suitable choice of material, far more.

We owe the existence of some of the oldest and greatest monuments, such as Stonehenge and the Easter Island statues, to the ability of the lever to turn anyone into Superman.

Musical instruments

Ur (modern Iraq)

The earliest instruments were simple objects such as bones or gourds, which could be persuaded to emit more or less musical sounds just as they were, or with a few holes drilled in them to produce different notes. Such pipes were often made of hollow swan wing bones. A flute carved from the leg bone of a bear was unearthed in 1995 alongside Neanderthal tools and an ancient fireplace. It is thought to be 50,000 years old.

The first musical instrument to feature some human involvement in its construction was probably the drum, made from the skin of cattle and stretched tautly over a bowl or tube made of wood, pottery or metal. There have been suggestions that the earliest drums, from around 12,000 BC, were made from mammoth skulls, but the earliest manufactured drums that we know of, found in the Czech Republic and Germany, were made of clay and date back to 3000 BC.

Unable to do much more than produce a few notes or generate a rhythm, these early instruments were limited in their musical abilities. By 4500 BC one solution to these limitations had emerged in the Sumerian city of Ur: the stringed harp. This is the earliest known instrument to be entirely manufactured.

The ability of a string to produce a musical note when plucked was probably discovered much earlier still, but the harp was the first instrument specifically designed to exploit the fact that altering the lengths of vibrating strings affects their pitch.

4500 BC Civilization emerges in Mesopotamia, the Indus Valley and the Yellow River delta as communities co-operate to deal with common threats such as flooding of rivers.

Below : Harp, late 18th-century, Iran; horizontal grand piano, 1826.

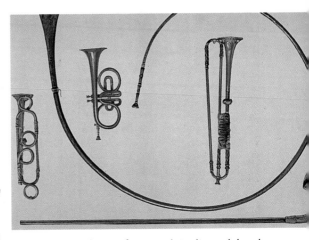

As the loudness of a sound is dictated by the amount of air physically set in motion, string instruments cannot, by their very nature, produce very loud sounds: the string simply doesn't move enough air. A whole range of new and more powerful musical instruments thus became possible only when the sound of strings was boosted using the resonant sound cavity of a drum or pipe.

A crucial discovery was that natural tubes, such as bamboo or animal horns, could make musical notes when the column of air they contained was made to vibrate by blowing across their top. The deep, resonant sound produced by such wind instruments gave them a special, almost magical status among non-literate peoples. Some Amazonian Indians, for example, keep giant flutes, over a metre long, in a shrine where they are worshipped.

Horns made of metal date back at least as far as the Egyptians. A hunting horn was found in Tutankhamun's tomb by Howard Carter and his team in 1922: the sound of its being played for the first time in more than 3,200 years is one of the eeriest sound recordings ever made.

By around the end of the Dark Ages, the basic instrument had transformed into the oliphant, a blown horn imported into Europe about 1,000 years ago from the Middle East. Made from an elephant tusk, the oliphant became a precursor of the

trumpet. Instrument makers noticed that lengthening the tube allowed more notes to be played, and by the Middle Ages horns were bent into more compact shapes for manageability. When the German instrument-maker Anton Weidinger introduced keys to the basic horn tube in 1801, the trumpet was born.

The ability of reeds to make musical sounds appears to have been discovered first in South East Asia several thousand years ago. So-called double-reeded instruments were first to appear, probably because they could be made simply by squashing a tube of cane to form a slit with edges that could be made to vibrate against each other using the player's breath. An early example of such an instrument is the aulos, a precursor of the bagpipe, which originated in ancient Greece and was still in use in Europe up to the early Middle Ages.

Single reed instruments came later, and took the form of a pipe over whose opening a vibrating reed was fixed. Around 1700 one such instrument, known as the chalumeau, was widely used in folk music in Europe when its extra potential was spotted by a German instrument-maker, Johann Christoph Denner. Modifying the basic design by doubling the chalumeau's length gave the new instrument a range of three octaves. Denner called the result the Klarinette – which ultimately became the modern clarinet.

From far left: Aulos (bagpipes), c.1340; Etruscan stick-shaped lituus and circular cornu, and later trumpet, cornet and baroque trumpet; German flute and oboe players.

A desire to sustain musical notes on wind instruments emerged perhaps as early as the 1st century BC. It led to the idea of connecting them to a bag of air, which the player could squeeze. The result was the bagpipes, mentioned in European manuscripts dating back to the 9th-century.

The drive for ever-greater musical range led the 18th-century Italian instrument maker Cristofori Bartolomeo to invent the piano. Originally a harpsichord maker, Bartolomeo sought ways of altering the force with which the instrument's strings could be made to vibrate. This led him to replace the plucking mechanism in the harpsichord with an arrangement of leather-covered hammers, which struck the strings with a force dictated by the player. The resulting sounds could be soft (*piano*) or loud (*forte*), and gave the instrument its name: the pianoforte.

While attracting little interest in his own country, Bartolomeo's design was adapted by German instrument-makers, who turned the pianoforte into one of the most sophisticated of all musical instruments, and one whose vast range of expression immediately attracted the attention of composers such as Mozart.

Another attempt to improve the sound of an existing instrument – this time the bass clarinet – led to the invention of the saxophone. Belgian-born Adolphe Sax, a gifted flute and clarinet player, came up with his eponymous instrument while he was a student at the Brussels Conservatoire. Patenting the device in 1846, Sax found great interest in the instrument among the bands of the French Army, prompting other instrument-makers to attempt to copy his ideas. The result was an endless series of legal disputes which helped push Sax into bankruptcy twice. A group of eminent composers, including Camille Saint-Saëns, lobbied the French government for a small pension for Sax, but it was scant recompense for his creativity, which had cost him so much. He died a bitter and disappointed man in 1894.

A 12th-century ivory oliphant, precursor of the trumpet, imported from the Middle East about a thousand years ago.

Metalwork

Egypt

4500 BC

The Stone Age ended around 8000 BC when Neolithic man first profited from his discovery that some stones did not chip or flake when worked but, due to their softness, distorted and carried an edge.

These stones were nuggets of copper. Hammered into sheets they could be used to make tools, jewellery and arrowheads, or beaten into rock cavities they could be made to form a shape. At about the same time, man discovered a material that retained a superior edge to copper. This was meteorite iron, which Egyptians called "black copper from heaven".

For around 3,000 years, man made do with this crude use of raw metal nuggets. But in about 5000 BC three discoveries heralded the true beginning of the metallic age.

The first was that lumps of virgin copper and iron would melt in a fire and could be cast into shapes by pouring the liquid into moulds.

The second was that pure metal could be extracted from the ore by smelting.

The third discovery was that when copper and tin were melted together a new material was created that had all the advantages of copper with few of its disadvantages. It could be hammered to a sharper edge and poured into moulds and it triggered a new age in man's development: the Bronze Age.

The ease with which bronze could be worked meant its impact was widespread. The short dagger used for stabbing grew into a long implement for slashing. The fish hook was transformed from a thick bone into something thin and barbed. Plates, nails, rivets and tweezers were invented.

Early Bronze-Age cultures travelled in search of ore, spreading the use of bronze from Egypt in 4500 BC and arriving in northern Europe by 1800 BC, by which time its heyday was almost over. Man had by now discovered a way of extracting iron from its ore to make wrought iron. The process required a reducing atmosphere (a continuous blast of hot air at a high temperature) and brute force to beat out most of the ore's impurities.

Iron revolutionized metalwork. The next stage was to add carbon to iron to give it a steel edge. With this harder and less easily corroded material, a myriad of new, tougher tools were made that could clear forests and cultivate previously unworkable land.

Bridges

Africa

We have all made bridges at one time or another: a log across a stream, maybe, or just building blocks spanning the gap between two rugs. And they usually aren't very good. They wobble about, they sag in the middle and they look clumsy. If there is one lesson to take from the history of bridge building, it is that the best, most efficient designs are also the most beautiful.

People seemed to have noticed this some time before 4000 BC. Up to then, if a gap was wider than the height of the local trees, bridging the gap was a problem. Then someone must have noticed how vines span big distances in the jungle canopy, and how strong they are. Woven together, they could make a bridge of any length. Getting to grips with the forces of tension and compression led to a breakthrough by around 200 BC. The Romans had found an improvement on the Greek pillar-and-beam system, which had demanded an incredibly strong beam, supported by a substantial set of pillars in pure compression under the load. Roman engineers realised they could transfer this load into the ground more effectively using an arch. The whole structure was then in compression and it no longer needed an incredibly strong pillar or beam.

With a beam bridge, you always have to worry where the load is, and what it is doing: two high-speed trains passing each other on a simple pillar-and-beam bridge would be asking for trouble. But an arch bridge spreads the load, making it stronger and safer.

Roman arch bridges also made excellent use of a convenient and common material: stone, which is extremely strong in compression. In an arch bridge the forces are all directed downwards into the ground, which also dispenses with the need for binding material.

After the fall of Rome, bridge design slowed dramatically and only picked up again during the High Renaissance. By the mid-18th-century the use of masonry bridges had peaked. Wood became the choice of bridge builders and this was only superseded by iron and steel in the early Industrial Revolution. In 1841 the engineer John Roebling, one of America's great bridge builders, developed a rope woven from iron wire, making possible the modern suspension bridge.

However, one stormy night in December 1879 bridge engineers learned a hard lesson: the great Tay Bridge near Perth collapsed and more than 70 people were killed. The engineers had ignored one elemental force: the wind.

The evolution of the bridge: from compression beam-and-arch bridges to the suspension bridge based on the rope bridge.

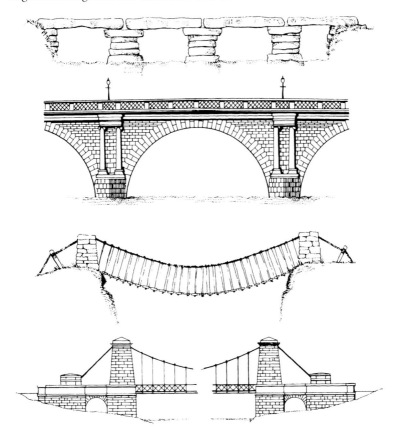

Cosmetics

Egypt

Egyptian women were the first to use rouge and kohl, made of lampblack, in about 4000 BC.

The remains of artefacts used to store eye makeup, scents and unguents indicate that cosmetics were used, at least by the nobility, as long ago as 4000 BC.

By the time of Christ, kohl (made from lampblack), rouge, scented bath oils, and abrasive dentifrices were widely used in the Roman Empire.

They fell out of favour after the fall of Rome until the Middle Ages, when European travellers and traders brought back cosmetics and perfumes from the Middle East.

The Wheel

Sumeria

It is nearly impossible in our post-industrial society to conceive of a world without wheels. From clocks to huge machinery and from cars to computer discs, everything employs cogs, wheels or other types of cylindrical components that spin on an axis. Yet the wheel took a relatively long time to be invented and several civilizations, such as the Incas, Aztecs and native North American Indians, reached a relatively high level of technological sophistication without it.

The most likely explanation is that for many years neither terrain nor climate suited the wheel. Until 10,000 BC, much of the world was in the grip of the last vestiges of the Ice Age. What was not under ice sheet was covered by jungle, desert or bog – conditions obviously unsuited for something like the wheel.

Also, man did not domesticate animals until around 7000 BC, when horses, oxen and zebu cattle were fitted with yokes and made to pull simple devices such as harrows, spikes and tree roots to disturb the soil ready to plant seed. Without an animal to pull it, the cart had little value.

By 3500 BC, when a Sumerian pictograph shows the earliest indication of the existence of the wheel, the world was a

Assyrian King Assurbanipal on his chariot, from 650-9 BC. The wheel brought many advantages including the mobiliza-tion of warfare.

relatively civilized and sophisticated environment, and man had begun to live near open grassland and steppes, or learnt to burn down trees to create land suitable both for agriculture and wheeled vehicles.

By this time, a manufacturing industry and a trade in products such as wines, oils, cosmetics, pottery, tools and hunting implements was growing. The world was ready for the wheel and it is likely to have evolved from the sledge, which is known to have been used by Stone-Age hunters.

Something that looks like a wheel was discovered by archaeologists in what used to be Mesopotamia. It was dated to 50,000 BC but the social and geographical conditions of that time make it unlikely.

Most experts agree that the wheel evolved from the fact that Neolithic man was familiar with moving heavy objects by putting a roller, such as a tree trunk, under the load. Such techniques were used to move the huge stone blocks to build the pyramids from about 2980 BC and probably Stonehenge, which dates from around 2000 BC.

Another technique for moving large, heavy objects was to place them on sledges and to put the sledges on rollers. In time, it is likely that the sledge wore grooves into the rollers and ancient man had a ratio – a small turn of the inner edge of the worn groove generated a large turn of the outer edge of the roller.

The next and final step in the invention of the wheel was to reduce the weight of the roller by cutting away the wood between the grooves, thus creating an axle with a wheel at each end.

The evolution of the wheel and of wheeled vehicles continued to be one of incremental development rather than sudden breakthrough.

Early wheels were sliced from the trunks of trees and then honed into shape. However, this method produced wheels of limited size and strength, so ancient mechanics pegged three wooden boards together and cut them into a round shape. As this produced a very heavy wheel, sections of it were cut out to reduce its weight, and so, in the second millennium BC, wheels with spokes were invented.

At last man could better indulge his passions for travel, speed and movement.

The wheel has been around so long and is so ubiquitous that it can be difficult to see exactly why it is such a brilliant invention. In the end, it boils down to the way a wheel exploits the mathematical properties of a circle.

The definition of a circle is a curve which keeps the same distance from a fixed point. So a wheel with an axle through its centre allows you to keep the same height over the ground as you roll. And, as tangents touch circles at only one point, the rim of a wheel is guaranteed to make minimal contact with the thing most likely to slow it down: the ground.

Constant height, minimal contact – geometrical perfection turned into mechanical perfection.

Then there's the fact that you never come to the end of a circle: you can just go round it forever. That makes it perfect for tasks needing to be done over and over again: precisely the kind of tasks we humans loathe.

Attempts to make the wheel better have focused chiefly on how best to connect it to the axle, and thus the load. At first, wheels were firmly fixed on the axle and the axle turned in the holes of the cart frame. Later the axle was fastened so that the wheels revolved on its ends, in an early form of bearings. This made the cart more stable around corners, as the wheel on the outer part of the curve, which had to travel further, could spin faster and thus keep up with its partner on the inside.

As speeds have increased, so has the sophistication of bearings. The lateral forces acting on wheels going round corners has led to sports cars being fitted with angled wheels and bearings capable of taking the strain.

Transport
Middle East

3500 BC

It was a small step once the wheel had been invented to add wheels to the ends of simple sledges, such as the travois, to create a two-wheeled cart of the sort seen in Sumerian pictograms of 3200 BC. This innovation galvanized production on the land and brought new wealth for those engaged in agriculture. The two-wheeled cart could easily be unhitched from an animal and tipped up to help loading.

Beasts such as the ox and zebu cattle were commonly used, and a simple yoke secured the cart to the animal. From this developed a more sophisticated harness, enabling lighter animals such as mules and horses to pull the cart more efficiently.

The first evidence of carts in Europe is a pottery cup in the shape of a four-wheeled wagon found in a grave of 2900 BC at Sziget-szentmarton in Hungary. Actual wheels from 2500 BC have been found in northern Europe.

Soon, those with military intentions took up the cart. It metamorphosed into the chariot – the first wheeled machine of war. Bronze spoked wheels appeared in around 2000 BC; they made the chariot light enough for horses to pull, and this turned the slow oxen or donkey-driven vehicle into a much more versatile carriage. In combination with the bow, the chariot became a symbol of military might.

Elevated transport: A Bronze-Age clay chariot from Duljija. Serbia. The wheels, spinning free of the axles, were a Celtic invention.

The ultimate chariot may have been the Persian eight-horse machine of the 1st and 2nd-centuries BC. This had blades sticking out of each wheel to slash any nearby infantrymen to shreds. Driven boldly, the chariot was able to outflank foot soldiers with ease and provided a stable platform for bowmen to rain down arrows on the enemy.

By the time the Roman Empire was criss-crossed with an extensive network of military roads, wheeled methods of transportation were common – but one more improvement remained to be made that to our modern eyes would seem obvious. Even in Roman times it was still common for both wheels on an axle to be fixed rigidly, so the whole axle an-two-wheeled assembly had to pivot and roll, often stressing the axle, the wheels, or all three. It was also quite common for a four-wheeled cart to have two axles both rigidly fixed to the cart.

It took the mechanical intelligence of the Celts to realize that disc wheels needed to rotate on the axle, each wheel free to find its own way round an obstacle, and with the front axle able to steer the whole carriage.

Carpentry

Egypt

3500 BC

The invention of metal-edged tools such as the plane, the saw and the adze gave an impetus to woodworking. By 3500 BC building construction, ship design and weapon development changed, and it brought about the flowering of elaborate furniture design in Mesopotamia.

Metal chisels, iron hammers and the handsaw, first made of copper, enabled carpenters in Egypt in around 2680 BC to build the oldest surviving furniture, richly gilded and carved. Mortice-and-tenon joints joined strip wood at right angles and the dovetail was used to make boxes, drawers and chests.

Following page: Joseph the carpenter, altarpiece by Robert Campion (1375–1444).

Silk

China

3200 BC

According to Chinese legend, it was the carelessness of a Chinese queen while drinking a cup of tea that caused the discovery of the ultimate luxury fabric, silk. Lei Zu, wife of Huang Di, legendary founder of the Chinese nation in around 4000 BC, is said to have accidentally dropped a silkworm cocoon into the hot liquid. By the time she removed it, it had dissolved into a mass of long, smooth strands. However, instead of throwing them away, she had the inspired idea of spinning the fibres into a thread.

Silk thread factory in Korea.

3100 BC Stonehenge I is
built.

3000 BC Athens, the
birthplace of Western
civilization, founded.

Phoenician sailors
discover glass after
lighting a fire on a
sandy beach.

While the story may be apocryphal, archaeological evidence confirms that silk was being cultivated in China as long ago as 3200 BC. Chinese farmers had a detailed knowledge of silkworm biology, selectively breeding the best silkworm caterpillars and cultivating the best mulberry leaves, their sole food source.

Keeping the silk-making process a secret, the Chinese traded the luxury fabric with the Indians in exchange for precious stones and metal.

The Indians traded silk with Greek and Jewish merchants, who sold it to the Romans. The demand for silk led to the opening up of the 7,000-mile Silk Road, bringing regular communication between China and the Roman Empire via northern India.

The reluctance of the Chinese to divulge the origin of the fabric paid rich dividends, for the Romans, who believed the Chinese harvested silk from trees, paid outrageous prices for this desirable high fashion product.

Written language

3000 BC ## Sumer

Sumerians developed the first written language, which evolved from their farmers' need to count sheep.

C apturing speech in written form presented problems to the inventors of writing which are now hard to appreciate.

How does one take the stream of information in speech and break it up into meaningful units? What should those units be? Should they be words, or syllables? Should they be individually distinct sounds (in linguistic parlance, phonemes)? What kind of symbols should represent these sounds? And why bother anyway? These were the questions facing the inventors of written language. Working independently in China, Mexico and Mesopotamia, they came up with a variety of solutions, largely in response to the demands of their agricultural economies.

For Western civilization, the crucial developments took place around 3000 BC in the Sumerian culture, in modern Iraq where a writing system, cuneiform, inspired all European systems. Even before the emergence of cuneiform, Sumerian farmers had used simple clay tokens for accounting purposes, such as recording numbers of sheep. Just before 3000 BC, new accounting signs and symbols appeared, along with flat clay tablets, which were scratched with pointed tools.

Conventions evolved; the writing was organized into horizontal rows and read from left to right, and from top to bottom.

The first Sumerian writing signs were logograms, with whole words represented by pictures of the object in question. For example, the word *su*, meaning "hand" in Sumerian, was shown by a picture of a hand. But the Sumerians realized that the hand symbol could also be used to write down any other word which, when spoken, contained the sound "su".

Egyptian hieroglyphics on a coffin at Petosiris, from the 'Book of the Dead', 4th-century BC.

This was the dawn of phonetic script, rather than drawing a picture of an object.

Early writing was a strictly practical business of recording agricultural products and the payment of rations to workers.

In time, the symbols changed so that they could be written more rapidly on papyrus. Eventually, the alphabet, which broke the syllable into its constituent vowel and consonant sounds, was invented, probably by Phoenician speakers of a Semitic language sometime in the 2nd millennium BC. This early alphabet used only consonants, but developed into a full alphabet with vowels when the Semitic script was adapted to the Greek language in about 1000 BC.

Candle

3000 BC

Egypt, Crete

The lamp is one of mankind's earliest inventions, dating back to Africa around 70,000 BC. It soon evolved from a rock bowl holding moss or some other naturally absorbent substance soaked in animal fat, to a shell, or pottery or bronze saucer, with a loose wick made out of vegetable fibre floating in oil.

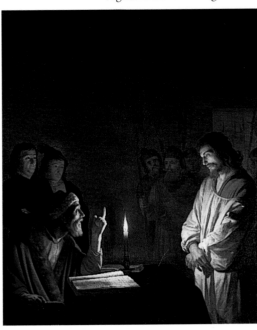

The next stage was to fashion tallow or beeswax into a solid stick with a wick running through its centre, and so the candle as we know it was invented, around 3000 BC. Little changed in candle design until 1825, when Michel-Eugène Chevreul, a French chemist, separated the fatty acid from the glycerin portion of fat to produce stearic acid, which could be used to make superior stearin candles.

Timepieces

3000 BC

Egypt

Instruments to measure the passage of time are almost as old as civilization. The shadow clock, a primitive predecessor of the sundial, was used in Egypt from about 3000 BC. Before that there were water clocks, possibly an invention of the Chaldeans of ancient Babylonia. Some of the oldest existing watches have been found in Egypt and dated to the 14th-century BC.

The water clock was a bowl with a small hole near the base. As the water leaked out, the level of the water, measured against graduated lines on the inside of the bowl, indicated the time elapsed since it was filled. Another design, also used by North American Indians, worked in reverse. An empty vessel with a hole in its base was floated in a bowl of water. Water trickled into the vessel and the passage of elapsed time was measured by how deep the vessel had sunk into the water, until it filled completely and dropped to the bottom.

A modern sundial from Carcassonne, southern France, shows the seasonal adjustment for the sun's movement across the skies.

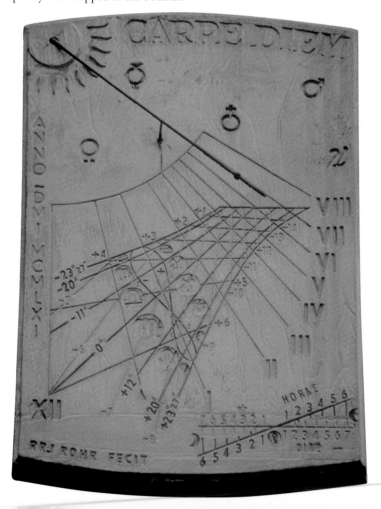

Water clocks seem to have been the most common form of timekeeper up to the end of the Middle Ages and at least one type of water clock was still in use as late as the 19th-century, most often in rural areas.

The Dark Ages saw the introduction of several other types of timekeeper. The invention of the candle clock is traditionally ascribed to Alfred the Great in around AD 878. He used it to regulate the time spent on various tasks. His candles were marked at intervals; the level to which the candle had burnt down indicated the time that had passed. Little is heard of candle clocks after about 1400, but a related device, the oil lamp clock, appeared in the 16th-century and survived into the 19th-century. The level of oil remaining in the glass reservoir indicated the passage of elapsed time.

Documentary evidence for sandglasses goes back to the 14th-century. They were particularly useful for sailors, who used them to measure the length of each watch as no other timekeeper worked properly on a ship at sea. They also had their uses on land, timing the sermon in church, and the sandglass survives today as the familiar egg timer.

A brass log glass, 1817. Filled with mercury and lasting just 28 seconds, it was used for reckoning a ship's speed.

Early sundials divided daylight into a number of intervals, 12 "hours" in the case of the Egyptian instruments, and four "tides" for Saxon sundials. The intervals varied in length according to the time of year and consequent length of passage of the sun across the sky, but these unequal hours sufficed for most purposes. Few people needed to know the precise time, except priests to mark the times of prayer.

By the 14th-century there was a gradual move from irregular intervals of time to the modern, equal hours, system of timekeeping. At the same time the first mechanical clocks were made, and sundials were designed to indicate equal hours. The trick with a sundial was to set the edge of the gnomon, the shadow-casting part, parallel to the Earth's axis.

It is easy to assume that these early timekeepers survived alongside mechanical clocks and watches only because they were simpler and cheaper, but that is not true for sundials. A sundial was the ultimate timekeeper for most people. Except at an astronomical observatory, the only way to set a mechanical clock was to use a sundial. This was the case until the railway telegraphs first provided a means of disseminating time signals in the mid-19th century. Standard time was not available everywhere in Britain until the BBC started broadcasting the "six-pips" time signal in 1924. Only then did the sundial finally become obsolete.

Soap

Babylon

2800 BC

Roman legend has it that soap was discovered by women doing their weekly wash in the River Tiber. According to the story, a mysterious substance deposited on the river's clay banks had a miraculous effect on the cleanliness of their clothes. The source of the substance was traced to Mount Sapo, where animals were sacrificed. A mixture of melted animal fat and wood ashes was washed into the river – and the result was soap.

The legend may explain the origin of the word, but the discovery that animal and vegetable oils mixed up with ash produces a cleansing substance goes back further. In 2800 BC, the Babylonians boiled fat and ashes in clay cylinders to make soap, and the Egyptians also produced a soap-like material.

Soap is created by a chemical reaction that takes place between neutral fat and a strong alkali, originally wood ash but now more often caustic soda or caustic potash. This process produces soap molecules. Their miraculous abilities come from the fact that they have two ends, one of which is hydrophilic – attracted to water – and the other hydrophobic,

2800 BC Rice is cultivated in India.

which is repelled by water but attracted to oil and grease. After the surface becomes wet, it absorbs the soap. The soap molecules then attract oil and grease from the surface and hold it in a suspension until it is rinsed away.

Soap chemistry remained unchanged until 1916, when a shortage of fats led to the invention of detergents.

Common everyday soap cannot be used in sea water because the animal fat, or talon, from which soap is still produced, is sensitive to salt and thus will not lather.

To get around this problem seafarers use what has become known as marine soap, produced using coconut oils instead of animal fats. These natural oils, not being salt sensitive, make it possible to produce a soap which will lather in saltwater.

From the Tiber to the tub.

Swimming pool

Egypt

2500 BC

There is archaeological evidence of recreational swimming pools being built as far back as 2500 BC in Egypt. The Assyrians, Romans and the Etruscans also constructed pools, and learning how to swim was part of the Roman martial tradition and an important part of elementary education.

Gaius Maecenas designed and built the earliest heated pool in the 1st-century BC, using hypocaust flues. These terracotta blocks were part of a central heating system that channelled warm air from heating furnaces.

Etruscan painting from the Tomb of the Diver from the southern cemetery at Paestum, southern Italy, 480–470 BC.

Ink

Egypt, China

2500 BC

The rise of civilization itself owes much to the invention of ink, which has played a crucial role in the spread of ideas, the process that underpins so much of human achievement. The first inks were developed around 2500 BC in ancient Egypt and China, and were made from lampblack, a fine black soot, ground up with glue or gums.

Now, 4,500 years later, the ascendancy of ink as a means of conveying ideas is being seriously challenged, with the advent of electronic documents, e-mail and the World Wide Web.

Parasol

2400 BC Mesopotamia

The umbrella began life as the parasol in Mesopotamia, where it was used only by nobles and held by bearers. The earliest indication of its use can be seen on a monument dated to 2400 BC in Iraq.

It soon spread around the Mediterranean and to the East. Mycenaean pottery and Indian statues of the 1st millennium bc depict nobles, pharaohs or Buddhas being shaded from the sun by a flunkey holding a parasol. The parasol became the umbrella in Rome (umbra is Latin for shade) where its construction came closer to the modern device. They stretched a cloth over a wooden frame, but it was still not the collapsible kind we use today.

During the Wei Dynasty, a water-repellent umbrella made of oiled mulberry paper was invented; red and yellow for the emperor and blue for his subjects. When the Chinese developed stronger silk in the 14th-century, its use for umbrellas was restricted to the royal family.

By 1340, the water-repellent umbrella had migrated to India, where the papal envoy John of Marignolli spotted it. He brought one back to Florence, describing it in a letter to the Pope as a "little tent roof on a cane handle, which they open out at will as protection against sun or rain".

By the 17th-century the umbrella had become a fashionable accessory in France, but only for women. Only when the philanthropist and man-about-town Jonas Hanway carried an umbrella publicly for 30 years from around 1750, did it become popularized for men in London and it was often referred to as a Hanway.

In 1852, Samuel Fox invented the modern umbrella when he stretched oiled canvas over steel ribs in place of cane or whalebone. He did it primarily to use up a surplus stock of stays made to shape women's corsets. The first all-umbrella shop, James & Sons, opened in London in 1830 and is still trading today.

Opposite: In the shadow of the master: the Spanish artist Pablo Picasso wields a parasol for his paramour, Francoise Gilot, in Golfe-Juan, France, 1948.

Irrigation & canals

2400 BC   Sumer, China

B y 5000 BC in Egypt primitive attempts were being made to channel water from the River Nile into the lands around it where man had begun settling in permanent habitations and growing crops. A way had to be found to raise water up beyond the flood plain of the river.

First attempts used a simple lever-and-bucket system, called the shadoof. Another method was the Persian Wheel, a waterwheel that dipped containers into a river, and tipped the water into the upper course as it revolved.

By 2400 BC the Sumerians had begun digging drainage, irrigation works and canals. However, they fell foul of one of the unseen problems of irrigation. Land had to be kept moist below root level and salts had to be flushed out from the topsoil. If this was not done, salinity built up and the harvest

The Pont du Gard in southern France is part of a Roman aqueduct that was 30-mile long and brought water to the town of Nîmes. Made of dressed stone without mortar, it has a gradient of 0.4 per cent.

was ruined. The once-fertile valleys of the Tigris and Euphrates in Mesopotamia began to suffer in about 2100 BC and by 1700 BC Sumer settlements had collapsed after harvests had failed because of this problem.

From 2400 BC the Chinese constructed several schemes, one of which, the 1,000-mile Grand Canal from Hangchow to Beijing – started in around 700 BC and completed around ad 1280 by Kublai Khan – still exists as the oldest artificial water course in the world.

The Romans had a particular problem gleaning enough water to supply Rome – the Tiber was a muddy, stinking ditch. The growing metropolis needed clean water and a route out for its sewage. Appius Claudius planned and built first the Aqua Appia, around 312 BC, then a series of underground and overground aqueducts, vaulting across the city. It was only with the invention of the arch that Roman architects could raise aqueducts to the height needed to span valleys. The Roman version of concrete was vitally important to constructing the new works. Patches of Roman concrete can still be seen all round the countryside. Eventually 11 aqueducts linked together kept taps and fountains running in Rome. Some are still used today.

Cartography

Mesopotamia

The earliest known maps, showing parts of northern Iraq during the reign of Sargon of Akkad in Mesopotamia, date from around 2300 BC.

The first serious attempts to map the whole world emerged in the work of Ptolemy of Alexandria in the 2nd-century AD. Greek astronomers had previously divided the globe into regular divisions of longitude and latitude, but Ptolemy took these and developed much more sophisticated projection techniques.

His eight-volume treatise on cartography, called *The Geography*, with latitudes accurate to within one-twelfth of a degree, remained the standard work in the West until the 17th-century.

2300 BC

Ptolemy of Alexandria refined mapping techniques that had been invented 2,500 years earlier..

The oldest topographical map still intact shows a gold mining area and basanite caves at Wasi Hammamet, Egypt. It was drawn between 1186 and 1070 BC.

2100 BC Stone henge II is built.

Locks

2000 BC

Egypt

The first primitive locks from Egypt were wooden. Each had a sliding bar that latched the door shut, preventing it from moving. Vertically arranged pins fell into slots in the bar. To raise the pins and free the latch bar, a key was inserted through the door. The key had dowels sticking up which pushed up the pins, allowing the bar to slide back and the door to open.

2000 BC The building of Stonehenge III begins, the remains of which can be seen today.

Primitive locks in ancient Greece were nothing more than a hole in the door and a sliding bolt with a slot in it. All that was needed to gain access was an appropriately shaped stick to poke through the door, engage with the slot and lever it along, opening the latch.

The next great leap in security was the invention by the Romans of metal locks. These iron locks were opened by bronze keys – some of them beautifully decorated or cast in the shape of mythical creatures such as dragons or serpents. From around 750 BC, these intricate mechanisms protected the garrisons and gold of the Roman Empire.

As well as using metal, the Romans devised a security measure to make sure only the right shaped key would enter the lock. This was called the ward. Wards could be passed only by a key of the right cross-section: even if a key could by some chance engage with pins correctly, wards would stop the entry of the key into the lock in the first place.

When caravans carried the riches of the East across central Asia it was vital to be able to keep the commodities locked in boxes or bags. Some ingenious Roman invented the padlock, a small portable lock, although it is thought possible that the Chinese evolved a similar invention at about the same time.

It wasn't until the 18th-century that further progress was made in the technological development of locks. Robert Barron invented the tumbler lock, which was hard to pick. It used a key which, when inserted, raised a series of levers. Each lever, or tumbler, had to move to exactly the right height, or else the lock would stay steadfastly bolted.

But the Barron lock was not impregnable, and in 1818 Joseph Chubb of Portsmouth improved on it. He invented a way of using tumblers and pins in combination to protect the latch. If an errant lock picker pushed a tool into the lock and managed to raise a tumbler, a sprung trap would capture the lever when the lock picker raised it too far out of position. The lock would then jam and alert the householder.

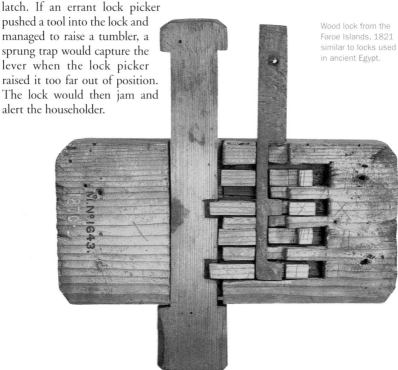

Wood lock from the Faroe Islands, 1821 similar to locks used in ancient Egypt.

These inventions were nothing when compared with the genius of Joseph Bramah, another English locksmith. In 1784 Bramah came up with a tubular key and lock mechanism of such complexity and security that he put it in his shop window as a challenge to lock pickers. He offered a reward of 200 guineas to anyone who could open it.

(It was 67 years before American locksmith A.C. Hobbs found a way to break in through the Bramah lock. At the Great Exhibition in 1851, Hobbs challenged British locksmiths to bring along any kind of lock they could come up with – the swaggering American said he could open them all. And he did. He also brought along his own Impossible lock, but none of the English locksmiths could find a way in. Unfortunately for Hobbs, his perfect lock was so complex and expensive to make that it was only ever used for bank vaults. It never made him rich. However, the company that Bramah founded is still making his locks, in slightly modified form.)

Manufacturing such locks as Bramah's took a long time and required intricate work to fashion. His own answer was to develop machines and jigs to mass-produce the lock to the quality he desired, and these were among the first machine tools ever produced.

The next landmark in the history of the lock is the work of the American Linus Yale. In 1848 he patented a new

Linus Yale junior's lock based on the Egyptian model.

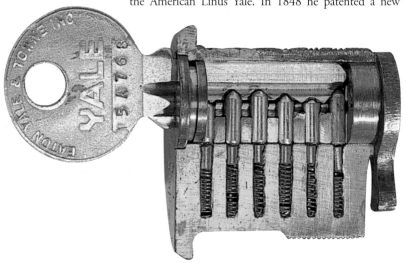

version of the ancient Egyptian lock, adding a pin-tumbler mechanism. His son further developed the idea into the Yale lock still in use today. A small flat key is pushed into the cylindrical lock, with a serrated edge pushing up five sprung pins. If all five pins are pushed to the right position, the barrel is freed and can be twisted to lift a latch on the inside of the door.

Locks demonstrate that evolution doesn't just apply to living things. From the earliest Egyptian security devices of 4,000 years ago, locks have been in a state of constant evolution, because no matter how well something is protected, someone will try to find a way to get at it.

This race for survival of the fittest was never faster than in the first half of the 19th-century, when the forerunners of today's pin-tumbler locks were in their infancy.

The race continues to this day, but in a different guise. Where once it was booty, or slaves, or a woman's chastity that was under lock and key, now it is information that people are keen to protect.

From encryption software like PGP to biometrics, which includes techniques such as voice recognition and hand-scanning, the race is to keep ahead of the new predators – hackers.

Using their ingenuity, hackers continue to keep multina-tionals on alert. Some hackers are then hired by companies to test their security.

Joseph Bramah's almost unpickable lock.

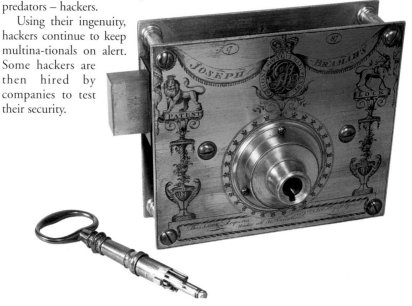

Sanitation

Crete

Polite breakfast room conversation may not often cover the question of efficient means of sanitation. To the Minoans of Crete, however, it was a subject of fascination. By around 2000 BC they had evolved the first known form of flushing toilet, and they went on to use a system of interlocking clay pipes, sewers and manhole covers to inspect the sewage flow.

The Minoan flushing toilet was simple; a stone cistern was gravity fed with water from a stream and the water released into the toilet by pulling a lever.

However, attempts to clean up human effluent began well before this. Archaeologists on the Orkneys, Scotland, found evidence of attempts to clear foul water that date to 4800 BC. Waste was simply directed into a passing stream.

The supremacy of the Roman Empire was founded on brick-built sewers and terracotta drain pipes. Gravity still did the work in their conveniences and it would be another 1,500 years before Sir John Harington rediscovered how to flush a toilet with a gush of water.

Paved roads

Crete

The earliest roads were little more than pathways – well-trodden trade routes like The Ridgeway in southern England, frequented by foot travellers and mounted traders and merchants.

The Ridgeway passes near Stonehenge and the ancient sites around it, and it is thought to have been used at least 5,000 years ago. But it is not the earliest track. Tracks have been found around Jericho that could date to around 6000 BC.

As cultures advanced, the volume of trade on the roads increased. Burgeoning populations needed to travel to and from market to trade in livestock, food, clothing and trinkets.

With increased trade between nations, networks of paths, such as the Amber Routes which crossed central Europe, grew into semi-surfaced roads. At well frequented crossings or boggy ground, timber was used to construct a raised surface.

Stone paved streets are known to have existed in Ur (in modern Iraq) in 4000 BC. A timber surfaced track of a similar age excavated at Glastonbury in Somerset was little more than a raised narrow path over a bog.

The earliest surviving proper road was built by the Minoans of Crete. It was a 30-mile limestone paved carriageway running from the city of Gortyna to Knossos. It was built around 2000 BC and has gutters and a raised crown in the middle.

Some of the grandest routes of the ancient world were built for ceremonial use – in Babylon in 615 BC the Chaldeans built brick processional roads linking the royal palaces to temples.

In Egypt, it was the River Nile that was the main trade route. But by around 2000 BC the enormous blocks of stone used to build the pyramids and temples were being transported along purpose-built tracks.

It was the Romans who truly understood the importance of good roads. To them, roads were a tool of military strategy that ensured the flow of trade and guaranteed lines of communication. Their pioneering use of concrete enabled them to build a 53,000-mile network of properly founded, well surveyed roads. Almost 30 main military trunk routes, or Viae Militares, allowed legions to subdue uprisings quickly.

Roman road builders started a route by digging two parallel trenches, about 12 metres apart.

A raised, well-drained road surface was made by using spoil from the trenches together with stone aggregate. The foundations of a Roman highway were around a metre thick, which is why some are in use today.

1850 BC First contraceptives in use in Egypt: suppositories of honey and camel dung.

Corsetry

1800 BC **Crete**

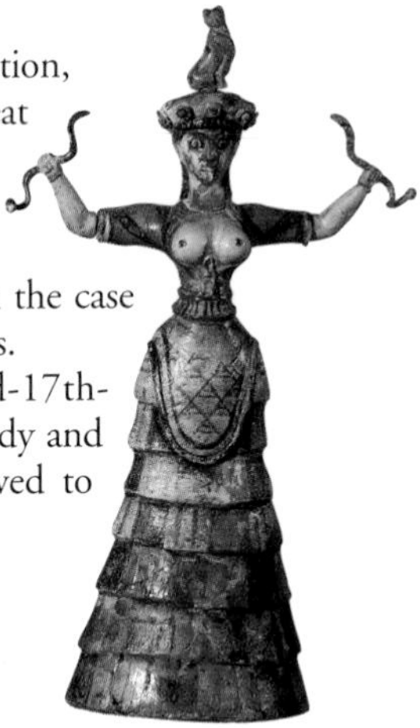

The well-corseted
Minoan Snake
Priestess from
Knossos in Crete, 7th-
century BC.

In the annals of fashion invention, the corset is a garment of great antiquity. We know from statues that Minoan women and, more strangely, men wore a corset to clinch the waist and, in the case of the women, to raise the breasts.

From the mid-16th to mid-17th-century it flattened the upper body and from 1660 the design was evolved to emphasize the bust.

The oar

1500 BC **Phoenicia**

Moving through water is hard work, making it a perfect application for that most simple of means for making hard work easier: the lever. That is what an oar is: a lever with the mass of water forming the load, the fulcrum on or near the edge of the boat, and with you providing the effort – boosted 300 per cent or more by the lever effect.

The paddle originated in ancient Egypt on the River Nile, and later on the sea, sometime around 3000 BC. By 1500 BC, the Phoenicians had added a fulcrum and extended the length of the shaft to create the oar, which enabled people to propel themselves across large expanses of water, and at great force over small ones – features that military commanders did not hesitate to exploit. The oar thus underpinned the construction of the first warships and led to nations extending their military power to the seas.

An umla oar from an
Arabian fishing boat,
the same shape as
modern cleaver racing
blades.

The struggle for naval ascendancy led to an arms race during the 3rd-century BC. Macedonians built ramming ships with crews of 1,800 arranged in banks of 18 on each set of oars, while the Egyptians had ships with banks of 20 or even 30 per set. Later Roman warships were bigger still, and carried mechanical artillery such as catapults.

Long after the fall of the Roman Empire, the oar was still at the centre of military technology, this time with the 9th-century Vikings, whose oared long-boats played a crucial role in their invasions.

Today we usually associate oars with more congenial pursuits of sport and leisure. From the 13th-century, when the watermen of the River Thames first began taking passengers around London, races started to take place. By the 18th-century there was a full programme of regattas.

These races prompted some major technical innovations, such as the sliding seat, invented in 1857, which allowed legs to add to the force. Today's racing oars are sophisticated, and built from carbon fibre to reduce the oar's weight while maintaining its stiffness and thus its ability to transfer as much of the oarsman's effort to the water as possible. In the early 1990s, so-called cleaver blades also started to become popular, their rectangular paddle boosting propulsion by allowing the oar to shift a bigger mass of water.

1500 BC Egyptians use water clocks to keep track of time.

Shoes

Egypt

When Otzi the Ice Man set out around 5,200 years ago to cross what is now the Austro-Italian border, he was wearing purpose-made shoes of waterproof deer hide and bear-skin soles, with soft grass inside. Even in the Stone Age, people knew the importance of comfortable footwear. The discovery of such sophisticated footwear stunned archaeologists who discovered Otzi's remains in 1991.

By around 1500 BC, the Egyptians knew it, too: they had well-made, good-fitting sandals, often highly decorated. This

1500 BC

Ancient Egyptian
sandals shaped for
each foot.

was the first footwear to be shaped to fit left or right feet, and is regarded as the first real shoe.

Footwear has long been expected to say something about who is wearing it. Which leads us to that long-standing icon – the trainer. In 1971 the American athlete Bill Bowerman poured molten rubber into a waffle iron and produced the grid-like sole still seen on trainers today.

In 1979 Bowerman helped set up Nike, which launched the innovative air-cushioning system and began the trainer's evolution from a simple sports shoe to a piece of high technology footwear.

Tap and hinge

1500 BC Egypt

The earliest known taps are Egyptian, devised in the 2nd millennium BC. A resource as precious as water needed to be carefully husbanded in the dusty lands of the pyramids. It was the unifying dynasty of Menes in 3000 BC that channelled the waters of the Nile into the surrounding countryside, and from this moment technologies were developed to store, transport and deliver water into the homes and palaces of the Delta.

In Egyptian plumbing, copper and lead were used widely and cast bronze and brass taps were relatively common. The Romans made extensive use of taps. In the most lavish Roman palaces and baths, taps fashioned from solid silver or gold were not unknown.

The hinge is believed also to have been invented in Egypt, where exceptionally dextrous carpenters adapted ball-and-socket joints used to hang doors. Hinges were initially used instead of sliding lids on boxes and for centuries they remained a luxury. Made of copper or gold, they appeared on items such as caskets found in the tomb of Tutankhamun.

Swords and armour

Egypt

The first swords forged from iron were made by 1200 BC. They had been developed to penetrate copper or bronze breastplates as if they were made from leather and were therefore a major step in the development of arms and armour.

Like all military hardware, armour had evolved as the result of an arms race; in this case one that began at around 3000 BC in Egypt. Up to that point, armour was little more than stout leather that provided a protective wrap around the torso and upper legs but did little to shield the warrior against the might of a stone axe.

The next inevitable development was the invention of metal armour. Mesopotamian armourers began to fashion helmets and breastplates from copper and bronze in about 3000 BC. Suddenly stone weapons, such as the mace, spear, sling and arrow, became useless.

Man's immediate reaction was to begin developing armour-piercing weapons to take their place. Heavy copper maces were invented to penetrate the bronze plates sewn on to leather backing like overlapping fish scales. These weapons were small and easier to mount on a haft or handle. With more concentrated mass than the stone versions they ripped through the soft bronze breastplates with ease. The mace evolved into a fearsome sharp-edged axe, which in turn led to the iron sword.

Like most early inventions, armour was invented several times in different parts of the world. We know that by 1000 BC Chinese warriors wore armour made of five to seven layers of rhinoceros hide. And it is possible that the Scythians, a

1200 BC

1200 BC Decline of Mycenaean civilization on Crete; beginning of Iron Age in Middle East and south-east Europe, spreading to China by 600 BC.

Helmet from Italy, 6th-century BC.

1,000 BC Chinese calligraphers are using ink brushes.

753 BC Rome founded.

warlike and fierce nomadic race who conquered much of the Balkans, had bronze armour as early as 5000 BC, but there are few traces of their advances.

In the foremost arms race, in the eastern and central Mediterranean, the spread of iron swords changed armour everywhere. Greek helmets evolved from the early, expensive one-piece hammered bronze works of art into cheaper, tougher, iron. Copper and bronze armour was superseded by iron plates sewn on a leather backing garment.

The last great advance took place in around 400 BC. At this time chain mail appears in Greek sculpture and friezes, although it may be of Celtic origin. The Romans adopted it widely and by 400 BC *lorica hamata* was used by many legions.

The problem with chain mail was that it did nothing to absorb the energy of a sharp blow, and so segmented iron breastplates, which could do so were introduced. By 100 BC *lorica segmentata*, the first practical plate armour, was clothing the soldiers of Rome.

After the collapse of the Roman Empire in the 5th-century there were no major advances in the design of armour until mediaeval times when other weapons began presenting problems to the armourer: how to keep out the crossbow bolt, the heavy shaft fired by the longbow, and the first lead shot.

False teeth

700 BC

700 BC Sundial invented in Egypt.

Etruscan

A round 700 BC a weary Mediterranean traveller with a tooth abscess would have pointed his horse, mule or camel in the direction of the Etruscan Empire in Italy. The Etruscans were Europe's most accomplished dentists and they were the inventors of false teeth.

Partial dental plates were made and gold bridgework was carried out. False teeth were carved from bone or ivory, or were recycled from other human mouths.

It is not known if permission was sought from the patients

or whether this was a form of punishment, but what is known is that the new dentures continued to decay.

Roman false teeth in gold setting.

Coinage

Lydia

640 BC

For many people, money is one of the most fundamental of all man's inventions. It is among the oldest of our social tools and at global, national and local levels, synonymous with power and the shaping of history in every generation.

"Every branch of knowledge has its fundamental discovery. In mechanics it is the wheel, in science fire, in politics the vote. Similarly, in economics, in the whole commercial side of man's social existence, money is the essential invention on which all the rest is based" – so wrote Geoffrey Crowther, former editor of *The Economist*, on the invention of money.

The first true coins date from around 640 BC but long before that, forms of preferred barter items became customized into primitive forms of money, some of which, such as cowries, remained popular right up to the second half of the 20th-century.

Alexander the Great's conquests gave rise to the first – and most successful – single currency.

Where the cowrie originated nobody knows, but it is probably the longest-lasting and most widely used of all so-called primitive moneys. It forms the Chinese character for money. And after Ghana gained its independence it chose, in 1960, to base its currency on the *cedi*, the Ashanti word for cowrie.

Other examples of early instruments of barter include amber, beads, cows, drums, eggs and feathers such as the quetzal used by the Aztecs and adopted as the modern currency unit of Guatemala. The ability to "invent" or add new forms to old uses increased their acceptance as money, for in essence money is anything that is widely used in making payments and accounting for debts and credits.

It would seem logical that banking, bills of exchange and the like developed long after the emergence of coinage. In fact quite sophisticated forms of banking were invented and developed in ancient Babylon a thousand years or

A Celtic coin probably used by the Venti of Asmorica, France, in the 1st-century BC. It shows a person guiding a horse with a human head which is leaping over a boar, an emblem of strength to the Celts.

so before coins. Writing, accounting and banking were born and grew up together in Mesopotamia. The earliest known texts are lists of livestock and agricultural equipment which come from the city of Uruk around 3100 BC and writing was invented primarily as a method of book keeping.

The royal palaces and temples provided security for deposits of grain, tools, fruit, cattle and, eventually, precious metals. Receipts testifying to such deposits gradually led to third-party dealing via written orders including bills of exchange. In this way the loan business originated and reached a high stage of development in Babylonian civilization. A document originating from 1792 BC, the first year of Hammurabi's reign, and preserved in the British Museum, authorizes its bearer to receive in 15 days eight and a half *minae* of lead.

The Money Changer and his Wife by Quentin Massys, 1574. At this time milled coins had just been introduced to prevent illegal "clipping".

Thousands of cuneiform blocks confirm that banking operations were everyday affairs in ancient Mesopotamia. Nevertheless the full flowering of money awaited the series of inventions in coinage which occurred most successfully in Lydia and Greece from around 700 to 640 BC.

The rivers of south-west Turkey, typified by the River Maiandros, silt up as they meander over the plain. It was from panning in these rivers that the Lydians and Ionians derived their amber-coloured electrum, a natural amalgam of gold and silver (and the philological father of e-money; the word electron comes from amber's characteristic of generating a charge when it is rubbed) from which by the 7th-century BC they produced their bean-shaped coins known as "dumps".

The improvement in the technique of coinage progressed from unstamped dumps pre-700 BC and dumps stamped on one side only, through to proper double-struck coins carrying the lion's head of the royal house of Lydia, clearly circulating by 640 BC.

The extraordinary characteristic of Greek coinage is the speed with which it developed so that the 5th-century saw the minting of the most beautiful coins ever made. At Sotheby's in July 1995, 200 such coins, including electrum and gold and silver examples, fetched £2,099,295.

The sword, as well as trade, rapidly extended this early coinage empire. The adventures of Alexander the Great played a key role in the monetization and diffusion of the vast, relatively dormant and centralized gold and silver reserves of the Persian empire. Alexander's insistence on a crystal-clear and simple 10:1 ratio between silver and gold greatly improved, simplified and speeded up payments.

Later Roman developments did not improve in quality terms on the Greek precedent, yet Rome played a key role in the development of money. For long periods in Rome, the primary function of the coins was to record the messages the emperor and his advisers desired to commend to the population. The Roman Empire established the world's longest-lasting and most successful single currency, and invented hyperinflation. From about AD 410, while the richer citizens could trade in stable gold *solidi* and the poorest were "impecunious", the majority were reduced to dealing in trashy *pecunia*, subjected to inflation which at its height made the debased *denarius* exchange at 30 million to one gold *solidus*, hastening the fall of the empire. Hyperinflation is not purely a modern phenomenon, but paper-based money systems make it a more prevalent and persistent evil. The development of modern money would have been impossible without the movable-type printing press invented by Gutenberg in about 1440 and further developed by him as a result of a loan of 1,600 guilders by his hard-headed and hard-hearted local banker, Johann Fust.

Europe took longer to divert these presses for banknote inflation than had China with her earlier block printing. But the printing press proved surprisingly adaptable and it was modified to produce coins. This lead to significant improvements in minting, a process which interested Leonardo da Vinci, who had worked for the papal mint.

The new minting processes, which gave a milled edge to the coins, led to a reduction in the illegal clipping of coins whereby bits of their valuable metal were nicked off the edges to be melted down. The technique spread from Augsburg via Paris to London by 1553 though it was not until the 1630s that the first substantial amounts of milled moneys were produced in London. By that time London goldsmiths were about to emerge as bankers.

With the founding of the Bank of England in 1694 and the early prominence it gave to note issue, paper money began to show its irreversible pattern of exceeding hard cash. For the next 220 years this inverted pyramid of paper money seemed to depend on a gold-based circulation of full-bodied coins backed by the meagre gold reserves of the Bank of England, a situation brutally ended in 1914. In the following decades most of the world lost its gold anchors and began to indulge in a prolonged orgy of paper (and later, plastic) inflation.

As well as a need for gold coins, the Industrial Revolution had created an unprecedented demand for low-denomination coins for the mass payment of wages. This was fittingly met by the application of steam power to the minting press. The concept of using James Watt's steam engine to mass produce coins has been seen as one of the most elegant ideas in the entire history of technology. Watt's gospel was spread worldwide, with new mints in Denmark, Russia, India, Brazil and five in Mexico.

But by the 20th-century, coins had become the small change of modern money supply. In the UK today they are only 1/300th part of broad money. The great bulk of the world's money now consists almost entirely of computerized accounts.

From the dawn of recorded history, the bankers, makers, minters and printers of money have been associated with the cutting edge of technology, yet we have still not achieved the goal of an instant global currency. We seemed near that goal when sterling ruled the international cables of the Victorian age. The euro, not quite on a parity with the US dollar, provides another stage for the emergence of universal money.

Lighthouse

The Mediterranean

600 BC

For the traders, warriors and fishermen of the ancient world, navigating the sea was fraught with danger. A light on the distant shore was always a welcome sight for exhausted sailors. In 1200 BC, in the *Iliad*, Homer made the earliest known reference to on-shore bonfires to guide the ships home. In the 7th-century BC, crude tower structures crowned by beacon fires were built all around the Mediterranean.

The most extraordinary lighthouse in history, however, must surely have been the Pharos of Alexandria, built in 285 BC. Considered by the ancients to be one of the seven wonders of the world, it is reputed to have reached 134 metres into the Egyptian sky. Said to have been built from blocks of marble bonded with seams of lead, it may have had a square base,

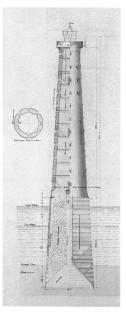

A 19th-century lighthouse design. The first lighthouses around Britain's shores were built by the Romans.

surmounted by an eight-sided tower, topped by a round column and crowned by a beacon fire that burned 24 hours a day.

By AD 400 the coastlines around the Roman Empire – in the Mediterranean, the North Sea, the Black Sea and the English Channel – were marked by a whole network of 30 lighthouses.

The invention of the multi-faceted glass Fresnel lens in 1822 by Frenchman Augustin Fresnel focused light further out to sea, leading to the remotely-monitored beacons that guard our shores today.

Anchor

592 BC **Greece**

Greek geographer Strabo (64 BC to AD 25) described the invention around 600 BC of the first iron anchor by Anarcharsis the Greek. It had curved

arms that fastened to the seabed. Greek coins minted just after the period, now in the British Museum, show the likeness of the Anarcharsis creation. It was much more reliable than the stones or blocks that had been previously used to secure vessels.

Catapult

400 BC **Greece**

Bottling up energy until it is needed is still a major challenge in engineering. Over the years people have come up with all kinds of answers, from batteries and flywheels to pumping water up hills; none of them is ideal.

But around 400 BC Greek engineers exploited one of the simplest and best means of storing energy: elasticity. Using the elasticity of rubber, even a child can make a catapult with a pent-up energy that can be unleashed to send decent-sized stones zooming at 100 mph towards targets 50 metres away. If the amount by which the elastic is stretched is increased, the range of the catapult increases dramatically – as the Greeks realized when they built their first ballista. Essentially a scaled-up bow and arrow, the ballista used an energy storage system in the form of a rope tightly twisted and stretched back using a crank. The elastic energy pent up in the rope was then ready for release whenever needed, and would send a spear hurtling towards the enemy.

A French policeman uses an ancient weapon during protests in Paris in 1970.

Simply dropping a rock on one end of a lever to launch a stone at the other makes a pretty good catapult, too. The energy stored this time is the potential energy of the dropped rock. This principle was exploited in the trebuchet, which was used extensively in the Crusades of the 12th-century. Using a huge 10-tonne counterweight as the energy store, the trebuchet could hurl 150 kg rocks around 350 metres, with devastating effect.

The catapult had a new lease of life in the 20th-century when its elastic energy principles found use in hurling fighter aircraft from sea-borne carriers – and for trapping them when they land. Which just shows that a great idea never becomes obsolete.

Mirror

Lebanon

400 BC

I n the 1960s, British archaeologists who excavated Çatalhüyük, a Stone-Age settlement in central Turkey, were astonished to uncover perfectly flat, unscratched mirrors.

These mirrors were fashioned from obsidian, a volcanic glass that is extremely hard to work. Apart from naturally occurring reflective surfaces, they are the earliest known mirrors.

Throughout the Old World, mirrors were made from highly polished stones and metal discs or pieces of reflective crystal. As well as indulging the human fascination in its self-likeness, bronze mirrors were used to cast light into dark corners.

The Chinese used bronze mirrors to invent the periscope during the Han Dynasty.

A wall mirror made of black or semi-opaque glass was found attached to a wall at Pompeii but, according to Pliny, the first real looking-glasses were invented in the Lebanese city of Sidon. These were made in 400 BC by applying gold, silver or copper leaf to a sheet of glass.

Ice cream

400 BC Persia

Many myths surround the invention of ice cream but the commonly held view is that it evolved from cooled wines and flavoured ices that were popular with Alexander the Great and later with Roman nobility.

285 BC The Pharos lighthouse is built at Alexandria.

The emperor Nero, according to Roman records, sent slaves to the Apennine mountains in 62 BC to collect snow to be flavoured with honey, nuts and fruit pulp.

Hydraulics; Water clock

c.270 BC Ctesibus

Greek engineers in the 3rd-century BC spent a lot of time worrying about ways of raising and delivering water. This precious commodity enabled cities and cultures to grow in hot climates at a time of expansion for the Greek Empire following the death of Alexander the Great. Culture, commerce and thinking flourished before Greece succumbed to the point of the Roman spear, shield and siege machine. Devices such as the piston, the Archimedes screw, gears and the endless chain were all developed by an ingenious and innovative engineering community to quench the thirst of the new populations.

A noted physicist as well as an engineer, Ctesibius was the son of a barber. Nothing remains of any of his many inventions, but anecdotal evidence, notably from the Roman architect Vitruvius, explained the principles behind many of his ideas. The engineering principles established by Ctesibius were the inspiration for generations of Greek and Roman physicists and engineers, such as Hero of Alexandria and Philo of Byzantium.

Like all great engineers, Ctesibius's inspiration came from his surroundings. While working on a way to raise and lower a mirror in his father's barber shop by counterbalancing it with a lead weight, he stumbled on a method of automatically closing the shop's door without its slamming shut. He ran a weighted line from the door over a pulley and into a pipe, which slowed the speed at which the weight dropped. As the door hissed away, opening and closing, he realized the weight

From using mechanical power to move water we have progressed to utilizing water power to generate electricity. Hero's force pump based on Ctesibius's designs.

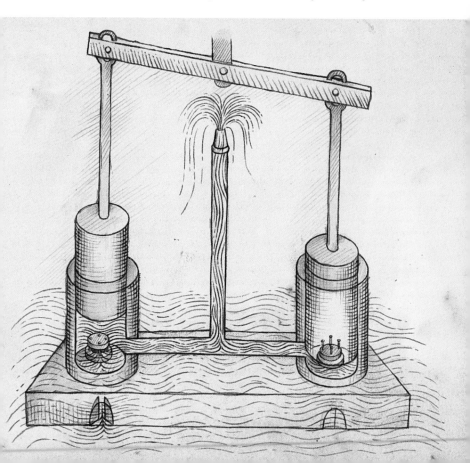

Opposite: 2,200 years on, the 230-ton centre-piece of a hydro-electric power station.

was acting as a piston, displacing air. This realization led Ctesibius to investigate methods of moving fluids along a pipe using a piston and to the founding principle of hydraulics.

In his force pump, stream water passed through a one-way valve and entered a cylinder. From there a piston, driven by human effort, forced the water through another one-way valve into the final delivery pipe. A second piston and cylinder were quickly added to the machine to work together with the existing assembly as a compound pump and the basic design proved to be sound.

Although it was not suitable for pumping very high volumes of water, the force pump played a vital part in many aspects of the Hellenistic culture. Among other uses, force pumps drained the bilges of the trading ships of the time, they were used to extinguish fires and they brought to life the fountains that graced Alexandria.

From the idea of the force pump, Ctesibius developed the Hydraulis, the first mechanical wind organ. As bagpipe players have always known, the problem with wind instruments is ensuring the supply of a consistent amount of air and plenty of puff at the right time. Ctesibius's machine was intended to provide a constant gust of air to a large bank of pan pipes situated on top of a wind chest.

The chest was connected to a conical air reservoir which sat in a tall bath of water. Air was forced into the cone from bellow like pumps. When the pressure dropped, as it does with rhythmically pumped bellows, atmospheric pressure on the water supplemented the human effort and a steady note was played on the pipes. Although it was invented in the 3rd century BC, the Hydraulis was still in frequent use in the 9th century AD.

Ctesibius pre-empted the obsession of 21st-century engineers to build reliable robots that can assist with everyday life but his interest was slightly more down-to-earth. He just wanted to build a maintaining machine that could power a timepiece on its own, with little in the way of human attention. Today these objects are called self-maintaining machines. Ctesibius dubbed his creation the Clepsydra, or water clock, and it worked ingeniously. Surrounded by a tower of supporting timber, a reservoir of water was fed by a stream. If the reservoir got too full, it just overflowed, so the pressure never became too great. The reservoir fed a lower vessel, which contained a float

that rose as the vessel filled, pushing up a lever that moved a geared hand around a clock face. At the end of the day, the float rose high enough to release a stopclock which released all the water and the cycle restarted. As daylight hours varied with the seasons, the length of time the clock functioned had to be adjusted. This was done by means of a tap that controlled the amount of water released from the reservoir into the lower vessel. The outflow was used to power all sorts of automata, such as ringing bells and moving figures.

As well as metering out water with machines, Ctesibius devised ways of heating water and using the expansive power of steam. His Aelopile, while perhaps just a decorative plaything, was the first steam-powered rotary object. It had a copper globe suspended on bearings at its top and bottom, and nozzles around its circumference.

The globe was heated from below. As the water boiled within, jets of steam were expelled from the tiny apertures, driving the Aelopile round on its bearings. It was a simple reaction engine. Escaping steam was action, movement of the globe was reaction. Modern reconstructions of the Aelopile have found that it was able to rotate at up to 1,500 rpm.

Ctesibius's pupil Hero refined his master's theories about steam. He devised an automatic door-opening and closing apparatus which was based on a primitive steam boiler. This he showed to most dramatic effect with a temple and altar tableaux. As steam gathered within the fire-heated altar, so the temple doors parted.

Buoyancy; the lever; Archimedes' screw

Archimedes

c.250 BC

The myths surrounding the greatest scientist and mathematician of ancient times and the first warrior-scientist are so good – and often so entertaining – that few choose to question them. Irrespective of the finer detail of the folklore, this famously absent-minded man (287–213 BC) discovered important principles of physics and, in the process, made a number of vital inventions.

Undoubtedly, the most celebrated story surrounding Archimedes concerns his discovery of the principle of

Opposite : Building the Tower of Babel. The use of cranes and pulleys was explained by Archimedes, who single handedly pulled a fully laden ship on to dry land.

buoyancy. Archimedes had recently returned to Syracuse from studying in Alexandria under Conon, a Greek mathematician, when Hieron II, king of Syracuse, asked him to determine whether or not his crown was made entirely of gold.

Archimedes was at a loss. The king had given him strict instructions not to damage the crown during his investigation, but Archimedes could not work out how to determine the volume of the crown without taking it apart. Then he stumbled upon the answer, as first described by Vitruvius, the Roman architect.

"While Archimedes was turning the problem over, he chanced to come to the place of bathing, and there, as he was sitting down in the tub, he noticed that the amount of water which flowed over its edge was equal to the amount by which

240 BC Halley's comet first recorded by Chinese astronomers.

Archimedes takes one of the most famous baths in history, from a woodcut of 1547.

his body was immersed. This showed him a means of solving the problem, and he did not delay, but in his joy leapt out of the tub and, rushing naked towards his home, he cried out with a loud voice that he had found what he sought. For as he ran he repeatedly shouted in Greek, 'Eureka! Eureka!'."

Since that day, the cry of *Eureka* ("I have found it") has become the clarion call for any discovery, but the story has an unhappy ending. When Archimedes measured the volume of water that the crown had displaced he found it was greater than the volume of water displaced by a lump of gold of the same weight. This told him that the goldsmith had substituted some of the gold in the crown with a less dense material, probably silver. The goldsmith was executed.

Archimedes's second discovery, which played a part in countless inventions, was the principle of the lever. It led him to make his second famous remark: "Give me a place to stand on and I can move the world." The remark referred to the ability of the lever to use distance from the fulcrum to magnify the weight that can be moved. Strato, a Greek physicist, had suggested the basic law of the lever, but it took Archimedes to apply to it the simple mathematics of inverse proportionality. It is the principle used when a child balances on a see-saw with a heavier child.

Provided Archimedes could find a lever long and rigid enough and a firm place to stand, he could, in theory move the world. It was a claim that Hieron II found hard to believe, so he demanded that Archimedes prove it by moving something impressively large. Archimedes built a crane with a series of levers and pulleys and sat himself comfortably in a chair to demonstrate his principle. Plutarch, the Greek biographer, described what happened next.

"He fixed accordingly upon a ship of burden out of the king's arsenal, which could not be drawn out of the dock without great labour and many men; and, loading her with many passengers and a full freight, sitting himself the while far off, with no great endeavour, but only holding the head of the pulley in his hand and drawing the cords by degrees, he drew the ship in a straight line, as smoothly and evenly as if she had been in the sea."

Archimedes, the greatest inventor of ancient times, demonstrated how one man could move the world.

Archimedes had such a low regard for his practical inventions that he left no written record of them. Instead, he left nine treatises in ancient Greek on theoretical mathematics. He regarded his work on the volume of a sphere enclosed by a cylinder as his best and requested that his tomb should be inscribed with a representation of the two and their mathematical ratio.

The invention of a screw to lift water is often attributed to Archimedes and the device is named after him. However, the device was without doubt in use in Egypt and a more accurate explanation is that Archimedes was the first to try to describe in mathematical terms the way it worked.

Archimedes's screw, from Vitruvius's 'De Architectura', 1511. As it turns, the thread of the screw raises water continually along its length.

Musical notation

Greece

250 BC

The earliest evidence of written music comes from the 1893 excavation of Delphi. Two stone slabs were found inscribed with hymns celebrating Apollo, the god of sun. Between the lines were markings and characters which were initially a mystery. Fortunately, in the 4th-century AD Alypius, an Alexandrian, had written of the Greek system of musical notation, listing and explaining its symbols. Using Alypius's notes, scholars were able to determine that the slabs were in fact the earliest evidence of fragments of a musical score.

Manuscript copiers kept Alypius's notes alive and the Western system of musical notation, established in the 17th-century, was largely inspired by the Greek system.

There are references to an Indian musical notation system in use in 700 BC but no original written examples have survived.

214 BC The Great Wall of China is completed.

Ravi Shankar plays the sitar: musical notation may have existed in India before it did in the West, but no written evidence remains.

Horseshoe

200 BC   Rome

The precise date when the horseshoe was introduced and who invented it is not entirely clear, but it obviously played a critical role in transforming the horse from an ancillary beast of burden, useful only for light duties, into a highly versatile source of energy in peace and war.

Horseshoes seem likely to have been in use by the 1st-century BC. Writing at this time, Catullus, generally regarded as the finest lyrical poet of ancient Rome, described a mule's loss of its shoe. But the invention of the horseshoe could date back even further.

Although the Bible has the first known reference to a blacksmith, Tubal-Cain, he is not believed to have shoed horses. In 1865, J.P. Mgnin published *De l'origine de la ferrure du cheval*, in which he claimed that the Druids invented the horseshoe around 500 BC. He wrote: "The Druids learned the structure of the horse's foot by the numerous sacrifices they made of this animal. Accustomed to the manipulation of metals, and their intelligence continually cultivated by study, they were marvellously disposed to be the inventors of shoeing by nails."

The horseshoe shape has remained a similar shape for thousands of years. These are all from the 18th-century.

Further evidence comes from examples of horseshoes found by archaeologists, including some from Silbury Hill in Wiltshire said to date from 200 BC, but the dating methods used are vague.

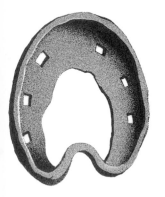  

The first written record of iron horseshoes appears in documents belonging to Leo VI, the Byzantine emperor who died in AD 912. Records of equipment to be carried by his cavalry include "crescent figured irons and their nails". But it is thought that the iron horseshoe was in use by AD 500, although there is no conclusive item of evidence.

The Koran, dated to AD 610, refers to "horses... which strike fire by dashing their hoofs against the stones..." and it is believed that the horseshoe was soon after introduced to Europe by invaders from the East.

Oscar Brown, an American, patented the first double or compound horseshoe in 1892. An upper shoe was secured to the hoof and a lower shoe attached to the upper shoe, which could be easily replaced.

Book

Greece

180 BC

If the definition of a book is a written instrument of communication, then clay tablets, papyrus and even microfilm would qualify. But if it has also to be portable, easily accessed at any point, and of considerable and flexible length, then the first bound volume, known as a codex, produced in the 2nd-century BC, is the father of the modern-day book.

Up to this point, papyrus rolls had been the most flexible form of "book". Some from 2500 BC have been found buried in the desert sands of Egypt, preserved by the extremely dry climate. The rolls were up to 35 metres long and included stories, myths, scientific writings and, most famously, the Book of the Dead, a collection believed to help the deceased in the afterlife.

By the 5th-century BC, significant papyrus roll libraries of philosophy, drama, history, oratory and poetry existed in Greece. Some of the most famous held several hundred thousand rolls, but the length of any one was limited to the maximum length of a papyrus roll.

The Chinese 'Diamond Sutra', showing scenes from the life of Buddha, dates to AD 868 and is one of the oldest surviving books. However, the first bound volume can be traced back to Ancient Greece.

The breakthrough came with the codex. Instead of papyrus strips interleaved to make a roll, it had leaves folded to make pages which were bound together on one edge. Although the oldest surviving codex has been dated to the AD 2nd-century, there are numerous references to those in existence in the 1st- century BC. Roman legal and Christian scholarship played a major role in the adoption of the codex and it revolutionized

the collation of biblical texts. A roll could hold only a single Gospel; a codex could hold all four.

The codex's easily-accessed pages were better than rolls for making comparative studies. Unlike the roll, a codex could be opened to any point in the text, the text could be on both sides of each leaf, and the length of the text was not limited by the length of the medium.

The final stage in the downfall of the papyrus roll began with the invention of parchment later in the 2nd-century BC.

Parchment came about when Ptolemy V of Egypt placed an embargo on papyrus to outdo his book-collecting rival Eumenes II of Pergamum.

Ptolemy's plotting was publishing's gain. Parchment was made from stretched animal skin and could be made in larger sheets than papyrus. By the 4th-century AD, the codex was dominant and the papyrus roll was seen as a pagan form.

Central heating

Rome

150 BC

Inventions often start out with someone thinking, "It doesn't have to be this way." The fact that central heating originated in warm countries suggests that it may well have been invented after people asked themselves whether they really had to endure the cold in winter. A basic system of diffused heating appears to have been invented by the Minoans and can be seen at the Palace of Knossos in Crete. But it took the Romans to use a fluid medium to convey heat and create a system close to central heating as we know it today.

Sometime around 150 BC, the Romans developed and engineered the hypocaust system. Mosaic tile floors were supported by columns, creating an air space beneath the floor where warm gases from a central fire could circulate and escape through flues in the walls. It was widely used in buildings in Britain and Germany – where no doubt many Romans were grateful during the grim northern winters.

One spin-off from these systems was a demand for well-made pipes. Early conduits for the heating fluid, made from wood or earthenware, were replaced by malleable lead, creating the job of plumber (from plumbus, meaning lead) in the process. The Roman plumber dealt with pipes, but also worked on roofs, gutters, sewers, drains, supply and waste.

The fall of Rome also brought the demise of under-floor central heating, which is a shame. Such heating is efficient: it heats the whole floor. Modern central heating, based on radiators, is not nearly as good. You are either too far away or too close to the heat, and because radiators are invariably fitted on walls, you get powerful draughts.

Screw press

150 BC Rome

100 BC Glass blow-ing invented in Syria, allowing the creation of hollow glass vessels.

The inventor of the principle of the screw is commonly held to be either Archimedes or, possibly, the Pytha-gorean philosopher Archytas in the 5th-century BC. But it took an unknown Roman to realize the screw's ability to exert pressure.

Invented in the 2nd-century BC, the screw press was initially used by both Romans and Greeks to press the oil from olives. Described by Pliny, a Roman savant and authority on scientific and technological matters, it rapidly evolved into one of the most utilized mechanical devices.

The Romans soon adapted the olive screw press to press the water out of clothes and we also know that the Chinese separately invented the screw press in the 1st century AD for pressing oil from seeds.

In the early 16th-century, Benvenuto Cellini, the Florentine sculptor and goldsmith, developed a hand-operated screw press to stamp images on to medals and to press coins for Italian princes.

The olive press: the principle was adapted to many other uses.

Astronomical calculator

Greece

82 BC

During Easter 1900, some sponge fishermen were blown off course while returning home to the Dodecanese. After taking shelter near the rocky islet of Antikythera, north west of Crete, they dived in previously unexplored water. Around 65 metres below the waves they stumbled across a huge ship wreck that held what has become the most exciting relic of advanced ancient technology.

The discovery was a complex mechanical computer that provided the earliest evidence of the invention of an astronomical calculator. It was probably built in 82 BC and was capable of calculating and displaying the relative positions of the sun, Moon and other planets in the solar system over four-year cycles, regardless of the erratic calendars then in use. Even more intriguing is the fact that the device could have been the father of all the clocks used today and its complexity requires us to rethink our attitudes towards ancient Greek technology.

Cicero documented the extraordinary sophistication of the Antikythera device.

The 1900 discovery was so at odds with the thinking of the time that many scientists refused to acknowledge the provenance of the device. Its clockwork mechanism was so advanced that one professor declared that it was impossible for it to have been invented by ancient Greeks.

Other artefacts found on the wreck indicate that the ship had been sailing in about 65 BC from Rhodes to Rome, so the scientists tried to explain the device's discovery by reasoning that it had been made in the Middle Ages and by coincidence dropped on to the wreck.

But the writings of the Roman statesman Cicero indicate it is likely the Greeks had invented one of the first, if not the first, mechanical computer.

Cicero wrote that his friend Poseidonius, a Greek philosopher, had "made a globe which in its revolutions shows the movements of the sun, stars and planets, by day and night, just as they appear in the sky". Conclusive evidence of the device's provenance came from Derek de Solla Price, the Avalon Professor of Science History at Yale University.

In 1951 Price analysed the fragments taken from the wreck.

It took him more than 20 years to prove that the lumps of bronze had once formed part of a clockwork computer.

By shooting high-energy X-rays into the lumps he was able to see inside the device. Price worked out that a complex series of geared bronze wheels had been housed inside a wooden box. One turn of the main wheel represented a solar year. Smaller cogs then rotated into the relative positions of Mercury, Venus, Mars, Jupiter and Saturn, the planets known to the ancient Greeks.

The front dial revealed its age. It had a scale which was out of phase with a slip ring by 13 and a half degrees to compensate for the fact that the Egyptian calendar did not take account of leap years. The only time when this discrepancy would have occurred is 80 BC, or 120 years earlier or later, but 200 BC and AD 40 are archaeologically unlikely. From another mark near the month scale, Price was able to work out that the device was probably first used in 82 BC, then used for two years and taken on to the ship within the next 30 years.

Several characteristics of the Antikythera mechanism link it to Islamic clocks of AD 1000 and hence to the clocks of today. Unlike most mechanical machines, the clock did not evolve from the simple to the complex. Instead, the oldest clocks were the most complicated because they began as astronomical showpieces that also indicated the time.

Islamic clocks described by the astronomer al-Biruni in about AD 1000 have the same 60-degree gear teeth, layout and wheels mounted on square-shanked axles as the Antikythera mechanism. The design can also be traced forwards to a 13th-century Islamic calendar computer preserved at the Museum of History of Science in Oxford.

"The Antikythera mechanism was no flash in the pan but was a part of an important current in Hellenistic civilization," Price said when he published his examination. "It is frightening to know that just before the fall of their great civilization, the ancient Greeks had come so close to our age, not only in their thought, but also in their scientific technology."

Opposite: Replica of the Antikythera device, a clockwork astronomical calculator so advanced that historians refused to believe that it had been invented by the Ancient Greeks.

Shorthand

63 BC   **Greece**

The Greek historian Xenophon, who used a self-devised system to record the teachings of Socrates, is generally regarded as the father of shorthand, but it took a Roman freedman to invent the first widely used rapid writing system.

In 63 BC Marcus Tullius Tiro invented *notae Tironianae* to record the Latin speeches of Seneca, Cicero and other members of the Roman Senate.

Within a few years, Tiro had collated a shorthand dictionary and his invention was widely taught. Julius Caesar and Titus became skilled practitioners but *notae* fell out of use in the Middle Ages, when it became seen as a form of witchcraft.

Thomas Becket reawakened interest in *notae* while Archbishop of Canterbury, but it took another 300 years and the discovery at a Benedictine monastery of Ciceronian notes and a Psalter written in *notae* to give Timothy Bright the impetus to devise the first modern shorthand system. In 1588, Bright published *Characterie: an Arte of Shorte, Swifte, and Secrete Writing by Character.* Another 13 English shorthand systems were devised in the 17th-century, one of which was adopted by Samuel Pepys when writing his shorthand diaries.

Isaac Pitman, knighted by Queen Victoria, is regarded as the leading proponent of the next development in shorthand – the change from alphabetic to phonetic systems with simple abbreviations to improve speed.

Invented in 1837, Pitman's Stenographic Sound-Hand is based on phonetic principles devised by Rev Phillip Gibbs. It became the world's most widely used system, adapted to dozens of other languages including Japanese and Gaelic.

Xenophon, the Greek historian, was the "father of shorthand" who recorded the wisdom of Socrates.

Calendar

Rome

45 BC

Calendars date back to the origins of civilization but it was only with the advent of the Julian calendar, instigated by Julius Caesar on the advice of an Alexandrian astronomer, Sosigenes, that they became reliable.

Early calendars, used for agriculture, regulating civil life, business and religious observances and the vital illusion of understanding and controlling time itself, depended on astronomical cycles, notably the day (based on the rotation of the Earth on its axis), the year (based on the revolution of the Earth around the sun) and the month (based on the revolution of the moon around the Earth).

But because these cycles of revolution do not comprise an integral number of days, and because astronomical cycles are not constant, early calendars were riddled with errors.

Julius Caesar invented the leap day and a calendar year with 365 days but no weekends.

The first practical calendar to evolve was Egyptian, and it was this that the Romans developed in 45 BC into the Julian calendar that served Western Europe for 1,500 years.

Sosigenes abandoned the lunar calendar, instead basing months on the seasons, and decided that the solar year was 365 days. Caesar then directed that the year should have 365 days and a leap day should be added every fourth year. To aid farmers, Caesar issued an almanac which showed when particular astronomical phenomena would occur on the new calendar.

There were several tweaks to this system, but it largely remained unchanged until the 4th-century, when Emperor Constantine I introduced the seven-day week. Until then, there had been fixed dates in the calendar for administrative and other matters, usually at the beginning of the month, but no weekends.

The Gregorian calendar, announced by Pope Gregory XIII in 1582, was a further improvement, taking into account that the 365 days year is an approximation that

Opposite: An early
19th-century Russian
calendar. Its biblical
theme includes a
descent into Hell.

needs readjustment every century, and a further tweak every 400 years.

Its adoption in Britain was fraught with confusion and unrest but it was universally taken up because it drew into one system the dating of religious festivals based on the phases of the moon and seasonal activities determined by the movement of the sun. However, even today there are thought to be around 40 types of calendar in use.

Chapter 2
Looking at new horizons
AD **1** to **1649**

M any misconceptions surround the development of technology during the period from the birth of Christ until the early Renaissance. The popular image is that of the glorious cultural achievements of the Roman Empire collapsing under a barbarian invasion in the fifth century to be followed by a period, often known as the Dark Ages, when innovation ceased. In fact there was little about this period that was dark, isolated, or backward. It was a period when ideas began to travel freely, when society became highly adaptive, willing to embrace new inventions and techniques. More technological progress was made than during the Greco-Roman era, marked though this had been by engineering of great ingenuity and aesthetic sensibility.

Following the collapse of the Western Roman Empire, the eastern civilizations of Islam, India and China became the new centres of technological and scientific innovation. In China, technical skills were highly valued by the mandarins, who prevented any new method or machine from being introduced unless it offered a clear benefit. The stability of their rule allowed paper, the lock gate, the abacus, the first mechanical clock and gunpowder to be developed. India and the Islamic world built on the scientific attainments of Greece and made significant breakthroughs in the fields of mathematics, astronomy, architecture and clock making.

The most profound influence in Western Europe was the incursion of the Teutonic and Viking tribes, colonizers who combined innovation and vitality. The marauding tribes are thought to have spread the use of stirrups, saddles and metal horse bits, which led to the invention of the lance, wielded in battle by the knight on horseback. The Vikings made several improvements in ship design. Later, the deep keel and triangular lanteen sail emerged to enable sailors to venture further, faster.

The Teutonic tribes improved upon the plough to turn the heavy soils of the forested lowlands of northern Europe, enabling new settlements to be established. By AD1000, new agricultural techniques, such as three-field crop rotation, had resulted in grain surpluses that led many peasants to abandon small, individual farming plots in favour of the successful medieval communal pattern of open-field agriculture or of artisan life in towns and cities. Within cities, there was a vigorous market for rope makers, coopers (barrel makers), metalworkers and leatherworkers.

Vigorous commercial urban life and an ordered society resulted in the establishment of patent laws, the application of scientific principles of investigation and observation in research and development, and the eventual adoption of engineering and other technological disciplines by the new universities.

The first significant invention of the period was gunpowder, which came to Europe probably from China and became an agent of a new spirit of nationalism, exploding the fortifications of feudalism and threatening the rule of the church. The second was Johannes Gutenberg's printing press, the spark and most lasting legacy of the European Renaissance, which promoted the spread of knowledge, liberated learning from the confines of church and court, and reintroduced many of the scientific, philosophical and technological teachings of the ancient world.

Also of importance was the harnessing of wind and water to power complex machines, developing man's knowledge of mechanical contrivances, which would be applied in mechanical clocks. The clock began the transition from a life governed by the daily course of the sun to one controlled by the artificial structure imposed by timekeepers.

But the most significant invention of the Middle Ages, and certainly the one that had the largest influence on the pre-Renaissance world, was the magnetic compass, which increased the safety and scope of sea voyage, opening up the Eastern hemisphere to Western travellers such as Marco Polo. With marine and navigation inventions, such as Mercator's maps, the compass probably had a more profound influence on man's perception of the world than even Copernican astronomical theories.

Heavy plough

Rome

Ploughing with oxen,
from the 14th-century
Luttrell Psalter.
Padded yokes,
developed in the
10th-century, gave the
animals maximum
pulling power.

The first farmers in the Middle East found it relatively easy to grub up their arid soil. They used little more than an old tree branch or root which they dragged by hand or pulled behind an ox or some such beast. In the wetter lands and heavier soil of northern Europe, a bit more engineering and ingenuity was required.

Pliny the Elder describes the use of a wheeled plough in the 1st century AD, and there seems little doubt that the Roman plough, the *carruca*, spread from the foothills of the Italian Alps to other provinces in Europe. By the 6th-century it was widely used in central and southern Germany.

The *carruca* established the basic operating principles of the modern heavy plough. The coulter, or blade, dug the earth and a mouldboard turned over the furrow. This device, pulled by horses or oxen, increased the productivity of any land it was used on. The development of the rigid, padded horse collar around the 10th-century added to the effectiveness of the heavy plough, enabling a draught animal to give maximum effort without the choking effect of the old yoke.

Dental Drill

Archigenes

If there is one reason people don't like going to the dentist, it's because of the dentist's drill. With its gleaming metal tip and high-pitched scream, it is so often seen as an instrument of torture. Yet it is truly an engineering miracle. Enamel is the hardest substance in the body and the dentist's drill can cut through it with exquisite precision at an incredible number of revolutions a minute.

Primitive dental drills were in use almost 2,000 years ago. Doctors must have recognized early on that simply yanking out teeth at the first sign of decay was a bit draconian. Much better to drill out the bad part and leave the rest of the tooth in place. Around AD 100 the Roman surgeon Archigenes developed a drill that was powered by rope; one can only imagine the look on his patients' faces as he approached them.

It took about 1,500 years for anyone to do much better. John Greenwood, an American dentist who made George Washington's dentures, used a spinning wheel to power a drill in 1788. For the next century and a half progress remained painfully slow.

A key development came with the launch by Dr Samuel White in 1944 of a flexible shaft drill, allowing the rotating bit to be powered from any angle. It gave the dentist more freedom of movement, and thus a lower risk of hurting the patient. Dr White cleverly spotted that his invention had applications beyond people's mouths, and he sold it for a tidy sum to Henry Ford, who used it for the speedometer cable in his motor cars.

100 Soap in wide use as cleaning agent.

In 1957, dental drill technology took a big leap forward with the advent of air turbine power. Compressed air squirted into the drill drove the bit to astonishing speeds of 200,000 rpm and more.

Such speeds greatly reduced the amount of time and effort, and thus pain, involved in drilling out decay. But they also made unprecedented demands of the bearings in the drills. To put their work load into perspective, an average domestic cordless drill will run at 800–1,000 rpm, while a very fast power-tool might reach 4,500 rpm.

Only recently have ball bearings been manufactured to a sufficiently high tolerance and standard of design to allow high-speed drills to operate without excessive failures. Yet such are the tolerances of today's precision ball bearings that even at these high speeds a modern dental drill requires a change of bearings only every 18 months or so.

Paper

China

105

The invention of wood pulp paper by the Chinese around AD 105 meant that writing materials became affordable. Until then writers had used expensive vellum, parchment, papyrus, silk, waxed boards or clay tablets.

Ts'ai Lin, a eunuch in the court of the Emperor Ho Ti, is credited with the creation of a paper made from the bark of the mulberry tree which was combined with bamboo fibres, hemp and flax. These were crushed to a pulp and mixed with water, then filtered through a bed of woven split bamboo strips and left to dry. The paper was an instant success but the skill of paper making did not spread from China for another 500 years, after which it seeped west, via Japan, central Asia and Egypt, where workshops were set up around AD 900.

In Britain paper manufacturing using a watermill was first recorded in 1495, shortly after William Caxton had started up Britain's first printing press, and other mills quickly followed. Mills in the West commonly used rags to make paper and by the 17th-century so much paper was being made that finding sufficient rag became a problem. This was solved in 1840 by

Dyed paper drying in Nepal. Mulberry bark, bamboo fibres, hemp and flax were used to make the first papers.

the development of a new pulping process that used cheaper and more plentiful supplies of wood.

Paper making was a laborious business, but a mechanized paper-making machine invented in 1798 by a Frenchman, Nicholas Louis Robert, speeded up the process.

Wheelbarrow

China

118

I t is difficult to imagine it now, but the wheelbarrow was once classified military hardware. Developed more than 2,000 years ago to allow the Chinese army to transport heavy or large loads quickly and easily, the wheelbarrow is indeed a clever design. So clever, in fact, that it can be hard to see beyond it, to how it might be improved.

It was only after spending days using a wheelbarrow while renovating his house in Bath that James Dyson became aware that the 2,000-year-old design could use some improvement. And, what's more, improvement of the component widely regarded as the greatest of all inventions – the wheel. The result was the Ballbarrow, which was launched in 1975. It did away with the wheel, using a pneumatic ball that didn't sink into soft ground or damage lawns.

This wasn't the first time the value of the wheel had been questioned. Introduced into Europe during the early 13th-century, the wheelbarrow took a long time to catch on, because merchants and tradesmen, in Covent Garden for example, found it harder to push heavy loads over cobbled streets than to use the two-man, stretcher-like device with a load-bearing frame known as a barrow.

The Chinese wheelbarrow had the wheel directly under the load, and came in two varieties: the Wooden Ox, with two handles at the front, and the Gliding Horse with handles at the rear. Either way, it was pretty unstable. Yet such was the perception of the advantage it gave over its primitive forerunner, that the design was kept secret for many years.

The great European innovation was to put the wheel in front of the load and the arms at the back. This made the wheelbarrow more stable. It also turned the wheelbarrow into

Plenty of room for improvement: the traditional wheelbarrow.

an effective type of lever, allowing loads to be tipped out with minimal effort. To this day, the builder's "navvy barrow" follows this old European design, while the domestic gardening type is more in keeping with the original Chinese design.

In designing the Ballbarrow, Dyson followed the more efficient European design and put the ball in front of the load.

Abacus

China

190

The ancestor of the modern calculating machine probably originated in Babylonia some time before Christ, but the first documented use of the modern abacus, with beads, is in a book of the Eastern Han Dynasty, by Xu Yue in about AD 190.

The name derives either from the Hebrew *ibeq*, meaning "to wipe the dust" or from the Greek *abax*, meaning "board covered with dust", and this gives an indication of its early origins. The first abacuses were boards on which Babylonians spread dust or sand to trace letters for writing purposes. As they evolved into something used solely for accounting purposes, counters on a drawn grid replaced the sand. The Romans replaced the grid with grooves to hold the counters, and by the time the abacus was in use in China, beads on rods replaced the counters on grooves.

Although the world has now entered the age of electronics, the abacus still enjoys undiminished vitality in China, where the largest known example is displayed in a pharmacy in Tianjin. It is more than 100 years old, has 117 rods on a frame 30 cm high and is more than three metres long.

In 1946 a two-day contest between Kiyoshu Matzukai, a Japanese abacist, and an electronic computer resulted in an unmistakable victory for the abacist.

In 1980, China developed an electronic abacus that deftly integrated native Eastern methods for addition and subtraction with modern Western technology for multiplication and division, with both methods used simultaneously. Practised abacus users should still be able to beat electronic calculators on basic addition and subtraction.

Saddle and stirrup

Asia

200

The horse was heading for extinction when it was domesticated on the Ukrainian steppes in around 4000 BC. It was an instant success.

By the 5th-century BC, the speed and versatility of the horse had elevated it to pole position among beasts of burden. The first "pony express" ran from Susa to Sardis in the Persian Empire, a distance of 1,600-miles covered in nine days, a time that was not bettered until the Napoleonic era.

Early horsemen rode bareback or on a blanket, using a simple rope harness to direct the horse. Mouth bits, used to pull the horse's head up to bring the animal to a halt, were invented around 4000 BC, but examination of horse teeth found at a Copper-Age site at Dereivka in the Ukraine show signs of wear that would indicate the use of a rope bit. The earliest metal bits, dated to around 300 BC, have been found at Celtic sites.

The invention of saddle and stirrup created a symbiosis between man and mount, making accurate shooting on horseback possible.

These early horsemen might also have been using stirrups (a crude stirrup appears on an Assyrian sculpture of 853 BC) but these are unlikely to have been much more than leather straps used when mounting the horse. The metal stirrup, probably initially for allowing rapid release when the rider came unstuck, was invented in China around 200 BC, and made its way to the Mediterranean via the horse-riding tribes of central Asia. The saddle is thought also to have been invented by these nomads and been in use by AD 200.

With the invention of stirrups and the saddle, symbiosis between man and mount achieved a precision that defied further improvement. Accurate shooting on the run became possible from the secure hold of a saddle. It made cavalry lances far more formidable, since a rider, by bracing his feet in the stirrups, could put the momentum of both himself and his galloping horse behind the thrust of his spearhead.

The widespread use of horses as military machines led to the breeding of much larger horses, which in turn further enslaved the animal to man's needs.

Algebra

Diophantus

275

400 First public hospital in Europe opens in Rome.

The earliest writing on algebra is a treatise by Diophantus, a Greek mathematician who was the first to substitute letters and symbols for numbers in equations. All that is known about Diophantus, who lived in Alexandria around AD 275, comes from a mathematical epigram which indicates that he married at 33 and had a son who died at 42, four years before Diophantus at the age of 84.

Diophantus also set out rules for simple multiplication and division, for the use of the minus sign, and for generating powers of a number. Negative quantities were fully comprehended for the first time in 1629 by Albert Girard, a French mathematician, leading to a realization that algebra was a "generalized arithmetic", a convention that Isaac Newton named "universal arithmetic" in 1707.

Algebraic problems that are solved with whole numbers are still known as Diophantine equations, but the true father of modern algebra was René Descartes, the French mathematician and philosopher. He introduced the ideas of using exponentials to indicate powers, letters at the end of the alphabet to denote variables and letters from the beginning of the alphabet to indicate parameters.

Modern numbers
Aryabhata

T he Indo-Arabic numeric system of the digits 0 to 9 that we use today was developed in India and systemized by a Hindu mathematician and astronomer, Aryabhata I, in around AD 500. When, a century later, a Muslim mathematician and astronomer, Muhammad ibn Musa al-Khwarizmi, introduced the idea of using a zero (known as the "positional zero") to denote powers of 10, our current number system was firmly established.

500 Persians smuggle silk-worms out of China, spreading silk production.

The Indo-Arabic system replaced the cumbersome Latin system of calling large numbers by their initial letters and it was soon universally adopted in the West. Much of modern science would be far more difficult, if not impossible, without it.

Astrolabe
Alexandria

T he theory of how an astrolabe might work may have been known in the pre-Christian era of Apollonius, the Greek mathematician known as the Great Geometer, and Hipparcus, who estimated the distances of the sun and the moon from the Earth and catalogued more than a thousand stars. But the first description of a working instrument was written by John Philoponos of Alexandria in the 6th-century AD.

The astrolabe was a type of celestial calculator that would be used variously as a navigational instrument, a surveyor's tool and also as the means by which astronomers could explore the heavens.

It was used initially to measure the position of an object above the ground, usually a planet or a star. But it was also used for calculating the heights of mountains or buildings. A basic astrolabe consisted of a disk, similar to a protractor, which was marked in degrees. Attached to the centre was a moveable pointer (the alidade). When the astrolabe was held to the eye, the baseline of the protractor was lined with the horizon while the pointer was aligned with the object and the elevation of the object could be measured in degrees.

An astrolabe from 1585. Its base was aligned with the horizon and the 'pointer' aimed at a terrestrial or celestial object (such as the sun), giving its degree of elevation. Astronomical tables were then used to work out the time of day and latitude.

More elaborate astrolabes were engraved with the positions of planets, and when aligned with a known star they pointed to other heavenly bodies. By finding the height of the sun, and referring to the correct astrological tables, it was possible to find the time of day, the correct latitude, the time of sunrise or sunset and the direction of Mecca, which explains the popularity of the instrument particularly in the Middle East.

By the 8th-century the Persians were using astrolabes, and the device spread to Europe. By the 14th-century astronomical scholarship using the astrolabe was common.

The major use for the instrument by the 15th-century was to help merchants, explorers and naval officers. The astrolabe remained the principle navigational aid, after the compass, up until the advent of the sextant in the 18th- century.

Chess

India

621

650 Windmill invented in Persia.

The debate still rages over the origins of chess, the most enduring of board games. Many believe it was derived directly from the Indian war game Chaturanga; others look to the well documented development of Chinese chess.

The most likely path to modern chess is the mixing of cultures along the Silk Road and other nomadic trade routes

A 16th-century depiction of chess playing at court in Persia. Under the name Shatranj, the game passed to the Arabs, lasting 1,000 years, with rules and pieces similar to the modern game.

which played a significant part in the development of several ancient Asian board games. Chaturanga, a 6th-century Sanskrit word, means "quadripartite" and there was at this time a four-sided version played with dice which evolved into a two-sided game of pure skill. The men were *rajah* (king), *mantri* (counsellor), *gaja* (elephant), *asva* (horse), *ratha* (chariot or rook) and *pedanti* (infantry or pawns).

The Persians played the same game, which they called Chatrang, and it was enthusiastically taken up by Arabs after their conquest of Persia in the 7th-century. Under the Arabic name Shatranj, it endured for 1,000 years, with rules and pieces similar to the modern game.

Abu Bakr Muhammad ben Yahya, born in 854 at Jurjan on the Caspian Sea, was the first player to write about strategy. His chess problems, now recognized as works of genius, have been solved only recently by computers. He was also the first to develop *ta'bi'ah* (openings) as well as middlegame strategy.

The 8th-century Moorish invaders of Spain brought Shatranj to western Europe, where it evolved into the game we know today. One of the first books published in English was Caxton's *Game and Playe of Chess*, printed in 1475.

A Spanish priest, Ruy Lopez, wrote one of the first major works on the game, which included the first recorded dirty trick of positioning the board in a manner that would force an opponent to face the sun.

By the end of the Renaissance leading players in Italy, such as Gioacchino Greco, were sponsored by the ruling classes. Many openings used today were developed by Greco, who travelled to France, England and Spain expounding his theories on the game. France subsequently became the centre for chess. The game was taken up by the middle classes, chess clubs were formed and the top players gathered at the Café de la Regence in Paris.

In 1843, Howard Staunton defeated the French champion Saint Amant, and London became the chess capital. Staunton developed chess notation, the method of recording games, as well as creating a standard design for chess pieces which bears his name today.

The first chess computer was "The Turk", a machine which appeared to move its own pieces. Operated by a chess player concealed inside, it was the first great cabinet illusion in the history of magic. It entertained and bamboozled spectators, including Napoleon, around the world for more than 80 years.

The first proper chess computer was developed in 1958 and the first world computer chess championship was held in 1974. The huge increase in computing power, which enabled the silicon players to retain vast amounts of knowledge, changed the nature and conventions of the game. The first to threaten human dominance was Deep Thought, renamed Deep Blue when the research team, led by Feng Hsu, received backing from IBM.

In 1996 Garry Kasparov, the world champion, defeated the computer in Philadelphia. A year later he lost a rematch. A team of Grandmasters had been teaching the powerful hardware every possible end-game position with five or fewer pieces.

Porcelain

China

Coveted for its translucency, porcelain originated in China around AD 850. Much imitated by European potters, the perfect mix of petsunte (a feldspathic rock) and kaolin was not matched until 1707 at the Meissen factory in Saxony. William Cooksworthy was the first Englishman to produce it, in 1768.

850

850 Earliest reference to gunpowder in Chinese literature.

868 First printed book when the *Diamond Sutra* is produced in China using wooden carved blocks.

Canal lock

Chiao Wei-Yo

983

Putting untrained people in charge of something that shifts thousand-ton masses sounds like a recipe for disaster. But that is exactly what we do on our waterways, where people insouciantly shunt around huge masses of water, plus their boats, using the brilliantly simple device we call the canal lock.

The most impressive thing about the invention of locks is merely thinking they might be possible. Nature offers no clues: when rivers want to go downhill, they just tumble down hillsides. And of course the law of gravity guarantees that they

The Caen Hill flight of locks on the Kennet and Avon Canal, which connects Westminster to Bristol. These mitre locks enabled vessels to travel 'uphill', and were essential to the construction of Britain's waterways.

never run uphill. It takes an original mind to ignore all that, and come up with a way of defying gravity and sending boats up hill and down dale.

The credit for thinking up the key idea goes to a 10th-century Chinese engineer, Chiao Wei-Yo. China had an impressive canal system dating back to before the birth of Christ, but the gradients weren't very demanding. So called "flash locks", where upstream water was allowed to build up behind a gate and then suddenly dumped downstream by lifting the gate, did a reasonable job.

But in AD 983 Chiao Wei-Yo realized that putting two flash locks together would create a "pound" of placid water, which could be filled and emptied under complete control. The same idea appeared in Europe in the Middle Ages: the first example of something approaching the modern pound lock is recorded as being built in Vreeswijk, Holland, in 1373.

One major problem with these early locks was that their gates were lifted vertically. This imposed a height limit on boats that had to go under them, a problem that occupied Leonardo da Vinci (see page 141). This was resolved in the 16th-century when the modern "mitre" lock gates emerged. Using two horizontal-opening, hinged gates – thus eliminating the height restriction – the mitre gate formed a "V" that pointed upstream when closed, cleverly using the pressure of water to keep it shut.

To raise the water level in the closed lock, sluices in the upstream set of gates were opened until the lock was full. With water at the same level on both sides of the gates, the pressure was off and the gates could be opened.

Canals became the dominant means of transporting freight in Britain, only to decline in the 19th-century after the advent of the railways.

Unlike other European countries, where canals thrive commercially, Britain has a geography that compels canal-builders to use a lot of locks, which inevitably slows things down. But as our roads become more crowded, has there been a better time to look again at this wonderfully benign means of transport?

Gunpowder

China

1000

Few inventions have had as dramatic and decisive an impact on the course of civilization as the invention of gunpowder. It occurred first in China, where it appears to have been associated with medicinal and life-enhancing procedures devised by Tao alchemists of the Tang dynasty around AD 1000. Lines of communication between China and the new civilization that was emerging in Western Europe at this time were extremely tenuous, but information about the explosive mixture percolated slowly and by the middle of the 13th-century there are several instances of European scholars experimenting with forms of gunpowder.

Roger Bacon was the outstanding British example of such pioneers (see right). A remarkably independent and inventive thinker, he experimented with various proportions of gunpowder ingredients (saltpetre, sulphur and charcoal) and was so impressed by the results that he described them in a code which was not solved until the 20th-century. But he did not manage to obscure the nature of the mixture and its explosive

Roger Bacon, England's first scientist, kept his recipe for gunpowder secret by writing it in code.

Norse and Greek mills (gearless water-driven pairs of grindstones) appear in Europe.

properties, which rapidly became general knowledge. By the end of the 13th-century gunpowder was widely produced.

The name gunpowder was given to a number of low-explosive mixtures that could propel a bullet from a gun or blast a gap through rock. The first gunpowder, or black powder, was a chemical mixture rather than a compound: the ingredients were thoroughly mixed together by grinding or pounding, but they did not combine until the moment of ignition. The proportions in which they were mixed depended upon the nature of the explosion required, but the standard traditional mixture came to contain 75 per cent saltpetre, 10 per cent sulphur and 15 per cent carbon. These ingredients were all available in Western Europe: saltpetre (potassium nitrate), a salt obtained from decaying organic material; sulphur, from volcanic deposits in Sicily and elsewhere; and carbon, from charcoal produced from carefully selected wood.

However, it became an urgent matter of policy for the new European nation states to secure themselves an adequate supply of these ingredients, for the enormous success of the application of gunpowder to warfare meant that weaponry was transformed.

In turn the growth of gunpowder weapons was to transform

Ornate Berber gunpowder boxes. The mixture transformed warfare.

European warfare itself. Fortifications which had been impregnable before the advent of gunpowder could be broken down by the heavy cannon, and the traditional mounted knights, pikemen, bowmen and swordsmen of medieval armies were powerless against the killing range of gunpowder weapons. Not that they were infallible. The weapons sometimes backfired: the gunpowder was fine and so fast-burning that it created massive pressures that could split barrels.

In the 16th century, European pyrotechnicians began manufacturing gunpowder in large, uniform grains to reduce the speed of burning. In the 1850s, Thomas Rodman, an American military academy graduate, invented prism-shaped gunpowder, with which the maximum expansion of gases occurred when the bullet was travelling down the barrel.

The effect of gunpowder was not limited to its military applications, although these were enormously important. It also became a valuable tool in mining, quarrying and civil engineering. But these non-military uses were developed after the military potential of gunpowder had been exploited, as were its other uses in fireworks for celebratory purposes.

Gunpowder was unchallenged as an explosive material for all purposes until the chemical inventions of the 19th-century made available new types of high explosive, called nitroglycerine and dynamite.

Fork

Italy

1000

E arly table forks have been dug up from ancient cities in Turkey, but the implement disappeared from use until Italian nobility reintroduced it some time in the 11th century.

The story goes that Domenico Silvio, heir to the Doge of Venice, married a Byzantine princess who brought a case of two-tined golden table forks to Venice. She outraged the populace and clergy by refusing to eat with her hands. The

practice was seen as blasphemous because it substituted God-given fingers. The fork did catch on, however, as a means of shaking sticky sauces from morsels of meat or fruit, before removing the morsel from the fork and popping it in the mouth. Although God-given fingers were used, this, too, was seen as lascivious and an immoral influence, particularly as courtesans practised it, and the Church soon banned the use of forks.

Bows of war

Wales

1000

The crossbow appears to have been invented at different times in various parts of the world. In Europe, it is thought to have been invented in around AD 1000, but ornately carved examples have been excavated from tombs in China dating to the 5th-century BC. Consisting of a wooden or bone-and-wood bow mounted flat on a headstock, the string of the bow was drawn back using either a lever or, later, a rack-and-pinion winding mechanism.

1085 Britain's first census as William the Conqueror orders Domesday Book.

Crossbows gave foot soldiers a chance to bring down heavily armoured men mounted on horseback. They were much easier to use than conventional bows, and their bolts, easily mass produced, packed a killing punch. There were significant disadvantages. One problem was that it took time to reload and cock the mechanism, precious seconds during which the crossbowman would be vulnerable to attack. The Chinese finally tackled this problem with elaborate bows equipped with up to 10 bolts and drawstrings, all ready to fire – the first machine-crossbow.

In the 12th-century the crossbow wrested some degree of military superiority away from the mounted, landed elites, and its popularity all over Europe and Asia did not begin to diminish until the 16th-century.

But the crossbow's major failing on the battlefield was its short range. At the Battle of Crécy, in 1346, Philip VI of

France took on the numerically inferior forces of Edward III of England. Philip's Italian mercenary crossbow corps were decisively outranged, and the whole of the French force was cut to ribbons by a small band of English longbowmen.

The longbow originated in the Welsh borderlands, where it was the key guerrilla weapon during the centuries of insurrection before medieval times. Like all the best guerrilla weapons, it was cheap to acquire and easy to conceal. About two metres tall, strung with hemp or linen twine, it could hurl an arrow and pierce the best armour of the day at least 200 metres away.

The usual missile fired was around a metre long (an English cloth yard or 37 inches – 94 cm) but a skilled archer using lighter shafts and better fledging could kill at 500 metres or beyond. Archers could rain down six arrows a minute on the enemy, a crucial advantage over the crossbow. It took a special kind of strength and nerve to wield a longbow. It needed a pull of 45 kg to launch an arrow and seasoned yeomen developed twisted spines, deformed bones in their forearms and calloused fingers from the effort.

The devastating effect of the longbow during the Hundred Years War against France built lasting superstitions about the English archer and his formidable bow – myths possibly magnified by the fact that the best bows were made from yew, which carried a significant religious and symbolic resonance.

At the Battle of Crécy in 1346, and at Agincourt in 1415, English archers slaughtered noblemen and footsoldiers of France, outranging and out-shooting them. The French were at a strategic disadvantage for many years.

The crossbow was cumbersome and took time to load. The longbow could rain down arrows, giving England the edge in key battles against France.

As major confrontations by massed ranks of bowmen became less frequent, the required strength and skill died out in England. Only a few hundred elite archers could be mustered by the 16th-century, when a new arms race powered by gunpowder had begun.

Mechanical clocks

China

1092

T he science of telling the time with a reasonable degree of accuracy turned a decisive corner early in the last millennium with the invention of the first mechanical clocks. Although no more accurate than water clocks, they could for the first time be built into towers and buildings for all to see. It meant that by the end of the 14th-century, the citizens of Milan and Salisbury could avail themselves of the time – albeit with an inaccuracy that made a minute-hand unnecessary. It is unlikely that accurate time-keeping was important to the typical citizen in those days, but these were the first tentative steps towards the chronologically-obsessed society we live in today.

The most elaborate early example stems from China and was a hybrid machine. In 1092 a monk called Su Sung described a complex time keeping mechanism driven by a water wheel. The movement of the wheel was regulated by an ingenious device – an escapement in modern terms – operated by the water. Its main purpose was to drive models of the heavens, and it was an isolated example. The clock had no influence on future developments, but thanks to detailed contemporary records, a replica was built in the 1950s and is now on display in Beijing.

Although there are references to mechanical clocks in Europe from the 13th-century, no detailed descriptions appeared for another hundred years. Richard of Wallingford, Abbot of St Albans, left drawings of a clock built for the Abbey in about 1330. Giovanni de Dondi completed an elaborate clock in Padua in 1364. These not only told the time but also showed the movements of the planets.

These early mechanical clocks had no minute-hand and some had no faces, just striking the hour instead. They used a simple mechanism of a weight attached to a rope that was wrapped several times round a drum. As the weight fell, the drum turned and this movement was transferred to the hour hand or striking mechanism. The rate of fall of the weight was governed by an escapement of a type known as a "verge". This was controlled by a balance – either a wheel or a weighted bar called a "foliot" – suspended by a cord so that it twisted back and forth to mark out equal intervals of time. These three elements – power source, escapement and oscillator – were present in one form or another in all clocks for centuries.

The first public clock of which we have reliable knowledge appeared on the palace of the Viscount of Milan in 1335. Other clocks are recorded elsewhere in Italy soon after. A famous astronomical clock was completed in Strasbourg in 1354. Three clock mechanisms survive from the 14th-century. All were converted 300 years later from foliot to pendulum control.

A reconstruction of Giovanni de Dondi's 1364 astronomical clock from Padua. Its complexity was unmatched for two centuries.

Smaller clocks were also made but their development is more difficult to uncover. The water-powered alarm which was used to tell the sexton when to call the monks to prayer became a mechanical alarm, and then a clock with a dial. The town watchman may have been the first lay person to need a clock, to mark the time of curfew.

Many domestic clocks were weight-driven, but more compact and portable spring-driven clocks appeared in the 15th-century. Early 16th-century portable clocks, the first watches, were seen as novelties.

By 1700, England was pre-eminent in watch and clock-making.

An early clock-making system with weights, from Istanbul.

A characteristic English style made in larger numbers was the simpler brass lantern clock with the bell for striking the hours on the top. The basic mechanism remained the same until the middle of the 17th-century, when the introduction of the pendulum led to radical changes.

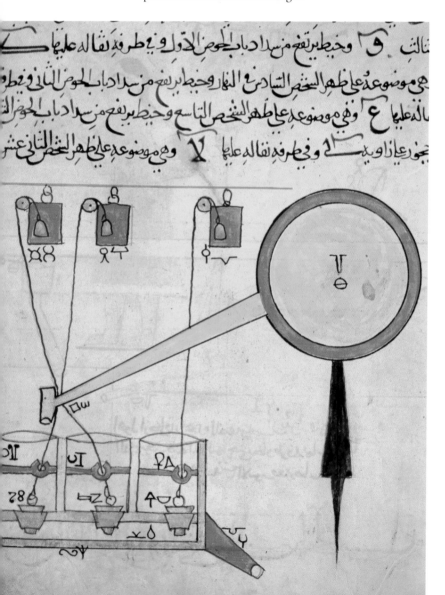

Navigational buoys

Baltic Sea

c. **1100**

The first navigational buoys are thought to have been sited in the Baltic Sea in around 1100 by founder members of what became the Hanseatic League, a federation of north German merchant communities that dominated commercial activity in northern Europe and Scandinavia until the 15th-century.

By the turn of the 13th-century, according to *La Compasso de Navigare*, a Spanish text, "there was a wooden buoy bobbing in the current of the Guadalquivir River, on the approaches to Seville in Spain."

Amsterdam introduced navigation buoys in its harbour in the late 13th-century. They were wooden casks, bound with iron bands and anchored to the riverbed by heavy stone blocks. In the 16th-century, levies called "light fees" were raised from mariners by port authorities to organize the maintenance of buoys.

First grants for the right to place buoys in a British harbour were made in 1503 to the mayor and corporation of Sandwich in Kent and the first navigational aids were positioned in the Thames in 1538.

Henry VIII granted a charter to the Guild of Shipmen and Mariners in 1514 to keep safe the seas around Britain by maintaining marks and buoys. This led to the founding of Trinity House in London in 1594, which still looks after the safety of Britain's seas.

1100 Construction begins on Chartres, Saint Étienne (Bourges) and Notre Dame (Paris) cathedrals in 12th-century.

Buoy-girl humour from the US. Buoys originated in the Baltic.

Flywheel

Western Europe

Repetitive mechanical operations were hard work in ancient times. The primitive kick wheel used by the potter, for example, needed frequent prods to keep it going. The answer was to connect a larger, heavier wheel, which with its greater inertia was slower to start but once turning steadily smoothed out the delivery of power. It was called a flywheel.

Although the earliest flywheels may have been used by potters, the first written evidence of them is from 1122 in treatises on the principles and physics of mechanical engineering and craftsmanship by Theophilius, a German Benedictine monk.

In 12th-century Britain, flywheels first appeared in blacksmiths' forges where they were used to grind or sharpen blades. Their heyday came in the 18th-century when the first

A 23-ton flywheel in the Palace of Engineering at the 1924 British Empire Exhibition. Steam engines used flywheels of enormous proportions.

steam-powered machines needed a way of smoothing out the action of the pistons. Flywheels on some early steam engines were 12 metres across.

When the first internal combustion engine was built in the last years of the 19th century, the flywheel was at its core, and it is still there today.

1170 The first English university is founded at Oxford by students barred from University of Paris.

Navigation

China, Europe

1190

To get from one place to another, a map is not enough. You also need some way of showing how this representation of the world relates to the real one – in other words, of orientating the map to your surroundings. For thousands of years, travellers relied on maps, natural cues such as landscape features and a knowledge of the direction of prevailing winds to get them to their destinations. Those living in the northern hemisphere were also able to exploit the existence of a bright star, Polaris, almost exactly at the north celestial pole.

But for long-distance travel, some method less dependent on weather and location was needed. It emerged during the 12th-century, when mariners first in China and then Europe independently discovered a naturally magnetic, iron-rich mineral that aligned itself with Polaris. This mineral – now called lodestone after the Old English for "journey" – became the basis of the first magnetic compass.

Gerard Mercator was the first person to work out how to represent the spherical world on a flat map.

We now know that the success of the compass is the result of a natural fluke: the approximate alignment of the north magnetic and geographic pole. Throughout recorded history, the alignment has held true, but the magnetic pole is around 12 degrees south of true north and the deflection will increase.

The mere fact that the magnetic compass kept the same direction was of key importance to navigators. By 1190, Italian mariners were using needles magnetized by lodestone and floating on bowls of water to act as ships' compasses. Since it could be trusted to bring them home, the compass gave mariners more confidence to venture further from the coastlines and their orienting landmarks.

Relatively accurate maps based on compass surveys began to emerge a century later, showing traders how to get from port to port. Yet while these so-called "portolans" were relatively accurate, they only covered the Mediterranean.

The cartographic expertise of classical times had been lost to the West after the destruction of the Library of Alexandria during the Roman civil war of the 3rd-century AD. For the next thousand years, Western cartographers were more concerned with biblical than geographic accuracy, marking their world maps with such places as the Garden of Eden.

From the 12th-century onwards, accuracy began trickling back into Western maps from the East, where Arab scholars had preserved and improved on the classical works. Using reports from traders, cartographers gradually began replacing dogma and guesswork with solid fact.

The journeys of the Venetian merchant Marco Polo between 1271 and 1295 had a huge impact on cartography. The sheer scale of his expeditions, which took him to the Pacific coast of China, and as far south as modern-day Singapore, brought in a wealth of information about hitherto unexplored lands. But at least as important was his account of the wealth and produce of China, which triggered huge interest in trade with this distant land and thus in accurate maps. The whole field of navigation then began to benefit from a form of positive feedback, where better maps led to bolder expeditions, whose findings in turn led to still better maps.

By the 15th-century, relatively accurate maps of large areas of the world were beginning to emerge. Classical concepts such as map projections and

An 18th-century compass.

longitude and latitude, lost for a thousand years, were re-discovered, and by around 1450, mariners had adopted the astrolabe to fix their latitude. Doing the same for longitude, however, remained problematic for another 300 years. Ironically, the most famous voyage of the era was based on biblical references and unreliable cartography. The Genoese navigator Christopher Columbus became convinced he could reach Asia by travelling west after making calculations based on faulty classical maps and information drawn from the Apocrypha. This led him to conclude that Asia was actually more than 6,000 miles closer than it was. Fortunately, the unreliability of the maps also saved him from disaster, as they failed to show the Americas, which just happened to be where Columbus was expecting to find India.

The ever-increasing scope of maps put pressure on cartographers to tackle the problem of projection. While small-area maps can ignore the curvature of the Earth, its effect becomes critical on world maps. Classical cartographers had

Gerard Mercator (1512–94), who devised an ingenious way of putting the round world on to a flat map, is pictured (left) with a fellow cartographer from the Low Countries, Jodocus Hondius. Using Mercator's maps and a compass mariners could navigate the seas.

made some attempts to solve the problem of projecting the curved surface of the globe on a flat map, but the resulting maps were far from ideal for navigation.

In 1569, a Flemish cartographer published a map that tackled the problem using a brilliantly simple projection method. His name can still be found in the margin of countless world maps: Gerard Mercator. Mercator's approach to the projection problem was to imagine a sheet of paper being wrapped around the Earth's equator in a cylinder, with a light then being placed at the Earth's core. The rays of light, projected on to the paper, then formed a grid on to which the world's countries could be drawn.

The resulting "Mercator projection" had its flaws, distorting the relative sizes of countries. No projection can be entirely distortion-free, however, and Mercator's projection had one huge navigational advantage: a course between two ports that appears as a straight line on a map really is a straight line. Mariners could simply read off from the map the direction they needed to go, and then follow that bearing on their ship, using their magnetic compass.

Despite its ingenuity, Mercator's map took a century to have a major impact. Its advantages were recognized after trade and colonization routes became so long that simple maps would no longer suffice.

Sawmill

1204

France

U p to the end of the first millennium AD, splitting, reducing and converting logs into timber was a laborious process, employing men standing with two-handed saws working long hours in dusty sawpits: one on top, one underneath. The lucky man who got to stand at ground level was the "top dog"; the unlucky sawyer, usually an apprentice, got the worst of it on the lower end of the saw down in the filth of the sawpit.

Water power changed much of this. By 1204, in Evreux, France, the first recorded mill driven by water was turning out timber far quicker than could be managed before.

1209 Cambridge University founded by students escaping friction between scholars and townspeople in Oxford.

The water-wheel, used from ancient times to mill flour, had become the principal form of power in many parts of the world, and in some places it still is. In 1068, the survey for the Domesday Book records 5,624 waterwheels in England alone, serving a population of two million.

As well as grinding corn and preparing timber they powered bellows in foundries, primitive weaving machines and the local blacksmith's hammer as well.

Button

Germany

1235

The Etruscans and ancient Greeks fastened their tunics using crude buttons and loops at the shoulder, but the button as we know it today did not come into use until the invention of the buttonhole sometime in the 13th-century.

The earliest evidence comes from sculptures in the cathedral at Bamberg in Germany. These are dated to the first half of the 13th-century and clearly show tunics with six buttons running from the neck down the chest of a secular benefactor of the cathedral.

It did not take long for the button to become popular and it soon replaced brooches, clasps, points and lacing as the method of holding garments together. By the 14th-century, sumptuary laws had to be introduced in order to put limits on the use of buttons, which were now being worn as decoration. Strings of precious metal, ivory, tortoiseshell and jewelled buttons, often running from the wrist or stomach to the neck, were used to denote rank, wealth or status.

In the 18th-century, Matthew Boulton, the manufacturing partner of steam-engine pioneer James Watt in Birmingham, invented highly polished steel buttons that became fashionable among those rich enough to afford them. At about this time, the first mechanized manufacture of buttons began, making them much more affordable.

Today, despite the invention of the zipper, the popper and Velcro, the button remains by far the most popular fastening for garments.

Screw lift

Low Countries

1250

1280 Spinning wheel, the first geared machine, migrates from India to Europe.

1305 When Edward I decrees that one inch is the length of three dried barleycorns, English cobblers devise standard shoe sizes for children based on barleycorn lengths.

1311 Professional cartography begins with a navigational chart, or Portolan, produced at Genoa.

I n the mid-13th century, the screw made the transition from a machine used to press things together (as in the screw press invented in around 150 BC to squeeze oil from olives) to a machine used to lift loads.

The screw lift, a crude and cumbersome machine, reeled in a rope and hook to lift loads. The mechanical advantage came from using a long lever to turn the screw.

By 1480, the screw jack had been invented by unknown artisans in the Low Countries and it came into its own with the invention of the cannon, when it was used to raise the tail to adjust precisely the weapon's angle of firing.

Cannon

Europe

1313

T here are several claims as to who first used a tubular barrel to fire a projectile at an enemy. It could have been the Chinese, it could have been the Arabs, it could have been a German monk called Berthold Schwarz.

Contemporary accounts describe Schwarz discovering black powder, or gunpowder, in 1313 and using it in a weapon. There is little doubt that the English, under Edward III, used cannon in a campaign against the Scots in 1327. These were small, primitive cannon, or *pot de fer*, used to fire darts that were little more than glorified arrows.

The English used cannon mounted on sleds at the Battle of Crécy in 1346. They needed to be immensely thick and, consequently, heavy to withstand blowing up and were generally used to lay siege to a castle. The main body of the cannon was cast as one piece of bronze or welded together from strips of iron. Firing such weapons was dangerous work. During the siege of Constantinople in 1453, the armourer Urban of Hungary was killed when one of his 14 giant guns exploded next to him.

Carved stone balls used as projectiles were replaced in the 15th-century by cast-iron balls. Siege guns at this time wrought destruction on a grand scale by using projectiles that weighed up to 400 kg in guns that weighed 19 tonnes. When an English naval squadron approached the Dardanelles in 1807, it was devastated by Turkish forces firing giant cannon dating from the 15th-century. One accurate shot from the Turkish battery was said to have killed 60 men.

It took the French to make the best use of the new weapons on the field of battle. Bernard Du Guesclin, Constable of France under King Charles V (r.1364–80) realized that new tactics were needed to beat the English. He issued orders to engage in a guerrilla war of attrition, avoiding pitched battles, and relying instead on artillery to break down forts and castles. Twelve years of defeat followed for the English, only halted when Du Guesclin died in 1380.

A cannon, or *pot de fer*, from around 1400. Early cannon fired arrow-like darts, but by the mid-15th-century, using cast-iron balls, they could wipe out fields of archers, changing the face of warfare for ever.

But those tactics were forgotten by 1415: at Agincourt another generation of Frenchmen were slain on the field of battle by the English longbowmen. The French artillery was ineffective, too heavy to move and incapable of following the action. It is thought that only a single Englishman was killed by French artillery.

By the Battle of Formigny in 1450, however, improvements in casting techniques led to lighter, more mobile guns. This was the first major battle won by the decisive use of artillery. The French drew the English in to a trap, opening up with light-weight culverins firing 3–4 kg iron balls at a range of 1,000 metres. In the end, 4,000 English longbowmen, footsoldiers and mounted troops lay dead.

The advent of the cannon had ensured that war would never be the same again.

Spinning Wheel

Britain

1325

For several millennia the spinning of thread to weave cloth on primitive handlooms had been achieved using a hand-held spindle. This was a simple device of a rod of bone or a wooden stick with a disc, usually of stone, attached towards the lower end. The start of a thread of linen or wool was attached to the spindle and the thread was teased out by the twisting and falling action of the spindle. The thread was then wound on to the spindle.

It is surprising that it took so long for the spinning wheel to develop. All it required was for the spindle to be turned on its side so that it lay horizontally on bearings and for the spindle to be attached by a belt drive to a larger wheel that was spun by hand. The inertia of the heavy-rimmed wheel produced a continuous, even motion. This was a simple concept but a significant step forward on the long path to creating a modern industrial society.

It is probable that the idea first developed in China with the rimless *charka* spinning wheel. From there it took hundreds of years to cross first to India, and then on to the Mediterranean and eventually to Western Europe. Along the route west it changed in design detail according to the

Spinning thread on to
a simple wheel in
India.

construction materials and methods available and the nature of
the local material to be spun.

There is enough evidence to suggest the so-called Jersey
Wheel was established in homesteads in Britain by the early
14th-century, spinning mainly wool and flax, and operated by
the women in the family. Several spinsters would be required
to supply the (usually male) weaver.

It was this technology that helped the great wool towns,
such as Sudbury, Worcester and Ashburton, develop in the
Middle Ages. The manufacture of cloth suited and
strengthened the family unit and, provided it was undertaken
as a part-time occupation, met the cloth needs of a family.

Around the late 16th-century, several refinements were
added to the machine. Chief among these was treadle power,
which left both hands free. Another improvement was the
"flyer", a device which wound the thread as it was spun. Later
mechanisms introduced after 1800 enabled the bobbin to be
wound evenly.

The spinning wheel in the 19th-century was gradually supplanted by industrial, mechanized versions and processes. It became the Boudoir Wheel, an over-elaborate and sometimes impractical toy often owned by the middle classes to enable their daughters to learn the skill of hand spinning which, by then, was already becoming lost to the factory system.

Blast furnace

1350

Britain

A modern Russian steelworks. Early blast furnaces were fuelled by charcoal, but the discovery of coke by Abraham Darby revolutionized iron-and-steel making and ushered in the Industrial Revolution.

The Romans introduced furnaces in Europe for producing wrought iron but these produced brittle iron with large amounts of impurities.

The invention of the blast furnace in the 14th-century allowed greater production of more useful pig iron, and expanded the uses of iron. It is hard to pin down exactly where and when blast furnaces were first used, but recent excavations at Rievaulx Abbey in Yorkshire revealed the foundations of a sophisticated furnace dating to around 1350. The Cistercians of Rievaulx were renowned for their inventive skills and analysis of ore samples from the site indicate the monks were able to produce very pure iron.

Early metal workers knew that blowing air through a furnace would increase the temperature. The blast furnace took this principle a stage further. The metal ore was mixed with extra fuel, probably charcoal, before it entered the furnace. As this mixture started to burn hot air was forced through it. This raised the temperature and purified the molten iron. The

iron ran out through a tap and the impurities were easily scooped from the surface.

In 1621, the English ironmaster Dud Dudley perfected the use of coal, and Abraham Darby refined the process in 1709. Darby's coke-fuelled furnaces led to an explosion in the use of steel and iron, building blocks of the Industrial Revolution.

Firearms

Europe, America

The early "hand gonnes" that appeared in the late 14th-century were nothing more than small versions of the cannon developed in the wake of the invention of gunpowder. An iron tube, the barrel, was blocked at one end and a small opening, the touchhole, was cut through the wall of the barrel. Powder was poured down the barrel and a ball of lead or iron rammed home on top. A pinch of powder, the priming, was placed over the touchhole and ignited by pressing into it the tip of the match, usually a smouldering piece of cord. The flame from the burning powder passed through the touchhole to ignite the main charge in the barrel, generating a massive volume of gas that forced out the bullet.

It took several evolutionary steps for this rudimentary hand weapon to become the pistol and rifle of today. The first changes were concerned with improving the method of lighting the priming charge. Matches were replaced with mechanical and chemical means of generating a spark. The final stage was to combine the priming with the explosive and the bullet in a single cartridge that could be loaded simply from the breech end of the rifle.

By the late 15th-century, the simple "hand gonne" had evolved into the matchlock musket when the barrel of the handgun was lengthened and fixed to a wooden stock. In place of the hand-applied burning match, the priming was ignited by the mechanical matchlock, a smouldering cord on an S-shaped pivoted clasp called a serpentine. Pulling the bottom of the serpentine up plunged the tip, with its match, into the flashpan, where the priming was ignited. The matchlock was refined with the use of a spring catch, called the snap matchlock, manufactured by locksmiths.

14th C

From top: a bronze hand gun from about 1500 with a hole in the barrel to insert a match, and a hook below the barrel to rest on a wall to take the recoil; the cheap and reliable flintlock pistol made by Robert Murdon of Charing Cross, 17th-century; the all-weather percussion pistol, Forysth Patent Gun Company, London, from 1809; cutaway of a modern self-loading pistol with metal cartridges.

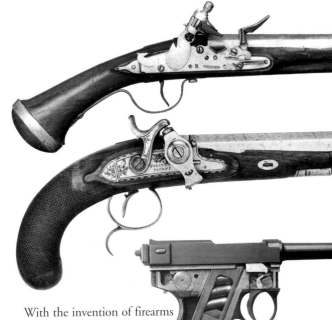

With the invention of firearms that could be fired by a lone gunner, knights in armour could be struck down by anyone with the briefest training. During the 16th-century, musketeers replaced archers, and although it was possible to make armour strong enough to stop a bullet, its greater weight made it too heavy for comfort. By the early 18th-century, most infantry and cavalry had discarded it completely.

Although the matchlock musket's ignition system had been mechanized, the burning cord match was still susceptible to wind and rain. Early in the 16th-century a mechanical system of generating a spark, the wheellock, was invented. This had

iron pyrites held against a serrated wheel that rotated when the trigger was pulled, generating a shower of sparks to ignite the priming. A burning match requiring constant attention was no longer needed. This made it possible to produce small self-contained firearms, such as the pistol, which could be loaded and carried ready to fire from a pocket or holster.

The wheellock was a major step forward but the mechanism was expensive and unreliable. In the late 16th- and early 17th-centuries a simpler system, the flintlock, was developed. Sparks to ignite the priming were produced by scraping a piece of flint down a steel plate. The flintlock was simple, cheap, reliable and could be made in any size. For two-and-a-half centuries it was the main firearm ignition system.

In 1805, Alexander Forsyth, a Scottish clergyman, keen amateur chemist and shooter, patented an even simpler system. In place of flint and steel he used chemicals, which, when struck, generated the sparks needed to ignite the priming powder. By the 1820s, the system had been improved by supplying the chemical deposited inside small copper caps that fitted over a short tube above the touchhole. These percussion caps were impervious to weather and needed only a simple mechanism to trigger the explosion. They opened up new fields of firearm development and made practical multi-shot weapons easy to produce. There had been attempts to produce revolvers using the earlier flintlocks and wheellocks but they were unreliable, cumbersome and expensive. In 1836, Samuel Colt, a Yankee, patented a simple, reliable revolver and by the mid-19th century multi-shot percussion revolvers were widely available.

Even with the percussion cap, the powder and bullet were still loaded in from the muzzle. What was wanted was a simple system of loading all three from the breech end. One early success was the needle gun invented around 1828 by Johann Nikolaus von Dreyse. It used a paper cartridge filled with powder and the bullet, and with the cap positioned in the powder. This cartridge was loaded into the barrel through a section that gave access to the breech. When the trigger was pulled a spring pushed forward a long needle that pierced the paper and struck the cap to ignite the priming charge, which lit the powder and fired the shot.

In 1857, the Smith and Wesson Company of America developed a metal-cased cartridge with the detonating chemical in the base. This was loaded into the cylinder of the revolver and as the hammer struck the base it exploded the compound. It was later improved by placing a percussion cap in the centre of the cartridge base, a system still used today. By the 1860s the rifle and the revolver as we know them today were with us.

Breech-loading metal cartridges increased the potential firepower of every soldier. In place of the three rounds a minute of the old flintlock musket, a First World War British infantryman could fire 10 rounds from his Lee Enfield rifle. So rapid was the firing that the Germans thought they were facing machine guns. The metal cartridge also made possible the development of self-loading pistols, which vastly increased the firepower of the handgun.

Carpenter's brace

Belgium

1400

The modern carpenter's reliance on expensive cordless electric power tools would be scoffed at by mediaeval Flemish craftsmen. Around 1400 they were commonly using the carpenter's brace, or wheel brace, and it quickly became the universal rotary tool for most woodworkers until the late 20th-century.

Consisting of a cranked handle and a swivelling chuck or socket, the brace was the first complete crank. It could be used almost anywhere, in a shipyard or on a construction site. By the 19th-century the brace was used to rotate a number of bits, such as an auger for boring long, wide holes, smaller bits for general drilling and screwdriver bits for fastening.

Spectacles

Florence, Italy

1451

Although precious stones and minerals ground to the shape of lenses are known from the ancient world, the earliest definite references to glass lenses being made to improve vision date from the 13th-century. One of the first comes from the writings of the Franciscan friar in Oxford, Roger Bacon. He described the magnifying properties of a convex lens, as well as the many distortions that could be achieved with transparent shapes of glass and other materials. But such feats were regarded with misgiving. If they were possible at all, they could not be natural and must therefore be illicit. Bacon fell under suspicion of sorcery or necromancy.

From the beginning, such optical devices were morally suspect. When lenses became more generally used through their application to spectacles, which were invented in Italy in the late 13th-century, some account of their moral integrity was urgently needed. One means of making them legitimate was to account for their operation within the science of optics – to reduce the sorcery to geometry. Another was to explain that the decay of human eyesight was a long-term consequence of the Fall in the Garden of Eden. Adam, created in perfection, had no need for spectacles. Spectacles could thus be associated with an impulse to recover Adam's state of grace.

The first spectacles had convex lenses, used to correct long-sightedness, and thus to prolong the possibility of reading and other close work beyond middle-age. Spectacles with concave lenses, the kind that correct for myopia by improving distance vision, appeared and spread with the introduction and widespread distribution of printed books. The earliest reference is of their sale in Florence in 1451, though they were probably not common in northern Europe until the mid-16th century.

A fanciful painting of Virgil. Spectacles were not invented until some 1,500 years after his death.

Printing Press
Gutenberg

I n modern Western civilization, the key means of communication, at least until the 20th-century, was the written word, and its instrument was the printing press. Traditionally, the invention of this seminal device is attributed to Johannes Gutenberg of Mainz, in Germany, who had produced a sufficient quantity of movable metal type to print the Vulgate Bible in 1454. While there is no reason to deny Gutenberg this honour, there are difficulties in the attribution.

In the first place, like so many of the great mediaeval inventions (windmill, compass, gunpowder) the idea had Chinese precedents, as block-printing had already been developed there around AD 500 and could have inspired emulation in the West, where it was being used by the 15th-century for playing cards and for "block books" compiled from pages printed from single wooden blocks.

By AD 868, the Chinese had produced the first printed book, the *Diamond Sutra*, and in 1041 Pi Sheng, a Chinese alchemist, invented movable type made of an amalgam of baked clay and glue characters placed on an iron plate coated with a mixture of resin, wax and paper ash.

The pictoriographic nature of Chinese writing did not make it easy to adapt to printing by movable type (a typical Chinese newspaper has 5,000 different characters) but Western scripts, with their limited number of characters, were able to exploit the idea.

It could also be argued that the success of printing involved merely the transfer of techniques already well established in other spheres. The printing press, for instance, was directly related to the heavy presses used for crushing juice out of olives and grapes. The oil-based ink used by printers had, moreover, already been made available by painters who had been experimenting with new pigments.

The casting of type received a great boost from metallurgical advances, although Gutenberg's choice of a lead-tin-antimony alloy for his type appears to have been a personal inspiration. And the invention of paper made from pulverized waste rags and water came at a convenient moment. Although

some books were initially printed on traditional parchment or vellum, the press could not have continued to operate satisfactorily without paper.

However, all great inventions depend upon the successful integration of good ideas developed for other purposes, and Gutenberg certainly deserves the credit for the inspiration which brought them together at the beginning of the European Renaissance. By setting the characters firmly in a "matrix" which could be brushed with a light covering of ink and then pressed down evenly upon a sheet of paper to produce a printed page, he devised a process which was capable of indefinite repetition and which, when complete, allowed the type to be broken up and re-used.

Gutenberg and his successors found an enormous market for their technique, so that it spread widely in Western Europe. By the end of the 15th-century there were almost 40,000 recorded editions of books printed in 14 European countries, with Germany and Italy accounting for two thirds of the total.

In England, the lead was taken by William Caxton, who acquired the new technology in Cologne and then introduced a press at Westminster in 1476. He printed almost a hundred volumes before his death around 1492, including the works of Chaucer and Malory and translations of works on religion, chivalry and chess.

Caxton's publications are not usually regarded as being of the highest aesthetic quality, but by the end of the 15th-century

Left: St Jerome, translator of the Vulgate Bible, below, the first book to be printed by Gutenberg. About 30 copies of this Bible were printed on parchment the rest on paper.

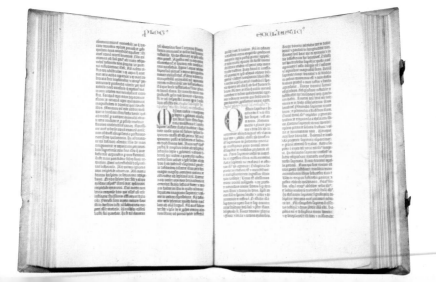

some books of very high quality were being produced, particularly in Italy by the Aldine Press of Venice and others, using an elegant version of "italic" type. By adopting a small format and establishing large print-runs of a thousand copies or more, these Italian printers made available to an increasingly literate public a wide range of classical and modern authors.

From the beginning of the printing press, illustration was provided by incorporating wood-cuts in the matrices of the type, and in some of the publications of the 16th-century, such as the German scientist Georgius Agricola's great work on metallurgy, *De Re Metallica*, published in 1556, these illustrations achieved considerable sophistication.

Thus equipped, the printing press enjoyed a long period of technical maturity. Leonardo da Vinci was involved in adapting it to produce coins and suggested a water-powered version which was later adopted. Otherwise it remained little changed – except for detailed refinements and the introduction of metal engraving techniques – until the 19th-century.

Gutenberg failed to capitalize on his invention, but he left behind one of the most influential of modern inventions, making the printed word an essential medium of communication in politics, social comment, religious argument and scientific progress. It also provided, through ephemeral publications such as pamphlets and news-sheets, a tremendously important means of disseminating information throughout society.

Johannes Gutenberg's attempt to invent the printing press is a classic tale of an idealistic inventor exploited and ruined by a venture capitalist. Born in Mainz in about 1398, Gutenberg (left) was the son of the city's Master of the Mint. As a boy, he often visited the Mint, where he watched goldsmiths stamp the coins. He also visited the city's monasteries, where he watched monks laboriously writing the Bible by hand.

In 1428, after serving an apprenticeship as a goldsmith, Gutenberg

Johannes Gutenberg invented movable type and printed the first books, but he was ruined by a venture capitalist.

moved to Strasbourg, where he began his experimentations with movable type and met Andreas Dritzehen, a German who financed him in partnership with two others. Dritzehen died soon after and Gutenberg abandoned his experiments when Dritzehen's partners threatened to steal his invention.

By 1450 Gutenberg was back in Mainz, where he met Johann Fust, a lawyer who lent him 800 guilders at six per cent annual interest in return for an insight into the secret of his invention.

Gutenberg set about making a press and manufactured a set of dies to make individual lead letters. However, he ran out of funds and could not afford the lead, ink, paper and vellum (for which the hides of 10,000 calves were required) to begin printing.

Fust seized the chance. He offered Gutenberg a further 800 guilders in return for a share of the profits. Gutenberg agreed and by 1448 he was ready to begin printing a Bible with two columns of 42 lines in Latin on each page.

At the moment the Bible was ready to be printed, Fust destroyed Gutenberg by calling in his loans plus interest. The sum amounted to more than 2,000 guilders. When Gutenberg was unable to repay his debt, Fust sued him and Gutenberg was forced to hand over his press, his tools and his business.

In 1455, Gutenberg, now under the employ of Fust, printed 300 copies of the 1,282-page Gutenberg Bible. This first book ever printed is now also considered to be the most beautiful, but its inventor received not a penny. Gutenberg, who never married, died in 1468, destitute and forgotten by his contemporaries.

Today the remaining copies of his Bible are more highly valued than any other books.

Toothbrush

China

1498

Chew sticks, twigs with one end frayed into bristles, have been found in Egyptian tombs dating back to around 3000 BC, but the first mention of a toothbrush with bristles at right angles to the handle is in a Chinese

encyclopaedia of 1498. The bristles, set into bone or bamboo handles, were plucked from hogs.

The toothbrush travelled to Europe in the 17th-century, where it soon became widely used. Americans, however, did not get into the daily habit until after the Second World War, when GIs were forced to brush their teeth regularly by the army.

The nylon toothbrush was invented in 1938, but remained too stiff to be used painlessly until the early 1950s. Since then, more than 2,000 kinds of toothbrushes, including mains- or battery-powered models, have been patented. The latest use sonic waves, which change colour when the user scrubs too hard, to protect the teeth and gums of over-zealous brushers.

The Romans invented toothpaste and mouthwash. They prized Portuguese urine as an ingredient; its ammonia content is an efficient bleach and is still used, albeit in the form of synthesized ammonium compounds, in modern toothpastes.

The American Rodius toothbrush.

Watch

Germany

1500

Pocket watches, made in gold and silver, were affordable only to the wealthy.

Peter Henlein, a Nuremberg craftsman, is usually credited with the development towards the end of the 15th-century of miniature, spring-driven, key-wound timepieces, small enough to be worn on a chain or in a pocket.

By the end of the 15th-century, improvements in spring-making had made watches possible, but such was the irregularity of power given by these early springs to the oscillating balance (which is the heart of any watch) that watches needed to be fitted with either a "fusee" or a "stackfreed". These devices evened out the power differences between a fully wound and an unwound spring, but even with

one of these regulators the first watches kept time to no better than within around 30 minutes a day.

Surviving watches from this early period are mostly drum shaped with gilt-metal cases, with just an hour hand on the dial and, if fitted, a pierced metal lid. Only with the invention of the balance spring was the minute-hand fitted to watches, although minute-hands had appeared on clocks a century earlier; their first recorded use being on the early precision clocks made by Jost Burgi for the Danish astronomer Tycho Brahe around 1577.

A pocket watch with a calendar inscribed in its lid, from 1635.

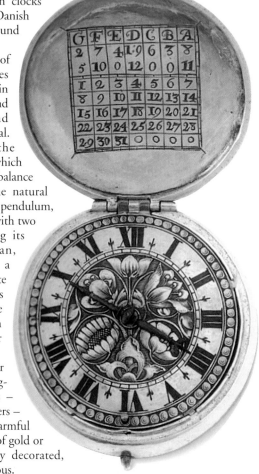

Following the invention of the balance spring, watches began to keep time to within 10 minutes or less a day and gradually the second-hand began to appear on the dial. The development of the isochronal balance spring, which controls the oscillations of a balance so that it operates with the natural evenness (isochronism) of a pendulum, became a contentious affair with two eminent scientists claiming its invention. An Englishman, Robert Hooke, designed a watch with a spring in the late 1650s but Christian Huygens of Holland was probably the first person to design a watch with a spiral balance spring, or hairspring, in 1675.

The watch became smaller and thinner as mainspring-making improved and cases – with rock crystal or glass covers – fitted better to keep out harmful dust and dirt. Usually made of gold or silver, and often elaborately decorated, watches became highly precious.

The 18th-century saw great strides in technology. The modern mechanical watch has its roots in the work of London-based makers of the period. Thomas Mudge invented the lever escapement in the 1750s. Building on the pioneering work of John Harrison, the chronometer makers John Arnold and his archrival Thomas Earnshaw developed the bi-metallic compensated balance during the 1780s and 1790s, which improved accuracy.

Division of labour within the trade meant that a watch, by the end of the century, contained the work of 30 or more specialist craftsmen and women. The result of these laborious efforts was exquisite, handmade watches that were still high priced and were now accurate to within a minute a day.

At the turn of the 19th-century, first Antoine Lepine and then Abraham Louis Breguet, both working in Paris, developed the style of the watch and the layout of the dial into what is familiar today. Many makers throughout Europe at this time were experimenting with keyless and even self-winding mechanisms in order to get rid of the troublesome, and eminently losable, key.

Steady industrialization throughout the 19th century, especially by American manufacturers such as Waltham and Elgin, saw timekeeping and dependability improve at the same time as the price of a watch steadily fell. By the end of the 19th-century, owning a reliable watch was within the reach of all working men due in large part to the firm of Robert Ingersoll, who was the first to market (via mail order in 1892) a watch costing the average day's pay of one dollar, coining the phrase "the watch that made the dollar famous".

This period also saw the watch move out of the pocket and on to the wrist. What was at first regarded as simply a "ladies fashion" gained masculine acceptance after being used so successfully by soldiers in the First World War.

The extensive and exhaustive war effort also marked the beginning of the end for both English and American watch production and the subsequent dominance of Swiss manufacture.

Etching
Urs Graf

Etching, a method of making prints by incising the design on to a metal plate, evolved from the practice of etching designs on to metal armour. Urs Graf, a Swiss artist, made the first dated etching in 1513. The technique remained popular into the 20th-century and was adopted by Picasso, Matisse and Chagall, although Rembrandt is regarded as certainly the first and probably greatest master of etching.

Woman Bathing (1658) by Rembrandt.

Warship
Henry VIII

Until the end of the 15th-century, most European warships had been adapted merchant vessels, equipped with one or two cannon, usually sited on deck. Most were clinker-built, strong and easy to construct, but there were limits to size, and cutting openings for gun ports was problematic as it weakened the shell.

These ships were often little more than sea-going castles; unsteady constructions with built-up wooden citadels to give protection to archers or primitive artillery. As the artillery became heavier and larger, the task of carrying the guns safely and firing them without sinking the ship became difficult.

The Naval Dockyard at Portsmouth was founded in the late

Henry VIII commissioned warships, such as the *Mary Rose*, whose design changed the balance of sea power.

15th-century by Henry VIII with the intention of developing powerful new fighting ships to rule the seas. It began 30 years or more of constant development, the first fruits of which were a sequence of successful new ships built by Henry VIII around 1513, including the *Mary Rose* and the *Henry Grâce à Dieu*. They had smooth carvel planking, built on a skeleton of sturdy timber frames. This soon became the way all great warships were built.

The new ships, for the first time in history, were built around their guns. The main gun decks were sited lower in the ship, for stability. The sides of the hull sloped inwards, from the gun deck upwards, an effect called "tumblehome", another attempt to lower the centre of gravity. The deep-draughted displacement hulls had stern-hung rudders, up to four masts and could sail much closer to the wind than older ships.

Cast-brass guns came into use at the turn of the 16th-century, giving the English advantages in firepower and accuracy. By 1547 a standardized range of guns had been developed with shot ranging in size from 15 kg down to 2 kg, a system that lasted for 250 years.

The main intention in earlier naval conflict was to get close to the enemy, degrade or disable him with whatever weapon that would reach, and then to board and fight for the ship. This is what the Spanish were waiting to do in 1588, and the development of the true cannon-armed warship by the English put paid to these ancient tactics for ever. As the enormous Armada which had been assembled by Philip II of Spain edged nearer to the English Channel, their complement of sailors and marines, priming their weapons, would have been expecting to get close in to the English fleet. Then they would board and cut the English to ribbons.

When the moment of conflict came, it developed into a numbing defeat, a pursuit around the British Isles and the loss of two thirds of the Spanish force, including tens of thousands of men.

The technological advances by the shipwrights and armourers of England allowed the nimble, heavily-armed carracks of Francis Drake to run rings around the Spanish ships, picking them off from long range with their guns before the enemy had a chance to reply. The Spaniards never got close enough to wield their swords and fire their musketry.

1519 Spanish explorer Hernan Cortés brings cocoa bean back from Mexico, beginning the spread of chocolate.

Coining mill

Augsburg, Germany

L eonardo da Vinci, who spent several years working in the papal mint, is said to have been closely involved in the adaptation of the printing press to mint coins, but the coining mill is thought to have originated around 1525 somewhere near Augsburg, not far from Mainz, the birthplace of Gutenberg's printing press.

The first coining mills, which used rollers to produce metal coins, were capable of minting coins only of copper, which had a lower melting point and consequently was softer and easier to work. Rolling mills would not be able to work with iron until the advent of steam power in the 18th century.

However, by sometime after 1583, when an English visitor to Spain observed a coining mill in action, the machine had evolved to incorporate separate processes of first rolling the metal, then stamping and cutting the coin.

"At the Mint of Segovia in Spain there is an engine that moves by water, so artificially made that one part of it distendeth an Ingot of gold into that breadth and thickness as is requisite to make coyn of it; it delivereth the plate that it hath wrought onto another that printeth the figure of the coyn upon it, and from thence it is turned over to another that cutteth it in due shape and weight," the visitor wrote.

The new minting processes, which gave a milled edge to the coins, led to a reduction in the illegal clipping of coins, though it was not until the 1630s that the first substantial amounts of milled moneys were produced in London.

Ceramic plates

France

1530

1530 The first
public lottery
with cash prizes intro-
duced in Florence.

The Romans had used plates for serving food but they fell out of use following the fall of Rome in AD 410. People either ate from bowls or used thick slices of bread, called trenchers, to carry their food until the ceramic plate was re-invented for the banquet celebrating the second marriage of Francis I of France to Eleanor of Portugal in 1530.

False limbs

Ambroise Paré

1536

Ambroise Paré, a
military surgeon, used
his knowledge of
anatomy to design a
range of artificial
hands and limbs.

Necessity may be the mother of invention, but compassion has been godmother to a fair few. Until the mid-16th century, little was done to improve the lot of those born without limbs, or who lost them in accidents or war. Then a French doctor showed that medics just hadn't been trying.

When Ambroise Paré began work as a military surgeon in 1536, aged 26, the techniques used were indistinguishable from those of a butcher. Badly damaged limbs were treated by amputation: the stump was sealed with boiling oil and patients were packed off to cope as best they could.

Paré decided this was not good enough. Following the teaching of Hippocrates to do as little harm as possible, he minimized the amount of surgery, and used hygiene, oils and ligatures to treat wounds more effectively and humanely. But he reserved his greatest innovations for those whose limbs he could not save. Using his knowledge of anatomy and an all-too-ready supply of patients, Paré designed a range of artificial arms and

A hand designed by
Paré.

hands. As his skill increased, so did the ingenuity of his solutions: one of Paré's prosthetic hand designs had fingers that could be moved individually via a set of gears and levers.

Paré's biggest contribution was not so much the devices he made – the sheer numbers left disabled by war meant that simple poles and hooks would have to suffice for most. Rather, it was by making disabled people a focus for ingenuity.

Only since the Second World War has that ingenuity been matched by technology to make prosthetics successful, however. In the past 20 years, medical technologists – many in military hospitals – have used lightweight materials, sophisticated mechatronics and electronics to create so-called myoelectric limbs. These detect nerve impulses sent to a limb by the brain, allowing the disabled person to activate a prosthetic like the original limb.

Paré would surely have been delighted by such miracles.

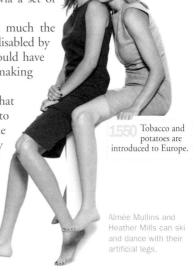

1550 Tobacco and potatoes are introduced to Europe.

Aimée Mullins and Heather Mills can ski and dance with their artificial legs.

Lead pencil

England

1564

The modern lead pencil (which got its name from the Old English word for "brush") became possible when an exceptionally pure deposit of graphite was discovered in 1564 at Borrowdale in Cumbria.

The following year, Conrad Gesner, a German-Swiss naturalist, recorded the first description of a writing instrument in which graphite, then thought to be a type of lead, was inserted into a wooden holder.

The breakthrough in pencil technology came in 1795, when Nicolas Conté, a French chemist, used a mixture of clay and graphite that was fired before it was glued into a slot in a thin cylindrical strip of wood that was sealed with a second, thinner wood strip.

Concrete

1568

Philibert de l'Orme

1582 The
Gregorian
calendar, as used
today, is introduced.

T he idea of mixing sand and gravel with a binder to make a tough construction material is about as old as construction itself: an early form of concrete was used to make the floor-slabs of huts built on the Danube more than 7,000 years ago.

The Egyptians developed a material similar to modern concrete for use on the pyramids, using lime and gypsum as binders, but it disappeared from use until Philibert de l'Orme, a Frenchman regarded as one of the great architects of the Renaissance, rediscovered it and published its composition. As a result, de l'Orme is regarded as the modern-day inventor of concrete.

Today, most concrete uses Portland cement. Its excellent binding properties were discovered by accident by a 19th-century bricklayer, Thomas Aspdin. Trying to make his own mortar, Aspdin over-heated a three-to-one mix of limestone and clay, which when mixed with water turned incredibly hard. Aspdin patented Portland cement in 1824, and it has been used ever since.

But while concrete is a byword for rock-solid toughness, it has to be used judiciously, as it is 10 times weaker in tension than in compression. This would rule out the use of concrete in huge bridges and dams – which bulge under load – were it not for an invention patented by a French gardener, Joseph Monier, in 1867. Looking for a way of strengthening flower-pots, Monier incorporated a metal mesh framework into concrete.

The metal in this "reinforced" concrete has the tensile strength pure concrete lacks, and led to concrete's becoming the material of choice in massive construction. Nowhere is its toughness more dramatically displayed than in the works of Fritz Todt, Hitler's construction chief, whose colossal structures such as the V2 launch complex at Wizernes, France, and the U-boat pens at La Rochelle have defied all attempts at demolition. They may, indeed, prove to be the only parts of the Third Reich to fulfil Hitler's dream of lasting for a thousand years.

Demolishing modern concrete structures is not something

Opposite: Cars set in concrete, an artwork from the César Foundation in Paris: though rock solid, concrete is 10 times weaker in tension than in compression, and has to be reinforced for building bridges.

to be attempted lightly, as many of them have been built using "post-tensioned" reinforcement. This is another means of boosting the tensile strength of concrete, using steel bars sheathed in plastic. Tightened with hydraulic jacks, these bars are under incredible tension. Accidentally releasing them during demolition could prove lethal.

Stocking frame

Rev William Lee

1589

Woollen stockings were a luxury afforded by few until William Lee, a clergyman from Calverton in Nottinghamshire, invented the stocking frame, the first practical knitting machine.

Lee is said to have invented the machine, in 1589, because his future wife was more interested in her knitting than in him. While his fiancée ignored him, he wondered why hundreds of small needles and hundreds of hooks could not be used to lift the loops over the wool in a single, simple action and thereby replace the two needles that were constantly working in his betrothed's hands.

With his first machine made, Lee forgot his calling as a cleric, and headed for London to request Queen Elizabeth I's patronage and the monopoly and patent rights to his invention. Disappointed that his stockings were coarser than silk ones from Spain, the Queen refused Lee a patent.

Lee returned to the Court a year later, when he had improved his machine to produce fine wool stockings, but fearing for the job security of tens of thousands of hand knitters, the Queen again refused him his patent rights. His new machine could produce material as fine as silk 10 times faster than hand knitters.

Lee headed for the Continent with his brother when he failed to secure funding from the City. With support from Henry IV of France, he manufactured stockings in Rouen, and prospered until Henry IV's assassination in 1610. Lee died in poverty and obscurity soon after.

His brother returned to England, where he found a sympathetic backer in the form of a merchantman from Nottingham. Against the opposition of the hand knitters they established the first frame-knitting factory in the north Midlands. It was a slow but sure success.

The principle of Lee's machine, with a so-called bearded-spring needle, is incorporated even in the fully shaped, or full-fashioned, stocking knitting machines of today.

Microscope

Van Leeuwenhoek

1590

The microscope was invented in about 1590 by Zacharias Jansen, a spectacle maker who combined a concave and convex lens, like the telescope in reverse, but failed to make any significant use of it. At about the same, Antonie van Leeuwenhoek, an insatiably curious Dutch linen merchant, succeeded in making a microscope that thanks to his glass-grinding skills produced a clear and impressively magnified image. Nevertheless, for the early part of the 17th-century microscopes were treated as toys and were often bought at fairs. This attitude changed around the middle of the century, not because of improvements in the instruments, but because of changes in assumptions about the world. The "mechanical philosophy" was gaining support and taught that all the phenomena we experience depended on the mechanical interaction of tiny particles or atoms. Since these could not be seen, the only hope of basing science on observation was to improve microscopes.

It was in this context that one of the most spectacular and influential books of microscopy was produced in England in 1665, the *Micrographia* of Robert Hooke. People were astonished to see an engraving of a common flea spread across a 35 cm folding plate. Blown up to this size, it seemed an enormous creature, with scales and hairs on its legs, unlike anything anyone had encountered before. This was, as Samuel Pepys recorded in his *Diary*, "the most ingenious book that I ever read in my life".

In Van Leeuwenhoek's microscope the object is held on the point of the rod and viewed through a minute aperture in two thin brass plates riveted together.

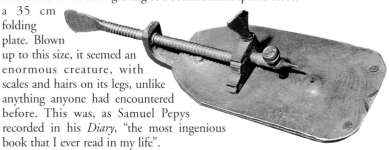

Not everyone was convinced: in the popular culture there was still something ridiculous about a man obsessed with the minutiae of his personal life, especially with fleas and lice. The dramatist Thomas Shadwell satirized Hooke in his play *The Virtuoso* as "one who has broken his brains about the nature of maggots... and never cares for understanding mankind". Hooke himself went to a performance; the audience, he wrote in his diary, were "Damned Dogs".

Only a little later Van Leeuwenhoek in Delft used tiny single-lens microscopes to show that invisible little creatures which he called "animalcules" were teeming everywhere – they seemed to be the alarming inhabitants of everything he examined. He also observed the connections between the arteries and veins, and gave particularly fine accounts of the microscopic structure of muscle, the lens of the eye, the teeth and other structures.

In a sense this sudden revelation of the invisible microscopic world and its inhabitants spelt an end for more than a century to what we would think of as the research application of the microscope. After all, if fleas were creatures as complex as anything in the animal world, and if the world was full of invisible little animals that had structure, motion, energy and life, there was no hope that microscopes would take us to the fundamental mechanical level of the atoms on which all this microscopic life, as well as our macroscopic world, was supposed to depend.

Thermometer

Galileo Galilei

1592

G alileo Galilei is generally credited with the invention of the thermometer, but the air thermometer that he created in 1592 was merely the crude trigger of a 150-year process of refinement that finally led to an instrument capable of accurately recording temperature.

The bulky and clumsy instrument invented by Galileo had a column of air trapped by water inside an open tube. A rise in temperature caused the air to expand and push up the level of the water against an arbitrary scale.

Despite the gross inaccuracy of the instrument (changes in air pressure affected the water level and made the water itself expand), Sanctorius, professor of medicine at Padua, applied it to measuring the warmth of the body by placing the bulb in the mouth of a patient, thereby inventing the clinical thermometer in 1594.

In the late 1680s, Guillaume Amontons, a deaf French physicist, refined the air thermometer by trapping the air with mercury instead of water. However, he complicated matters somewhat by keeping the volume of air fixed and then adjusting the amount of mercury in the device to read the temperature as a function of air pressure. Nevertheless, it was considerably more accurate than Galileo's thermometer.

Galileo Galilei invented the first air thermometer, and helped to develop the telescope.

Flushing toilet

Sir John Harington

Widely, though incorrectly, attributed to the 19th-century plumber Thomas Crapper, the seat of civilization – the flush toilet – was actually invented by Sir John Harington in 1597. Harington, a godson of Elizabeth I, maintained the fine English tradition of toilet humour by naming his creation the Ajax – Jacks or Jakes being a popular 16th-century term for the convenience.

Queen Elizabeth, who famously took a bath once a month "whether she needed it or not", was said to be delighted with the new Harington installed for her use at Richmond. She was rather less impressed by an accompanying tract on the subject. In his *Metamorphosis of Ajax*, Harington's prose was more Rabelaisian than mechanical, and he was banished by his godmother.

Alexander Cumming invented the "S" bend in 1775, and the first patent for a modern toilet belongs to him. Cumming's simple, sliding-valve

1597

John Harington, a godson of Elizabeth I, invented the flush toilet and the queen was delighted with it.

workings were bettered three years later by locksmith Joseph Bramah, who patented a privy with two hinged water valves. An original Bramah is still in use in the House of Lords.

The Public Health Act of 1848 saw £5 million invested in sanitary research and engineering in Britain. The brass and metal of the Bramah were gradually replaced with ceramics: Josiah Wedgwood, Thomas Twyford and John Shanks all played their part in toilet history. Indeed, Twyford is credited with the design of the one-piece toilet, the Unitas, in 1885.

The vice

1600

Europe

No workshop worthy of the name is without a vice, acting like an extremely strong but uncomplaining assistant who is willing to hold anything you're working on – and not yell if you miss.

The earliest vices were pretty basic: two leaves of metal hinged at the middle, one of which was fixed to the workbench, the other able to come over the top of the component being worked on, with a simple nut-and-bolt arrangement to grip it tightly.

By around 1600, the nut had been replaced by the decent-sized T-handle, which gave greater torque and allowed craftsmen to grip components very tightly. The force applied to the handle can easily be magnified by a factor of 100 or more.

Variations of the same theme were developed and adopted by different trades over the years, each evolving to particular demands. But it took the genius of the inventor Ron Hickman to see that even after 400 years the vice still hadn't reached its full potential. One of the best features of his world-famous Workmate, launched in 1971, is that it can be turned into a big, portable vice. Despite this, Hickman had huge difficulty finding a manufacturer.

English dictionary

Robert Cawdrey

The publication in 1755 of Samuel Johnson's *Dictionary of the English Language* is justly regarded as a landmark in the attempt to bring order to a living, evolving language.

Commissioned nine years earlier by five London booksellers seeking a definitive guide to English words, the two-volume work included more than 43,000 words – often wittily defined ("Dull: To make dictionaries is dull work"). It was also remarkable for its use of 118,000 quotations to illustrate the precise meanings of the words.

Johnson's novel approach was highly successful. His dictionary went into a number of editions, and for many decades it remained a benchmark for the English language.

Yet Johnson cannot be regarded as the absolute inventor of the concept of the dictionary. A short word list survives from ancient Mesopotamia dating back to the 7th-century BC. The ancient Greeks compiled several dictionaries, and by the Middle Ages many Latin dictionaries were in circulation.

The compilers of the first English-only dictionaries were motivated partly out of a desire to bring the word of the Bible to all – and partly out of sheer frustration over the variety of spellings in use. Schoolmasters were especially keen to bring some sense to the "disorders and confusions" in spelling.

The English dictionary was created to rationalize spelling.

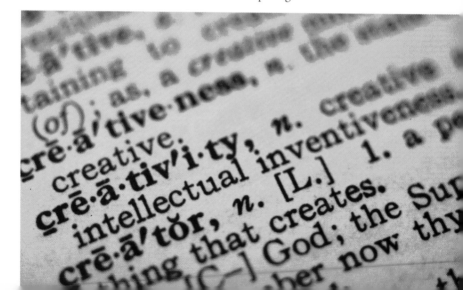

In 1604 Robert Cawdrey, a Coventry schoolmaster, published his "Table Alphabetical, conteyning and teaching the true writing and under-standing of hard usual English words". This is now regarded as the first English dictionary. However, containing only a few thousand words, its scope was far less than those available for continental languages, and it prompted a demand for something much more comprehensive.

With Samuel Johnson's dictionary, that demand was certainly satisfied.

Telescope

1608

Lipperhey, Jansen

1609 The Relation, the first modern newspaper, is published in Strasbourg.

B y combining convex and concave lenses at either end of a tube, the first telescopes were made. Galileo first heard of the device in 1609 as the work of "a certain Fleming" and modern scholarship has focused on two residents of Middelburg: Hans Lipperhey, who applied for a patent in 1608, and Zacharias Jansen, who may have an earlier claim but who may have been working from an even earlier Italian model. The precise origins of the telescope will probably never been discovered, but this is not surprising. The telescopic effect of early combinations may not have been very striking, but if anyone considered it worth exploiting, contemporary patent legislation would have afforded scant protection for an invention that was easily copied. It would have been more promising to have sought private patronage and market it as a "secret", the more so as these early instruments were thought to be applicable not to astronomy but to warfare. Galileo himself first sought to promote his telescopes on the basis of the advantage they would give over an enemy.

The fact that optical devices were still associated with a shadowy world of rumours, disputes, exaggerations and secrets was not a promising backdrop for Galileo when he sought to offer the instrument as the reforming tool of the ancient science of astronomy. He was not the first to apply a telescope to the heavens – so far as is known, that honour belongs to the Englishman Thomas Harriot (1560–1621) – but Galileo did marshal his observations in a powerful polemic for the

Copernican theory of the world, which has the Earth in motion round a stationary sun, and thus he placed his work at the centre of the cosmological debate of his time.

Galileo also turned his discovery of four moons of Jupiter into a gift to the Grand Duke of Tuscany, calling them the Medicean Stars, and thereby securing a prominent position at the court in Florence. His other discoveries included the mountainous surface of the moon, the phases of Venus, the strange "companions" to Saturn, the spots on the sun, and innumerable previously invisible stars.

In a sense Galileo was raising the moral stakes. Military gadgets and secret weapons were one thing, but how could Galileo presume to claim that a device that worked by deception – the nearness was, after all, an illusion – could enable its user to see more clearly into the nature of things? If the telescope showed stars never seen before, it was more likely that these also were illusions.

Hans Lipperhey of Middelburg applied for a patent for his telescope in 1608, though its precise origins are unsure.

Those who refused to believe Galileo's evidence are usually characterized as ignorant bigots, but the case is not so clear-cut. The supporters of the telescope might show the sceptics a distant object on Earth and then take them to examine the object at first hand, but contemporary Aristotelian philosophy taught that nothing about what happened on Earth could be extended to the fundamentally different world of the heavens.

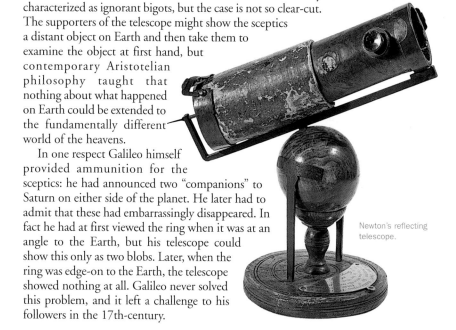

In one respect Galileo himself provided ammunition for the sceptics: he had announced two "companions" to Saturn on either side of the planet. He later had to admit that these had embarrassingly disappeared. In fact he had at first viewed the ring when it was at an angle to the Earth, but his telescope could show this only as two blobs. Later, when the ring was edge-on to the Earth, the telescope showed nothing at all. Galileo never solved this problem, and it left a challenge to his followers in the 17th-century.

Newton's reflecting telescope.

Surveyor's chain

1620

Edmund Gunter

The surveyor's chain, made of 100 links, was easy to carry.

W e have a former Rector of St George's Church in Southwark and Gresham College professor of astronomy to thank for the way in which until recently most of the English-speaking world divided up its land.

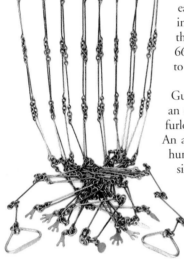

In the 16th-century, the mile was redefined and divided into eight furlongs, each of 625 feet. Then, in 1593, a statute introduced a shorter English foot so that the furlong, divided into four rods, was 660 ft long. The mile thus gained 280 feet to become 5,280 feet.

Into this confusion came Edmund Gunter. His masterstroke, which made him an early metric visionary, was to divide the furlong by 10, and to call that length a chain. An area of one chain by one chain was one hundredth of an acre, while a square having sides of ten chains each made an acre.

Gunter constructed an actual chain of 100 links that surveyors could easily carry and use to measure ground. It made perfect sense and caught on quickly.

Submarine

1620

Cornelius van Drebel

M ediaeval manuscripts claim that Alexander the Great used submersible vessels to defend his ships from divers attacking their anchor chains in Tyre harbour in 332 BC.

Whatever the truth, the first serious proposal for a genuine "submarine" capable of manoeuvring underwater surfaced in 1578. The British mathematician William Bourne envisaged a leather-coated craft propelled by oars and capable of submerging by shrinking in volume using a series of hand-vices. This made the vessel more dense, causing it to sink. When the vices were released buoyancy was restored.

Almost 50 years later, in 1620, the Dutch inventor Cornelius van Drebel built a craft along the lines proposed by Bourne. It successfully dived to depths of up to five metres in the Thames, an astonishing feat. James I reputedly took a trip in the vessel.

Despite this early British interest in the submarine, the Americans were the first to recognize its military value. During the American War of Independence in 1776, the British warship HMS *Eagle* became the first victim of a submarine attack, when it was damaged by a mine planted by an American submarine, the *Turtle*, in New York harbour.

The Royal Navy proved to be the most reluctant of all the major naval powers to adopt submarines. In 1900, one senior Admiralty figure described them as "underhand, underwater

Submarines, such as this one in Pearl Harbor, are key components in modern defence.

and damned un-English". Even so, the following year a small number of submarines, built to the design of the American inventor John Philip Holland, went into service. The effectiveness of submarines was demonstrated in the First World War when the German U-boat U-35 sunk 324 merchant ships.

In the late 1940s Hyman Rickover, a US Navy chief, pushed through the adoption of what proved to be the single biggest advance in submarines: nuclear power. The first nuclear-powered submarine, the USS *Nautilus*, set sail in 1954. Using its reactors for propulsion and to make its own air via electrolysis, it was able to travel submerged at 20 knots for months on end.

Because of their virtual undetectability, vessels such as Britain's Vanguard class of nuclear submarines, armed with Trident ballistic missiles, became a key part of Cold War strategy for the next 50 years.

Slide rule

William Oughtred

The work of John Napier, a Scottish mathematician, helped to give birth to an icon of science.

Before the advent of cheap electronic calculators in the 1970s, no self-respecting scientist would be seen without the icon of the boffin's craft, the slide rule. Able to perform multiplication, division and much more, simply by moving one part of the rule against the other, the workings of the slide rule often seemed mysterious to outsiders. Yet the basic principle, first exploited in 1621 by the English mathematician and cleric William Oughtred, is easily demonstrated.

Place two rulers on top of each other, so that their centimetre scales run in the same direction. Choose a number on the lower ruler – say, 11 – and slide the zero of the top ruler's scale along to it. Now the result of adding, say, 6 to 11 can be found by noting the number on the lower ruler that lies opposite 6 on the top ruler. Subtraction can be done the same way, after placing the rulers so their scales run in opposite directions.

Oughtred realized that by replacing conventional scales with logarithms, invented by the Scottish mathematician John Napier some years earlier, it would be possible to turn this adding and subtracting ability of scales into multiplication and division.

For some reason Oughtred failed to publicize his first slide rule, circular in design, until 1633, and only then after becoming involved in a bitter priority dispute with a former student, Richard Delamain, who claimed to have invented a circular slide rule in 1630. While both mathematicians appear to have invented the circular slide rule independently, historians give Oughtred sole credit for inventing the familiar straight version.

The slide rule. and a machine for calculating logarithms designed by John Napier in 1617.

In the years that followed, many special-purpose slide rules exploiting the same idea were developed. During the 1950s, engineers at the American aircraft company, Northrop, invented a circular slide rule the size of a coffee-table to use for aerodynamic calculations.

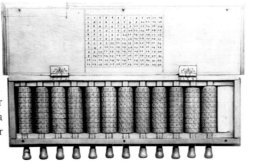

Calculator
Wilhelm Schickard

1623

I n 1935, Franz Hammer, a German historian who was then sifting through some old manuscripts, made a discovery that immediately overturned three centuries of academic thinking. Until then, historians of science had credited the brilliant French thinker Blaise Pascal with the invention of the world's first mechanical calculator some time around 1642.

1623 First modern
patent law
passed.

But while studying correspondence between the great astronomer Johannes Kepler and some of his 17th-century contemporaries, Hammer came across a small drawing of a gadget sent to Kepler by his friend Wilhelm Schickard, a German mathematician. Hammer realized that the drawing was of a calculating machine built by Schickard in 1623, almost 20 years before Pascal's effort.

Tantalizing hints that Schickard had beaten Pascal to the invention had been found in other manuscripts, but the details were assumed lost in the chaos of the Thirty Years War, which engulfed Germany in the late 1620s.

Hammer's discoveries made Schickard's claims of priority much stronger than previously thought. Not until 1957 did Hammer publicize the evidence, however, by which time he had found another drawing of the machine, together with a manuscript containing mechanical instructions.

Any doubts about Schickard's rightful place in history were finally dispelled in 1960, when Dr Bruno Baron von Freytag Loringhoff, an expert on mediaeval mathematics at the University of Tubingen, used the manuscripts to construct a working version of Schickard's device.

A replica of the first calculating machine, designed by Wilhelm Schickard. Six vertical cylinders convert numbers to logarithms.

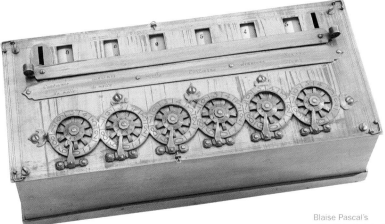

Blaise Pascal's
calculator, invented 20
years after Schickard's.

Looking like a miniature barrel-organ, the machine is really two devices stacked on top of one another. A set of six vertical cylinders converts any given number into their corresponding logarithms, with a set of gears adding or subtracting the result, thus allowing multiplication and division. Schickard developed an ingenious set of gears to deal with the tricky problem of "borrowing" and "carrying" when the machine was performing long division or complex multiplication.

In comparison, Pascal's calculating machine seems little more than an elegant toy. Essentially a collection of gears connected to dials, it could only efficiently perform addition; any other arithmetic function required considerable effort.

Even so, the so-called Pascaline was immediately hailed as a stunning display of ingenuity, having demonstrated that at least simple tasks performed by the human mind, such as arithmetic, could be performed by a machine. This proved to be the Pascaline's most important achievement. However, its unreliability led to its commercial failure, with perhaps no more than a dozen being sold.

Thirty years later, the third and most successful of the earliest mechanical calculators was unveiled by one of the few intellectuals in the same league as Pascal: the German mathematician Gottfried Wilhelm Leibniz.

During a trip to Paris in 1672, Leibniz pondered the ability of pedometers to count the steps taken by a walker and convert them into distance.

"It occurred to me at once that the entire arithmetic could be subjected to a similar type of machinery," Leibniz later recalled, "so that not only counting but also addition and subtraction, multiplication and division could be achieved by a suitably arranged machine easily, promptly and with certain results."

Leibniz set about constructing what became the Stepped Reckoner, a device whose capabilities far outstripped those of Pascal's machine. Consisting of a complex arrangement of gear, axles and dials held in a long wooden case fitted with a handle, it was designed to deal with numbers up to 12 digits long and to perform all four arithmetic functions. A major advance was Leibniz's invention of a gearing system able to perform multiplication mechanically.

As such, the Stepped Reckoner was the first mechanical calculator to approach the capabilities of its modern counterpart. Unfortunately, it also suffered from a design flaw which made it unreliable when numbers had to be "carried".

Some historians now doubt that Leibniz ever succeeded in getting his Reckoner to work properly. But over the next 150 years, many others built upon Leibniz's basic design to produce calculating machines that certainly did work.

These culminated in 1820 in the first commercially successful device, the Arithmometer, invented by a French insurance company owner named Thomas de Colmar. Using Leibniz's gearing ideas, de Colmar succeeded in making a machine that, while tricky to operate, was reliable and powerful enough for general use. Many hundreds were sold to businesses over the years, and the machine went on to win the ultimate accolade when the term "arithmometer" became the generic name for calculating machines, which in the 19th-century became smaller and less laborious to use. This continued until about a century ago, when they began to fall out of use.

The development of electronic data processing systems in the 1950s heralded the beginning of the end for mechanical calculators, but not until the 1970s, and the advent of cheap electronic calculators, did the mechanical devices finally disappear from general use.

Condom

England

1640

Although the condom has been in constant use for at least 3,000 years, the reasons for its use have changed – and in ways that cast interesting light on the changing priorities of society. Originally made out of animal gut, condoms were used by the ancient Egyptians. However, the oldest surviving remains of one come from the foundations of Dudley castle and date to around 1640. This is not the first condom, but it is the first we can be sure existed.

Early users of the condom were less interested in its ability to prevent pregnancy than to protect against disease. Even in the 16th-century, the Italian gynaecologist Gabriel Fallopius was advocating the use of condoms as a means of protecting against syphilis rather than pregnancy.

Evidence for its use as a contraceptive dates back to the 17th-century, which is also when the condom seems to have acquired its name. In one story, it occurred when a courtier named Condom explained its use to Charles II.

Of course, it is entirely possible that this dual use of a condom was widely exploited much earlier. Or perhaps the early condoms were not reliable enough to merit the title of contraceptive. Even so, one can't help suspecting that the emphasis on avoiding disease reflects the male-centred view that catching VD was worth worrying about but getting a woman pregnant was her problem.

Condoms made from animal intestines were still on sale in Britain in the late 1960s. They were surprisingly good as well, as I can personally attest, having once bought one made from sheep's intestine at a back-street shop in Bath.

The invention and patenting of vulcanized rubber by Charles Goodyear in 1843 did not just revolutionize motoring: it made the condom a cheap and reliable means of contraception.

It remained the most widely-used form of contraception in the West until the 1960s, when the Pill came in. Even so, the proportion of people relying on this simple method remained pretty much constant. While it may not be utterly perfect, it has minimal side-effects, something which has boosted its popularity among women concerned about the long-term effects of the Pill.

Since the early 1980s, the condom's disease-prevention role is again being emphasized, because of Aids. The history of the condom has come full circle.

Barometer

1643

Evangelista Torricelli

Galileo is believed to be the first man to predict the existence of atmospheric pressure – the pressure caused by the weight of gas above our heads, which, like all objects on the Earth's surface, is being pulled down by gravity. However, it was one of his pupils, Evangelista Torricelli, who in 1643 invented a device for measuring it. The instrument was christened the barometer a few decades later by the Irish chemist Robert Boyle.

Torricelli's first barometer was a glass tube, closed at one end and filled with mercury. When the tube was hung with the closed end at the top, the mercury fell, creating a vacuum above it.

His spark of genius was to realize that this vacuum would support a column of mercury when it became equal to the pressure outside. With the open end of the tube submerged in a cup of mercury, the height of the mercury column in the tube gives an accurate measure of the atmospheric pressure. Millimetres of

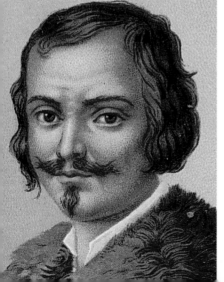

Evangelista Torricelli, a pupil of Galileo, invented the first device for measuring atmospheric pressure.

mercury are still used today as a unit of atmospheric pressure, although bars and millibars are more common.

Torricelli had created the first sustained vacuum and it was thereafter named the Torricellian vacuum. Four years later, the French mathematician and philosopher, René Descartes, realized that there could be a relationship between atmospheric pressure and the weather. He applied a scale to a Torricelli tube and wrote on 13 December 1647 to his colleague Marin Mersenne: "So that we may also know if changes of weather and of location make any difference to it, I am sending you a paper scale two-and-a-half feet long, in which the third and fourth inches above two feet are divided into lines; and I am keeping an exactly similar one here, so that we may see whether our observations agree."

Average atmospheric pressure will support a column of mercury of about 760 millimetres. In fact, any fluid can be used to fill a barometer but mercury, being 13.6 times heavier than water, is the most common. A barometer made by the French physicist Blaise Pascal used red wine, which, being lighter than water, needed a tube 14 metres high. He also forced his young brother to climb the summit of Puy de Dôme in the Auvergne mountains. It proved that the column of mercury dropped as the atmospheric pressure became less with altitude.

Set fair : a modern barometer

Air pump

Otto von Guericke

1650

1650 Controlled ageing of wine becomes possible with mass production of glass bottles and cork stoppers.

I t was the phrase "nature abhors a vacuum" that intrigued Otto von Guericke, an engineer who became mayor of Magdeburg in his forties, and set him on the path of inventing the first air pump in order to test the axiom based on Aristotle's belief that a vacuum could not exist.

Discussions about the possibility of vacuums had raged since the Greek philosopher came up with his law of motion. It stated that an object under a constant force would move faster through a less dense medium. The implication of this was that a body in a vacuum would move at an infinite speed, which Aristotle thought impossible and therefore declared that a vacuum could not exist.

This intrigued von Guericke, so he set about creating a vacuum and in the process became one of the most blatant showmen among scientist-inventors.

In 1650, in the first of many experiments, he constructed an air pump with all the joints sealed to be airtight. He then evacuated a vessel using the pump and demonstrated that within the vacuum a ringing bell could not be heard, animals would not live and candles would not burn.

Opposite: In 1654, in front of Emperor Ferdinand III, Otto von Guericke evacuated the air from two four-metre diameter iron hemispheres and demonstrated that the vacuum was stronger than two teams of eight horses, which were unable to pull the hemispheres apart.

Fig. III

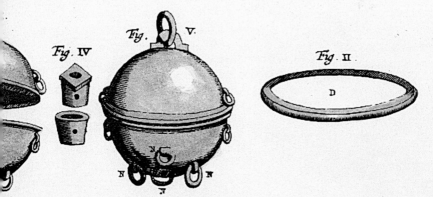

Fig. IV

Fig. V

Fig. II

D

N   N   N

Chapter 3
The age of industrial power
1650 to 1829

The period from the late Renaissance to the early Industrial Revolution was one of such massive technological upheaval that its effects can be clearly appreciated in statistical form: before 1600, twelve famines swept through England every 100 years; in the next century, thanks to agricultural innovation, there were only four. In 1750, the journey on horseback from London to Edinburgh took 10 to 12 days; by 1850, it took 17 hours by steam train.

These fundamental changes to the way in which we lived were the result of invention on an unprecedented scale. The scientific revolution that accompanied the Renaissance had borne technological fruits and the evolution of a self-conscious technique of innovation: what has come to be called "the invention of invention". With a political, commercial and social environment that valued inventors and invention, Britain led the world in applying scientific enquiry to technological improvement, which it used to expand its world influence. The prime examples are Robert Boyle's and Otto von Guericke's experiments with gases and vacuums, which led to the invention of the steam engine by Thomas Newcomen in 1712. This had repercussions throughout industry and society, not least because steam power, developed to pump water from mines, created a demand for fuel that in turn prompted the coal industry to expand and innovate. A product of this was Richard Trevithick's invention of the steam train to ferry coal from the pithead.

That conditions in Britain in the early 18th-century were ripe for innovation was partly the result of the invention in 1701 of the seed drill by Jethro Tull, an English farmer. Often seen as the beginning of the Agrarian Revolution, this event was in fact the trigger of the Industrial Revolution. In the early 18th-century improvements in cultivation techniques, the wider use of horses as draught animals, the policy of enclosure and the mechanization of agriculture begun by Tull, meant that for the first time the majority of men were liberated from the daily struggle for subsistence. As a result, large portions of the rural population moved to towns (London's population grew from some 100,000 in 1600 to more than one million in 1800). This stimulated an even larger requirement for more efficient food production and provided the unskilled labour needed to work in the factories that emerged following James Watt's development in 1781 of rotary steam power to drive the wheels of industry. Steam power transformed the metal industry, increasing supplies of coal to smelt iron and produce steel, and enabling new techniques of working with metals. The improvements in iron manufacture can be seen in the progression from the cast-iron bridge, known as the "Stonehenge of the Industrial Revolution" and completed in 1779 at Ironbridge in Shropshire, to the elegant glass-and-iron Crystal Palace exhibition hall of 1851.

With the harnessing of steam, technology truly released the brake of history, leading to an extraordinary quickening in the rate of change in manufacturing that far outpaced the late 20th-century innovation in electronics and computers. Capital and banking systems were established to initiate investment in mass-production techniques that saw more than 100,000 power looms with 9,330,000 spindles put into service in England and Scotland between 1790 and 1830.

The unpleasant consequence of the Industrial Revolution was a devaluation of the craft skills of artisans, as cottage industries gave way to mechanized manufacturing in the "dark satanic mills". The creation of the modern factory, reliant on unskilled labour undertaking tasks of mind-numbing tedium, introduced the concept of payment by wages in place of remuneration according to the quality of the finished goods, a change that discouraged further innovation.

Pendulum clocks

Christian Huygens

Christian Huygens, a
Dutch scientist, made
clocks more accurate
by introducing a
swinging pendulum.

Until the 17th-century clocks varied by as much as 15 minutes a day. Set in motion by a falling weight that was attached to a rope wrapped around a drum, they were controlled by a foliot, or balance wheel. These regulators, affected as they were by every imperfection in the running of the clock, did not make for good timekeeping. It was not until the pendulum replaced the balance wheel that a more accurate clock could be made.

Galileo was the first to realize that a pendulum was a natural timekeeper. In 1581, he noticed a swinging lamp in the cathedral in Pisa, and timed it with his pulse. He found that each swing, whether large or small, took practically the same length of time. Galileo went on to design a pendulum-controlled clock mechanism shortly before his death in 1642, but nothing came of it.

Fifteen years later, Christian Huygens, a Dutch scientist, designed the first successful pendulum clock and licensed Saloman Coster, a clockmaker in The Hague, to produce it. The following year Ahasuerus Fromanteel started making pendulum clocks in London, having sent his son John to learn about them from Coster. Other London clockmakers quickly followed suit. The superiority of pendulum clocks was obvious: they varied by no more than 10 seconds a day.

The pendulum was soon complemented by the "anchor" escapement. The escapement was the mechanism which came between the power source and the regulator, alternately checking and releasing the train and giving an intermittent impulse to the regulator. The anchor escapement replaced the verge escapement that had been used to control the previous power source of mechanical clocks, the falling weight. The earliest known anchor escapement clock was made in London by William Clement in 1671, although this invention is often attributed to Robert Hooke, Curator of Experiments at the Royal Society. The importance of the anchor escapement lay in the fact that it required a smaller swing of the pendulum – and a pendulum swings at a more constant rate if the angle of swing is small. Moreover, it worked best with short swings of a long pendulum, which was well suited to the fashion for long-case clocks that developed in London in about 1660.

Pendulum clocks were not portable, and Huygens was responsible in 1675 for a separate innovation that made portable clocks and watches much more accurate. This was the use of a spiral spring to control the motion of a balance wheel. Hooke was probably the first person to control a balance wheel with a spring, but he could not agree terms with potential backers and failed to perfect his ideas until too late.

The long-case clock, or grandfather clock as it was called in the 19th-century, with a 39-inch (99-cm) pendulum beating seconds, endured for almost 300

1658 Robert Hooke develops a clock regulated by a balance spring.

1660 1660 The cheque book is introduced by a London firm of bankers.

1665 Isaac Newton develops calculus, discovers white light is made up of all colours and formulates his three laws of motion over the next 20 years.

1666 The Great Fire of London leads to the introduction of fire insurance.

A 19th-century model of Galileo's design, the first known attempt to apply the pendulum principle to timekeeping.

This clock, from Augsberg, southern Germany, shows solar and lunar time and is typical of mechanical clocks in vogue at the time when the pendulum was invented.

years. Spring-driven movements with short pendulums were used for more compact "bracket" clocks, and the balance wheel and spring were used in portable clocks. Whatever the type, both movement and case were often beautifully made and decorated, for a good clock was a treasured possession.

Precision timekeepers such as the regulator clocks used in astronomical observatories continued to improve, with better escapements and features such as temperature compensation. From the 19th-century, electrically-operated clocks appeared, but they had little impact except as master clocks controlling a number of slave dials.

The best possible mechanical timekeeper is a pendulum swinging free of any interference. William Hamilton Shortt's "free pendulum" clock of 1925 achieved the closest approach to this ideal. It was the ultimate mechanical timekeeper, accurate to about a second a year. But within 20 years it was superseded by quartz clocks and then atomic timekeepers. The end of the long reign of the mechanical clock was in sight.

Champagne

Dom Pérignon

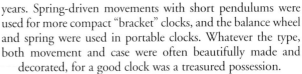

1670

Creating a wine with bubbles in it was, ironically, the last thing that Dom Pierre Pérignon, the inventor of Champagne, wanted to do. Indeed, this 17th-century Benedictine monk spent years trying to find ways of keeping bubbles out of wine, which were regarded as a sign of poor wine-making.

Pérignon's abbey, Hautvillers, sold large quantities of wine to the French court. As the abbey's wine-maker, Dom Pérignon was keen to service the aristocrats' preference for white wine, and some time around 1670 he invented a method of pressing white juice from the black grapes, which are easier to grow in the Champagne area.

But this created a problem. Lying to the east of Paris, Champagne has a relatively cold climate, and thus a short growing season for vines. The grapes were picked and turned into juice late in the year, so the fermentation process did not have time to finish before the winter set in. Only with the return of spring, and warmer temperatures, could the process begin again – by which time the juice had been bottled.

Dom Pérignon found that the result of this "double fermentation" was a wine packed with carbon dioxide bubbles, which he tried but failed to eliminate.

His aristocratic clients did not seem to mind: indeed, the new Champagne wine proved a hit with both the French and English courts. Dom Pérignon realized his success, and told prospective clients it "tasted like stars".

1675 William Woolcott invents a method of water filtration.

Dom Pérignon admires his handiwork in Hautvillers abbey. The invention of Champagne came about when he stopped trying to get rid of the bubbles.

He then focused on preventing the wine triggering explosions in cellars. He found the solution by switching to stronger English bottles, stopped with best Spanish cork.

In 1996, Jacques Pitoux, a Champagne-loving advertizing salesman, invented a safety cork to make opening a bottle less fraught.

Universal joint

Robert Hooke

1676

Mention the phrase "universal joint" to most people, and they'll probably think of the ball-and-socket arrangements that allow movement about all three axes. But for engineers, it is the term used for a much more clever invention: a joint that allows not just movement in any direction, but also the transmission of mechanical power. Try using a simple ball-and-socket joint to do that, and all you'll get is the ball spinning around while the socket remains resolutely stationary.

The 16th-century Italian mathematician Gerolamo Cardano is usually credited with thinking up the idea of such a joint. But as with so many Renaissance figures (most notoriously Leonardo da Vinci) Cardano never actually got around to building the thing. The credit for that goes to the 17th-century English scientist Robert Hooke, whose brilliance has long been overshadowed by the towering genius of his contemporary, Isaac Newton. In 1676, Hooke published a paper on an optical instrument that could be used to study the sun safely. In order to track the sun across the sky, the device featured a control handle fitted with a new type of joint, rather like two stirrups locked together, which allowed twisting motion in one shaft to be passed on to another, no matter how the two shafts were orientated.

Hooke's invention only really came into its own with the development of the car more than 200 years later. The challenge then was to transfer power from the engine drive shaft to the axle. Sprockets, chains and propeller drive-shafts were all tried, but none of them was reliable. In 1903, the American Clarence Spicer came up with a sealed universal joint system - heavily inspired by Hooke's ingenious solution - and it rapidly became the industry standard.

Pressure cooker

Denis Papin

I n 1675 Denis Papin, a French physicist, moved to London to work as an assistant to Robert Boyle, the chemist and physicist famous for determining the inverse relationship between the pressure and volume of gases. Boyle had discovered that if the pressure on a gas is increased sufficiently, it will condense to a liquid, or if a liquid is heated under pressure, its boiling point will rise.

In 1679, Papin used Boyle's Law to invent the pressure cooker. He heated water inside a sealed container with a pressure regulator so that he could adjust the temperature at which the water would boil. He discovered that bones softened and meat cooked quicker at the higher temperatures enabled by higher pressure. The following year, Papin cooked meals for King Charles II and the Royal Society in his pressure cooker; it earned him membership of the Society.

Papin's invention went on to become a major contribution to the Industrial Revolution. He realized that the steam that lifted the lid of his cooker could be used to drive a piston. His designs for a steam-driven machine proved to be impractical, but others took his ideas and developed the first steam engines.

1698 Condensation steam engine patented by Thomas Savery.

1699 Daumier Duperrier invents the portable fire pump.

'Auto-thermos' French pressure cooker, 1929, with spring loaded piston to regulate pressure.

Agriculture

Jethro Tull, Eli Whitney, Rev Patrick Bell, John Deere

Nasty, brutish and short – that was the farm worker's life until the first stirrings of agricultural invention some 300 years ago. The price of this progress has meant massive loss of employment in agriculture. In 1900, around half the rural population were engaged in farming; today it is just a few per cent.

The steps in mechanization began with the English farmer Jethro Tull in 1701. Concerned by the loss of seeds broadcast by hand on to the land, he developed a seed "drill". It placed the seeds below the soil surface in parallel rows, safely hidden from the birds, and planting in rows allowed weeds in the spaces in between to be controlled by hoeing, another step towards enhanced yield.

In the US, cotton production benefited enormously from Eli Whitney's cotton gin, patented in 1793, which removed the seed from cotton fibres.

Maximizing crop yield was further improved upon in the mid-1800s, when experiments at Rothamsted, Hertfordshire, by Sir John Lawes and Joseph Gilbert led to the use of artificial fertilizers such

Better sowing, harvesting and ploughing with Tull's seed drill (bottom) and Whitney's cotton gin.

as ammonium nitrate. Until then, spreading manure from housed cattle had been the only option for fertilizing crops.

The increasing yields from crops planted regularly in well-defined rows turned the attention of inventors to mechanization of the harvest. In 1826 the Reverend Patrick Bell, a Church of Scotland minister, invented a reaper with reciprocating knife and a rotary sweep. This cut the crop and laid it in wind-rows from which it could be gathered by hand into sheaves, to be placed in stooks to dry before storage. However, what was needed was a means of rubbing the grains from the ears, leaving the straw behind. So in 1836, the combined harvester-thresher, or "combine", was patented in America. By the 1870s in California, combines pulled by teams of 40 horses had a cutting width of up to 10 metres and harvested and threshed cereal crops of about one tonne per hectare. (The latest self-propelled combines, costing up to £200,000, have satellite navigation systems that record variations in crop yield as soil and other conditions vary across the field so that fertilizer application to subsequent crops can be metered accordingly.) The increasing use of combines forced farmers to find ways of drying large quantities of grain in bulk rather than on the stalk before threshing. Various kinds of drier were developed from the 1930s, with suitable buildings to house them and conveyors and equipment for handling grain in bulk.

For centuries the earth had been prepared for planting by wooden implements pulled by oxen, horses or mules, or in more primitive conditions manually with digging sticks. In 1837, the American blacksmith John Deere invented a steel plough. With steel shares to cut and steel mouldboards to turn the furrows, it was less given to wear and offered less resistance to the soil. Such ploughs enabled the untapped fertility of the Great American Plains to become a breadbasket for America and began the mass export of grains.

Towards the end of the 19th-century, mechanical power began to replace draught animals on farms, allowing the use of larger, heavier and more productive machinery across a longer working day. Steam-powered traction engines were used for cultivations or for stationary work such as threshing. At the end of the 19th-century, tractors with internal combustion engines were invented to pull ploughs, cultivators or carts. In

A combine harvester drawn by 30 horses cuts wheat in Moro, Oregon. The 'combine' was invented in America in 1836 and, together with the steel plough invented the following year, it opened up the Great American Plains and made the country, prosperous.

1904, Dan Albone, an English pioneer of tractor design, produced the Ivel, the first commercially viable British tractor. At least one working example still exists.

Development progressed rapidly. An importation in 1917 of American-built Fordson tractors was the first major move towards mechanization on British farms. The early tractors had steel wheels and were bone shaking to operate on hard surfaces; the introduction of pneumatic tyres for tractors and farm vehicles in the 1930s was a great improvement. The greater work capacity of even these early tractors reduced the British farm labour force as the number of working horses that

needed tending on British farms fell from more than a million in 1939 to about 50,000 by 1950.

Tractors were heavy, cumbersome and designed only to pull implements hitched behind, until the mid-1930s when Irish inventor Harry Ferguson developed a new concept: the tractor and implement designed as an integral unit. His "Ferguson system" used hydraulic power – oil under pressure – to control the functions and working depth of implements or machines mounted on the tractor. This enabled a light tractor to operate relatively heavy implements.

The next step was to link a trailer to the tractor instead of towing it behind. Theo Sherwen, an engineer employed by Ferguson, developed a pick-up trailer hitch that was subsequently copied worldwide. This enabled single-axle trailers with the wheels at the rear to be picked up hydraulically and locked to the tractor. The weight of the front end of the trailer was transferred to the tractor's rear wheels, greatly increasing wheel grip and traction. (It was such an arrangement of small Ferguson-type tractors with hydraulically coupled trailers that provided the only practicable means of transport for Kosovan refugees to escape across steep and difficult mountain routes in 1999.)

The importance of grass as a feed crop led to the development of machinery to harvest it as silage instead of hay. The Hampshire dairy farmer Rex Paterson invented the

Ferguson's tractor.

simplest and cheapest of these in 1948–49. His "buckrake", shown at the first National Silage Demonstration at Shillingford, Oxfordshire, in 1950, was a giant horizontal comb fitted to the back or front of a tractor to rake up the hay and transport it to storage in a silage clamp.

For livestock, the milking machine, invented in the early 1900s by the Swedish firm Alfa-Laval developed in efficiency until hand-milking virtually ceased in Britain. Enormous milking parlours became commonplace, with refrigerated tanks to hold the milk in bulk for daily collection by the dairy companies.

Animal welfare considerations led to the development of controlled ventilation for livestock transport vehicles and even to a machine that directs housed poultry into transport vehicles more gently than human hands.

Hot air balloon

de Gusmão, Montgolfier

1709

Joseph and Étienne Montgolfier organized the first manned flight in a hot air balloon over Paris.

On August 5 1709, Bartolomeu de Gusmão, a Brazilian Catholic priest, demonstrated his "lighter-than-air" device to the Portuguese Court when a small balloon housing a burning candle rose above the king's palace.

In 1783, Piâtre de Rozier and the Marquis d'Arlandes, a science teacher and an infantry officer, rose into the skies above Paris to become the first men in history to leave the Earth's surface for any length of time and survive. Their names, like that of de Gusmão, are all but forgotten, but those of the two brothers who made their achievement possible are still celebrated: Joseph and Étienne Montgolfier. The French word for hot air balloon is la montgolfière.

The sons of a wealthy paper merchant based near Annonay, in southern France, the Montgolfiers had access to both the money and technology needed to make the dream of manned flight a reality.

The first manned balloon takes off in November 1783, designed by the Montgolfier brothers, sons of wealthy paper manufacturers. A teacher and an army officer were the two volunteers on board.

Their inspiration came from watching small pieces of debris float gently upwards from the kitchen fire. They experimented by placing paper bags over the fire, which duly rose even higher. Fascinated, they tapped their family's paper-making expertise. By summer 1783, they had made a paper-lined cloth bag of more than 30 metres diameter.

On June 4, they took their huge balloon to the market square at Annonay and lit a fire beneath it. Together with an amazed crowd, they watched it sail 300 metres into the sky and land a mile and a half away.

They repeated the feat at Versailles on September 9, with a larger balloon, sending a duck, a rooster and a sheep on an eight-minute flight watched by Louis XVI and Marie Antoinette. It made the brothers famous.

On October 15, de Rozier made a successful tethered ascent, clearing the way for the first full manned flight with the Marquis on November 21. It lasted almost half an hour and reached an altitude of more than half a mile.

The Montgolfiers' invention attracted huge interest, not least from military commanders, who realized that balloons would allow them to watch distant enemy troops. Napoleon used them as observation platforms.

The brothers' triumph inspired others: 10 days later, the first manned flight of a hydrogen balloon was directed by the French chemist Jacques Charles.

Tuning fork

John Shore

1711

1711 John Shore, a trumpeter at the court of Queen Anne, invents the tuning fork.

John Shore, an instrument maker in London, invented the tuning fork in 1711. For the first time, musicians were able to tune their instruments to a known pitch. Because they produce constant wave frequencies, tuning forks can be used as accurate timing devices in laboratories and in 1923, American physicist Albert Michelson used a precision tuning fork to help measure the speed of light.

Steam engine

Thomas Newcomen, James Watt

1712

The invention of the steam engine was crucial to the industrialization of modern civilization. For almost 200 years it was the outstanding source of power for industry and transport systems in the West. It prepared the way for the development of more sophisticated heat engines and for the large-scale generation of electricity, which together effectively displaced it from its supremacy in the 20th-century.

From the introduction of the first viable steam engine by Thomas Newcomen at Dudley Castle coal mine in 1712 to the massive successes of the internal combustion engine in aeroplanes and automobiles at the beginning of the 20th-century, the steam engine was the inspiration of industrialization and its major work-horse.

James Watt realized the power of steam and built the machines that drove the Industrial Revolution.

The steam engine was a complex invention that underwent a process of incremental development which incorporated many important innovations. It was made possible by the increasing understanding of the atmosphere and the nature of a vacuum (see, pp. 162-3, 164-5). When Denis Papin unrented the pressure cooker and observed what was effectively a piston making a single stroke in a cylinder (see p.173).

This international effort culminated in the work of two Englishmen: Thomas Savery, who took out a patent in 1698 for a pump which worked by condensing steam in a cylinder to draw up water with the consequent suction; and Thomas Newcomen, who had the inspiration to combine the work produced from atmospheric pressure by condensing steam in a cylinder, with a piston and rocking beam which activated the pump rods in a mine. Savery had called his machine "the Miner's Friend", but it was the combination demonstrated in Newcomen's engine of 1712 which created the first efficient steam pump, because the working "stroke" of the engine could be repeated indefinitely, provided that the source of steam could be maintained.

By the late 17th-century, a national shortage of wood had created a lively demand for coal as a substitute fuel. Britain possessed huge reserves of coal, but the extraction of these required deep mines, and the curse of mines was their tendency to fill up with water. Newcomen supplied the coal industry with a tool for colossal expansion by providing a reliable pumping machine, and one, moreover, which consumed the "small coal" that was a waste material at collieries. This had tremendous ramifications for the British economy and led, among other things, to a rapid transition to coal fuel by British industries.

Despite its brilliant conception, Newcomen's machine remained for some decades a rough-and-ready sort of apparatus, assembled by local millwrights and blacksmiths, and much of its energy was wasted. A succession of refinements gradually improved its efficiency as a pump, and

James Watt's highly efficient beam engine became the driving force of the Industrial Revolution.

made it possible to use the steam engine economically even in districts where coal was comparatively expensive, such as Cornwall and Devon where the tin and copper mines urgently needed better pumps.

The main credit for these improvements belongs to James Watt, a Scottish instrument maker who in the mid-18th century had the inspiration that what he called the "elastic" property of steam would enable it to move from the working cylinder to a separate "condenser" to create a partial vacuum.

Strictly speaking, Newcomen's machine was an "atmospheric" engine rather than a "steam" engine, because it engaged the weight of the atmosphere to produce the working action, by successively admitting and condensing the steam in a single working cylinder. Watt recognized, in his patent of 1769, that the two functions could be separated, allowing the working cylinder to be kept permanently hot and the condenser cool, with much consequent saving of energy. The engines which Watt began to produce in 1775, in partnership with the Birmingham industrialist Matthew Boulton, were still atmospheric engines, but they achieved a spectacular improvement in efficiency.

Within a few years, Watt had gone on to design a genuine steam engine, by sealing the cylinder at both ends and allowing the expansive power of the steam to replace the action of the atmosphere. Then, by introducing steam both above and below the piston in the cylinder, he made it into a double-acting engine, so that the piston produced useful work with both strokes. His engines were still beam engines, producing a simple reciprocating action ideal for pumping, but with few other applications.

Urged on by his partner Boulton, who wrote to him in 1781 saying that "the people of London, Manchester and Birmingham are 'steam mill mad'", Watt then devised a method of producing rotary action by equipping his engine with a fly-wheel driven from the beam by a set of "sun-and-planet" gears. This system was rapidly adopted by industrialists, such as the cotton textile manufacturer Richard Arkwright, who had hitherto depended upon water or animal power to drive their heavy machines.

With Watt's added refinements of the steam governor to control the speed of the engine, and the parallel motion device which kept the piston upright in the cylinder, this engine was a huge success. Boulton and Watt produced about 500 engines altogether, from 1775 to 1800. They transformed the machine from a simple pump into a sophisticated and versatile prime mover which could be applied to a wide range of industrial processes.

The steam engine continued to evolve after 1800. The Cornish engineer Richard Trevithick demonstrated that steam could be safely used at much higher pressures than the very low pressures adopted by Watt. So it was Trevithick rather than Watt who stimulated the development of locomotive steam engines, the first of which he built for the Penydarren tramroad in South Wales in 1804. Trevithick also pioneered the highly efficient steam pumping-machine

Newcomen's atmospheric engine erected at the York Building Waterworks on the Thames around 1730.

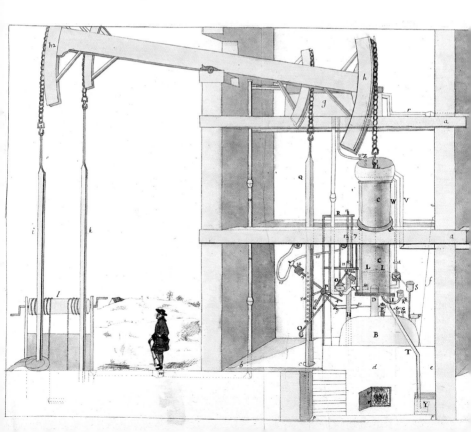

which cut off delivery of steam early on each stroke to exploit the expansive properties of high-pressure steam. This became known wherever it was used in mines and public services all over the world as the Cornish Engine.

Another development was the change from a vertical cylinder to horizontal and other arrangements, when it was realized that the position had no effect on cylinder wear. This led to more compact designs and also to some very large machines.

The steam engine became a varied engine, adopting a wealth of incremental improvements and local preferences. By the middle of the 19th-century, it dominated land and sea transport and was an essential component in every manufactory of any size in the industrialized world. The steam engine thoroughly deserves its status as one of the most successful and useful inventions of all times.

Temperature scales

Gabriel Fahrenheit

1714

Galileo is credited with inventing the thermometer in 1592 but his was a crude device. Thermometers became the instruments we know today thanks to Gabriel Daniel Fahrenheit, an Amsterdam manufacturer of meteorological instruments, who invented the first temperature scale and made other improvements on the basic instrument.

Fahrenheit's key advance came in 1714 when he found a way of cleaning mercury so that it would not stick to the narrow capillary walls of alcohol thermometers, by then in common use. This gave more accurate readings and all that remained was for him to devise a scale.

Isaac Newton had proposed in 1701 that the difference between the freezing point of water and body temperature be split into 12 equal divisions and used as the starting points for a temperature scale. Fahrenheit adopted Newton's proposal, but because he wanted to avoid negative temperatures on cold days, he added salt to ice water to obtain a temperature that he was sure would be below that at which water normally froze. This he called zero.

He then devised a 96-degree scale between zero and body temperature, which he later adjusted so that the boiling point of water was 180 degrees above its freezing point, which made body temperature 98.6°F.

Using his extremely accurate thermometer, Fahrenheit went on to make several fundamental discoveries, including the boiling point of different liquids.

Diving bell

Edmond Halley

1717

1719 Jakob Christof Le Blon of Frankfurt invents full-colour printing.

1730 John Hadley invents the quadrant. Until surpassed by the sextant, it is the primary instrument for measuring the position of the sun.

1732 London clockmaker Christopher Pinchbeck invents brass when he develops an alloy of copper and zinc.

Designs and devices used to explore the ocean depths have been found dating back to the 4th century BC. Alexander the Great is said to have dipped beneath the waves of the Bosphorus in a specially prepared barrel. At about the same time Aristotle said, "One can allow divers to breathe by lowering a bronze tank into water. It will not fill up, but keeps its air."

These early bells were crude and it was not until 1717, when Edmond Halley designed one of the first practicable diving bells, that life beneath the waves began to be fully investigated.

A diving bell operates on the principle that when an airtight vessel is lowered into water the air inside will not escape, even if the bottom of the vessel is open. As the bell descended, the pressure inside increased, but a pocket of air always remained. With early bells, when all the air inside was used up the bell had to be withdrawn.

It was Halley, renowned for his study of comets, who developed a method of extending the duration of the dive. Wooden casks containing air were lowered into the water and connected to the bell with leather pipes. When extra air was required more casks could be attached. In later bells air was pumped to the bell from above the water to extend endurance still further.

In 1899 Charles William Beebe became the first person to observe marine species at depths that could not be reached by divers. He used a spherical steel vessel, called a Bathysphere, that was lowered from a ship on a cable and in 1934 he reached a depth of 923 metres. The drawback with this type of vessel was that if the cable broke there was no way of returning to the surface.

Later designs of underwater craft incorporated methods of flotation and no longer needed to be connected to the surface. Diving bells are still widely used for under-sea exploration because of the length of time that divers can spend submerged in them.

Alexander the Great is lowered in a specially designed barrel beneath the Bosphorus 2,000 years before the first proper diving bell was invented.

Spinning machines

John Kay, James Hargreaves, Richard Arkwright, Samuel Crompton

The development of spinning and weaving machines in the mid-18th century was at the centre of the Industrial Revolution and established factory culture. Following the series of spinning and weaving inventions that transformed the industry, textile workers could no longer work at home; instead they had to work where there was power for machines.

In 1733, John Kay, who was born near Bury in Lancashire, invented the flying shuttle, the first step in the automation of the loom. Instead of having to pull the shuttle through the loom manually, a process that on a wide cloth required two weavers, the flying shuttle returned automatically. It allowed one weaver to work faster and make a broader cloth than two weavers had managed before, but it upset the balance of the arrangement between spinner and weaver. The rate at which yarn could be supplied became an issue; the only way forward was to mechanize spinning.

James Hargreaves, a poor, uneducated spinner and weaver from Lancashire, invented the mechanical spinning machine. It was named the "spinning jenny", supposedly after his young daughter Jenny knocked over a spinning wheel and he saw that the spindle, on which the yarn was spun and wound, continued to rotate in an upright position. This made him realize that he could construct a machine with several vertical spindles side by side and use a clamp as a substitute for the action of the spinster's fingers feeding the yarn on to the spindle.

Hargreaves made his first machine in 1764 and then went into production. Soon afterwards, hand spinners, fearing redundancy, broke into his house and destroyed several jennies, prompting Hargreaves to

Richard Arkwright's machinery brought textile manufacture under one roof, and started the factory system.

The Industrial Revolution devalued craft skills in favour of mechanized manufacturing, as in this cotton mill.

move in 1768 to Nottingham, which became one of the centres of the British textile industry.

Richard Arkwright's machine, patented in 1769, was based on the more complicated spinning wheel used for flax. In place of spinsters, who would carefully feed fibres through a rotating flyer that twisted them into a yarn, Arkwright's invention had a series of roller pairs. Each pair ran faster than the previous pair, to tease the fibres out, and several yarns could be spun simultaneously. It was not a new idea, having been tried by Paul and Wyatt in 1738, but Arkwright succeeded in developing a workable combination. The experimental machine in the patent drawing spun four threads at once.

Soon Arkwright was constructing machines with many more spindles, grouping them together to take advantage of substantial sources of power. In the first commercial venture, a horse drove the machinery, but Arkwright began driving his machines by a water-wheel, which is why his spinning machine became known as the water frame.

The simplicity and low cost of Hargreaves's jenny made it appropriate for cottage industry. Arkwright's machine, in

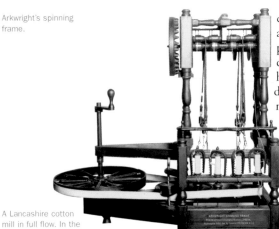

Arkwright's spinning frame.

A Lancashire cotton mill in full flow. In the 19th-century much of the world's cotton was spun and woven in the county. The industry was taken out of the home and installed in factories after Hargreaves invented the spinning jenny.

contrast, was complicated and costly, and he made a point of exploiting it differently. A key feature of his system was his development of further machinery for the preliminary processes of textile manufacture. The operation might be described as vertically integrated, having all the stages, from the opening of the cotton bale to the production of finished yarn, under one roof, and all the machinery driven from one water-wheel.

As well as adopting these economies of scale in his own factories, Arkwright forced others to capitalize similarly large installations; when he licensed the use of his invention he dealt in units of 1,000 spindles, setting the pattern for factory production.

In Bolton, Lancashire, in the 1770s Samuel Crompton developed a third machine, called the "mule" because it combined features of the other two machines. The mule could spin the finest yarns, such as those needed for the fashionable muslin.

Crompton's mule, which could spin fine yarns needed for muslins.

Developed into a "self-acting" machine (requiring less intervention and less skill on the part of the worker), the mule was the basis of the huge expansion of cotton production and exports during the Industrial Revolution. The value of British cotton good exports grew from £46,000 in 1751 to £5.4 million in 1800 and £46.8 million in 1861.

A modern reconstruction of the Spinning jenny.

Ball bearings

Philip Vaughan

1749

1750 Jigsaw puzzles invented in England as teaching aids for geography.

The entire industrial world runs on ball bearings, and would quite literally come to a grinding halt without them. Their importance was starkly demonstrated in October 1943 when the Allies committed almost 300 USAF bombers to the destruction of the ball-bearing factory at Schweinfurt. Two-thirds of the aircraft were lost or damaged through Germany's determination to keep the factory churning out its vital product.

The basic ball bearing design has remained unchanged since 1749, when Philip Vaughan patented it for use on carriage wheels. Two circular grooved rings known as "races", one fitting inside the other, provide the contact between the two components moving relative to one another. Metal balls fitted between the races keep the friction of the movement as low as possible.

The success of ball bearings boils down to the same simple geometrical fact that is exploited in the wheel: a circle gives the least amount of continuous contact with its surroundings. If the load carried by the bearings becomes too large, the balls are

Ball bearings being assembled between the grooved 'races'.

replaced with rollers; these have lines of contact with the races rather than just points and thus a larger load-carrying capacity.

Keeping the balls in alignment and keeping dirt out is crucial to the correct operation of ball bearings. Friction can generate enormous amounts of heat at the speeds at which many bearings operate. Any sideways pressure (thrust load) exerted by the axle on to the races can also cause trouble. Bearings with specially shaped races and roller arrangements have been developed to cope.

The traditional all-steel construction is being replaced in high-performance bearings with incredibly resilient silicon nitride ceramic balls fitted into races made from fabric-reinforced phenolic resin, giving substantial improvements in performance.

Lightning conductor

Benjamin Franklin

1752

Born in 1706, the 15th child of a poor Boston candlemaker, Benjamin Franklin went on to become one of the most remarkable figures of the 18th-century. After George Washington he was probably the most famous American of the period.

He was an author, printer, publisher, inventor and scientist who became a diplomat, playing a key role in representing the colonies in negotiations with the British Crown and Parliament. He had a hand in writing the Declaration of Independence, the American Constitution and the treaty that gave the 13 former British colonies their independence.

As an inventor, he devised bifocal spectacles and the "Pennsylvania fireplace", a wood-burning stove that could generate a greater heat than a conventional iron stove by using a grate for the wood and controlling the flow of air by means of sliding doors. It warmed millions of farms and city homes, and is still being made today.

Franklin's biggest achievements came in the field of electricity, which by 1745 had become a fashionable subject on the East Coast of America.

His foolhardy demonstration of the electrical effects of lightning impressed the Royal Society so much that they made him a member.

Franklin used this discovery to invent the lightning conductor, suggesting that pointed metal rods should be built on tall buildings and earthed on the ground. By 1782, Philadelphia alone had installed 400 of his conductors.

Franklin conducted many other electrical experiments, coined terms associated with electricity such as battery, conductor, negative and positive, and published a well-regarded book on the subject.

Marine navigation

John Harrison

1759

I n 1675 the Royal Observatory was founded at Greenwich with the sole remit of solving what was the greatest scientific problem of the age, the finding of longitude at sea. Every year men were losing their lives and many tons of valuable cargo were being lost because navigators were unable to establish their exact position when out of sight of land.

John Harrison won the Board of Longitude's prize for his timepiece after 50 years of trials and frustration.

Longitude is a measure of position east or west around the globe, usually determined by knowing the precise difference between the local time of that place and another, significant place, usually a home port. Determining one's own local time is accomplished relatively easily by measuring the altitude of the sun in the sky. The great problem in 1675 was how to determine the home time at that same instant, when thousands of miles away in the middle of an ocean. The simple answer was to take a clock, set to home time before leaving. But no clock had ever been made which could withstand the rigours of ocean travel.

The Observatory was thus founded to solve the problem by developing an astronomical system:

the lunar distance method, which used the night sky as a kind of celestial clock that could be read with accurate star charts.

In 1714, the government passed the Longitude Act which offered £20,000 (equivalent to about £1.5 million today), to be administered by the "Board of Longitude", for finding the longitude to half a degree on a six-week voyage to the West Indies. If a clock was to win this prize it would need to keep time to within 2.8 seconds a day throughout the voyage, a challenge that most people thought impossible.

The story of the winning of the prize is one of the greatest tales of a

1761 An electric harpsichord, the first electrical instrument, is invented in France.

man's triumph over adversity. Even more surprising than that the solution turned out to be a portable timekeeper, was that the inventor was not a member of the scientific establishment but a "rank outsider", John Harrison, a carpenter's son with no formal education, who grow up in a village in Lincolnshire.

Harrison had been interested in clockmaking from his childhood in Barrow-upon-Humber and had been trained as a joiner by his father. He taught himself the fundamentals of mechanics and in the early 1720s, with his brother James, constructed a series of precision long-case clocks, mostly of wood. These compensated automatically for changes in temperature, required no oil, and kept time to within one second a month, a fantastic achievement by any standards.

In 1726 Harrison heard about the great Longitude Prize and decided to create a portable version of his clock; in 1730 he journeyed to London with design drawings, to seek support. His first port of call was Greenwich Observatory, where the Astronomer Royal, Edmond Halley (of comet fame) recommended he visit the great clockmaker

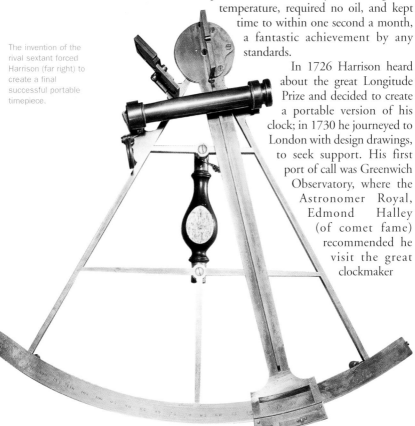

The invention of the rival sextant forced Harrison (far right) to create a final successful portable timepiece.

George Graham for advice. Following the meeting, Graham lent Harrison money for the project and he constructed the first of a remarkable series of five prototypes which was tested in 1736 on a voyage to Lisbon. It evidently did not go well enough to win the prize, but it gave Harrison hope. Design improvements led to his extraordinary third Timekeeper, H3 of 1740. In this he included his original invention of the temperature-sensitive bimetallic strip (still used today in thermostats, car direction indicators, and in toasters to ensure the toast pops out when done) and the caged roller race, the forerunner of all machine roller and ball bearings.

These timekeepers were necessarily large but now, encouraged by improved performance from his own pocket watch, made in 1753 to his design by John Jefferys of London, Harrison made a breakthrough. He would make a small, cheap and portable timekeeper as a supplement to his larger machines. Completed in 1759, it was only intended as a "hack watch" for carrying time from land to the ship, and from cabin to deck, but he found to his astonishment that his watch went far better than he expected. H4, as this new watch became known, was soon out-performing H3 and was tested with outstanding success on two voyages to the West Indies, in 1761 and 1764, with Harrison's son William on board. These trials should have proved, beyond all reasonable doubt, that the problem was solved.

At about the same time, however, the sextant was invented. In 1730, John Hadley had invented the reflecting quadrant, later known as the octant. It enabled the latitude, and also one's local time, to be determined with considerable accuracy. In 1757, at the suggestion of the experienced navigator and naval officer John Campbell, the angle marked off by the reflecting quadrant was increased to 60 degrees so that greater angles could be measured and the local time determined with increased accuracy. This new instrument, called the sextant, also made the lunar distance method of finding longitude – the rival to Harrison's method – more workable: the battle was not over.

John Harrison.

Harrison had been awarded only half the prize money and was told, in his mid-seventies, that he must make another watch. After successfully testing the fifth timekeeper, he received a bounty for the remaining prize money just three years before he died, on March 24 1776, his 83rd birthday.

The modern marine chronometer, developed as a simplified version of H4, enabled the British Navy to explore and chart the world for the next two centuries. The prosperity and safety of the world's ocean-going navies would not have been possible without it.

Sandwich
Earl of Sandwich

1762

A demand for a snack during a marathon bout of gambling brought lasting fame to the 18th-century politician John Montagu, the fourth Earl of Sandwich.

Despite his distinguished career as a Secretary of State, Postmaster General and First Lord of the Admiralty, it was as a gambler that Montagu achieved immortality. During one session at the gaming table in 1762, he called for some sliced meat and bread. Putting the greasy meat between two slices of bread, he was able to eat while playing.

Montagu's aristocratic title lent the humble meal a certain cachet, and it soon became highly fashionable.

Paper patterns
F.A. de Garsault

1769

The first paper clothing patterns are believed to have been produced by F.A. de Garsault, the author of a seminal work, *Art du tailleur* (Art of the Tailor), published in 1769 and containing patterns for such garments as short gowns and bed jackets.

In 1859, Ebenezer Butterick, an American manufacturer, invented standardized tissue-paper clothing patterns and went on to found the *Ladies' Quarterly Review of Broadway Fashions* and a fashion magazine, *Metropolitan*, to promote their sale.

The
Flat Paper Pattern
Of this
Complete Costume
Price 1/0½
Post Free
Can be obtained from
"Fashions for All"
290 Regent Street
London.

Steam Wagon

Nicholas-Joseph Cugnot

1769

The first experimental steam-powered vehicles were built by a Frenchman, Nicholas-Joseph Cugnot, in 1769 and 1770. They were a fearful sight, even for the metropolitan sophisticates of Paris. The frightened reaction of Parisians showed they were not ready for the forerunner of the motor car, and when Cugnot's second machine collided with a wall it confirmed the perilous nature of the invention.

Equipped with a single wheel at the front and two at the back, the snorting, hissing, high-pressure, steam-driven artillery tractor was finally impounded in the city's Arsenal to prevent it causing further harm.

Born in 1725, Cugnot was a former artillery officer turned military engineer. He took inspiration from the pioneering work of a fellow Frenchman, Denis Papin, rather than from the low-pressure steam theories of British engineers such as Watt and Newcomen. Only high-pressure steam and multiple-driving pistons could power a vehicle, and although the experiments of Cugnot were ultimately unsuccessful, others such as Watt's foreman, William Murdock, and the Cornish engineer Richard Trevithick, made more progress.

The Cugnot carriage was an ungainly looking device with a giant steam boiler situated precariously above the front driving wheel. The boiler supplied two pistons to power the wagon

forwards at about three miles an hour. Its limitation was range; the steam carriage soon ran out of water and wheezed to a halt.

Nevertheless, steam-powered road vehicles soon multiplied, and steam-coach services were popular in Britain many years before the spread of steam railways in the 1830s and 1840s.

The invention of the combustion engine failed to dampen any further developments in steam cars. The ultimate steam-powered car was the Stanley Steamer, an American automobile that held the world land speed record in 1906 reaching an incredible 127.66 mph.

Porcelain false teeth

Alexis Duchateau

1770

G old teeth, bone teeth, green teeth, multi-coloured agate teeth and, finally, dead men's teeth down the years have all been tried with varying degrees of success to fill the mouths of fashionable society with acceptable dentures.

The problem of replacing missing teeth was of such moment that it was discussed at length by Pierre Fauchard in his book *The Surgeon Dentist* of 1728, a work which resulted in the expansion of dentistry throughout Europe.

The use of gold by some early cultures such as that of the Etruscans had now been forgotten. Instead, 18th-century dentists resorted to ivory. This looked fine for a while, but was attacked by the acids in saliva, tasted awful and soon rotted. Other forms of bone were tried – canines made from Hippopotamus bone were popular for a while – until they, too, decayed, turning green. No expense was spared: mother of pearl was tried and Lord Hervey, the politician and wit, had a full set made in Italian agate.

Another popular measure was the acquisition of a full set of human teeth. Dead men's teeth became the vogue, and were in plentiful supply in times of war. (After 1815 there was a glut of "Waterloo teeth".)

In 1770, Alexis Duchateau, a French apothecary, pioneered the use of porcelain. After several failures he produced a practical set of dentures that were very popular.

1775 James Lind, a Scottish doctor, invents the pressure anemometer to measure wind speed.

1777 Samuel Miller invents the circular saw, which does not come into widespread use until sawmills become steam-powered.

1778 Robert Barron, a London locksmith, invents the mortise lock.

1779 A bridge made of cast iron – the first major structural use of iron material – is completed at Coalbrookdale.

When a process for producing vulcanized rubber was discovered in 1851, dentists were quick to use the new material. This was moulded against a cast of the patient's mouth and porcelain teeth attached. It reduced the cost of dentures, their use became more widespread and halitosis less common.

Steam rolling machine

Henry Cort

1783

Henry Cort, a Royal Navy Agent, could see that the iron and steel produced in British mills was inferior to foreign-produced stock. It was an important strategic failing: Britain was importing many tons of expensive Russian iron.

In 1775, after experimenting for several years with different processes, Cort left the Navy and established his own forge and foundry at Portsmouth. His researches led to the granting of patents to two key processes in 1783 and 1784. The first improved the quality of iron by beating out the impurities using a giant hammer while keeping the metal at a constant temperature, then repeatedly passing it through a rolling mill to iron out further impurities.

Critics of his process at the time were said to have pointed out that the rolling process just trapped the carboniferous impurities and slag within the rolled iron. He therefore devised an improved "reverberating" furnace, capable of better smelting and puddling of iron to produce a purer raw iron for rolling. These two discoveries revolutionized the iron-making industry. Over the next 20 years British iron production quadrupled. By 1820, there were 8,200 Cort-style furnaces smelting in Britain, and by the end of the 19th-century more iron was smelted in Britain than in the rest of Europe put together.

But Cort's invention did not make him rich and he sold his business in 1789.

Steam boat

Marquis Joffroy d'Abans

1783

I n 1783 on the River Saône at Lyons in France the Marquis Joffroy d'Abans slipped the mooring ropes of his steam boat, the *Peryscaphe*, and successfully navigated the hissing, wheezing vessel upstream for about 15 minutes – before it shook itself to bits.

1784 Benjamin Franklin invents bifocal lenses for spectacles.

Ten years of failure and despair led to this moment of relative success, the maiden voyage of the world's first steam-powered boat. In 1776 the Marquis's first boat failed when the flailing, lever-driven, duck's-feet paddles thrashed themselves comically to a standstill in moments.

His breakthrough was the use of paddle wheels to push the boat, and this led to paddle-driven boats crossing the seas of the world for the next 150 years. But the turmoil of the French Revolution halted his work before he could enjoy the fruits of his success, and he died in penury.

In America, engineers such as James Rumsey and John Fitch were making progress, too, and by 1790 Fitch had built a boat to provide a regular service on the River Delaware (below). Robert Fulton, the most famous American marine engineer, launched his vessel, the *Clermont*, in 1807. This became famous as the first passenger-carrying paddle steamer in the world, plying between New York and Albany on the Hudson.

John Fitch's first steam boat plying the River Delaware.

Semaphore

Claude Chappe

Before the electric telegraph, a message could travel only as quickly as it could be carried by messenger or, in the case of jungle drums and smoke signals, as far as it could be heard or seen.

That changed in 1791 when Claude Chappe, a Frenchman, refined the idea of the smoke signal, a sophisticated form of early telegraph, to create the first true telegraph system which we now know as the mechanical semaphore. The word comes from the Greek *tele* and *graph* meaning far off and writing, and it is used to describe any means of communicating by light, sound or gesture.

Instead of puffs of smoke, Chappe devised an ingenious system of signalling by semaphore between specially constructed towers. The person receiving a message from one tower relayed it to the next, and so on. In 1794, the news that Austria had captured the town of Conde-sur-l'Escaut reached Paris, 90 miles away, in less than an hour.

Despite limitations, it worked well, and by 1852 a 3,000 miles network of 556 Chappe semaphore towers extended throughout France, with a further 250 towers across the rest of Europe.

Chappe's early mechanism consisted of two synchronized one-handed clocks with 10 numbers on their dials, and an iron casserole dish which was beaten when the continually moving hand of the sender's clock swept by the number to be sent. Because the next post had to be in hearing distance, it limited its range and Chappe soon changed to a purely optical (and dish-less) system. This was a rotating bar, pivoted at its centre, with two smaller rotating bars suspended from either end. Codes could be devised from its 98 positional combinations.

Chappe's telegraph – he had initially called it *tachygraphe*, or fast writer – drew great interest from governments across Europe, primarily because of its potential military value. Chappe had hoped for a much more commercial use of the telegraph and he became disillusioned. In 1805 he drowned himself in a well. His tombstone depicted a telegraph tower indicating the "at rest" sign.

A German Nevy cadet signals the letter Q using semaphore flags.

The telegraph's commercial potential was instantly recognized in America. In 1790, the first stock exchange in North America opened in Philadelphia. Shortly after, a group of Philadelphia brokers set up a system of signal stations between Philadelphia and New York, specifically for the relaying of stock prices between the two cities. This system used heliographs, mirror devices for flashing messages by reflecting the sun's rays, in an early form of Morse code. This enabled information to travel from one city to the other in just 10 minutes, weather permitting.

Semaphore flags, developed from Chappe's methods, are still kept aboard naval ships, in case of emergencies or power failures, and rudimentary semaphore still has to be learned by naval recruits.

Guillotine

1792

Joseph Ignace Guillotin

It was the unfortunate legacy of Dr Joseph Ignace Guillotin, a French physician and penal reformer, to be associated for ever with one of the most brutally efficient machines of execution even though he did not directly invent it.

Although their precise origin is unknown, guillotine-like devices were used in Germany, Persia, Scotland and Italy long before France. An illustration held by the British Museum shows the first known use of a "guillotine" at the execution in 1307 of Murcod Ballagh in Ireland. And a replica has been built of the Halifax gibbet, a primitive pre-guillotine with a narrow, stubby blade that until 1650 was used to execute 53 men and women in the Yorkshire market town.

During the French Revolution Dr Guillotin was a member of the French Assembly who sought to banish the death penalty, at that time administered by quartering or on the rack in front of a square full of townsfolk. In 1789, Guillotin proposed to the Assembly that as an interim to banning capital punishment, the penalty should be made swift, painless and private for all classes.

In 1791, the Assembly rewrote the penal code, providing that "Every person condemned to the death penalty shall have his head severed." The next year, Tobias Schmidt, a German harpsichord maker, was paid 960 Francs to build a prototype for the necessary contraption.

The machine was first tested on sheep and calves, then on human corpses at an old people's home. On April 25 1792, Nicolas Jacques Pelletie, a highwayman, became the first person to be executed with Schmidt's devices, at first called the Louisette, but soon to become known as la guillotine. To make matters worse for Guillotin, his attempt to make the death

Swift, painless and classless... a condemned man approaches the guillotine in Tours, France, in 1920.

penalty private ended with thousands of public beheadings in a macabre carnival that went on during the French Revolution.

In 1854, Joseph Tussaud bought a used guillotine from Clément Sanson, the executioner who beheaded Pelletie, and installed it in his wax museum in London. The guillotine's last outing in France was on September 10 1977, in Marseilles, when a murderer, Hamida Djandoubi, was beheaded.

The machine is still in use in a handful of the many countries that adopted it.

Ambulance

Dominique-Jean Larrey

1792

1795 The metric system is developed in France in the wake of the Revolution and eventually is adopted worldwide.

One of Larrey's first field ambulances: they were much in evidence in Napoleon's campaigns.

Horrified by what he had seen on the battlefield as a military surgeon, a young Frenchman started the first ambulance service in 1792. Dominique-Jean Larrey introduced properly sprung horse-drawn carts to evacuate casualties quickly into newly established mobile field hospitals. From that moment on, surgeons had a reasonable chance of operating on the worst wounds before it was too late.

The success of Larrey's approach earned him the position as Chief Surgeon of the Guard in all Napoleon Bonaparte's major campaigns from 1805 onwards. Ten years later he was present at the Battle of Waterloo, tending casualties from both sides until he himself was wounded.

Vaccination

Edward Jenner

1796

Infectious diseases have claimed countless millions of deaths throughout history. The Black Death pandemic in Europe in the mid-14th-century alone left 75 million dead, while the flu outbreak of 1918–19 killed another 22 million.

Yet over the centuries, stories emerged of certain people who somehow remained free of the disease, and various explanations began to circulate to account for their miraculous escape. A common theme was that people exposed to weak doses of the disease were somehow protected against its worst ravages.

By the early 18th-century, some European doctors were advocating the idea of combating smallpox by deliberately injecting people with infectious matter taken from those with a mild case of the disease. The risks of such "inoculation" were very high, however, and those who were supposedly protected often fell victim to the full-blown disease.

Edward Jenner proved that inoculation was effective in an experiment that today would have landed him in prison.

During the 1770s, a young Gloucestershire surgeon named Edward Jenner began pondering another curious claim: that people who fell victim to a relatively harmless disease known as cowpox somehow subsequently became immune to smallpox. Jenner wondered if exposure to cowpox might provide a safe way of inoculating people against smallpox.

In May 1796, he decided to find out in an experiment so outrageously unethical that it would today undoubtedly land him in prison.

Jenner inoculated an eight-year-old boy named James Phipps with matter extracted from the finger of a dairymaid who was infected with cowpox. Six weeks later, the boy had suffered only a mild fever. So, on July 1, Jenner deliberately gave the healthy boy a dose of lethal smallpox.

Fortunately for the boy, Jenner proved to be right. We now know that the inoculation had trained the boy's immune system to recognize and destroy both the cowpox and the smallpox viruses.

Highly unethical today, but by inoculating a healthy eight-year-old boy, James Phipps, first with cowpox, then smallpox, Edward Jenner proved that disease could be prevented by building up the immune system.

While this was unknown to Jenner, he recognized the significance of his discovery about cowpox, which he named vaccination – after *vacca*, the Latin for cow. It went on to save the lives of millions, and it remains the primary defence against infectious diseases.

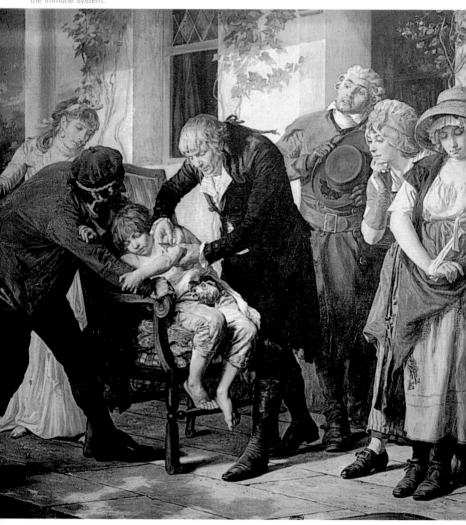

Parachute

André Jacques Garnerin

1797

Oddly enough, there is a link between parachutes and the bags in vacuum cleaners: in order for both to work, they have to be permeable.

In the case of the vacuum bag, tiny pores in its walls let the air out while trapping the dust inside – in theory, anyway. In practice, the pores quickly become clogged, and the suction drops dramatically.

If parachutes are to work properly and give a smooth descent, they too need to allow air through them, either via pores in the material or via purpose-made vents. Leonardo da Vinci famously proposed the idea of a parachute, with a pyramid-shaped canopy made from linen, around 1485. He claimed that by using it "anyone can jump from no matter what height without any risk whatsoever". As usual, however, he never actually got around to putting his grand claim to the test.

Similarly, dubious credit has been given to the Frenchman Louis-Sebastian Lenormand, who in 1783 made the first parachute jump. This was hardly death-defying: he leapt just four metres. The real credit should surely go to André Jacques Garnerin, a French physicist and aeronaut who in 1797 put his

Madame Garnerin joined her husband in airborne stunts.

research to the test by jumping out of a balloon at 1,000 metres. Now that takes real courage.

Garnerin was not content with merely surviving. After his jump, he took the counter-intuitive but crucial step of cutting a large hole into the centre of his parachute, and jumped again to prove that the greater permeability made the descent more stable

Lithography

1798

Aloys Senefelder

The lithograph (from the Greek for "stone drawing") was invented in 1798 in Munich by a playwright and printer, Aloys Senefelder, who was originally from Prague. Senefelder wanted to produce multiple copies of his play scripts. He discovered the process by accident one day while scribbling a laundry list in grease pencil on a stone slab in his studio, and he spent the next two years refining it.

Lithography is based on the principle of the antipathy of grease and water. An image is drawn on to stone with a greasy substance. The stone is then dampened and ink is applied with a roller. The ink sticks only to the grease. After a chemical treatment, the finished stone is placed with the paper on a bed that carries it through a press.

The resulting expressive prints were favoured by Goya in the 1820s and later by Toulouse-Lautrec, helping to legitimize lithography as a high art form.

A lithograph by Toulouse-Lautrec, a great exponent of the art that was invented with a laundry list.

Battery
Alessandro Volta

I n 1800, Alessandro Volta used metal strips to connect a series of bowls containing brine and unleashed an invisible force without which many subsequent inventions, including radio, telegraphy, electronics, computers and electric light would not have been possible.

Volta had invented the battery and with it the first steady supply of a dynamic electric current. It was an invention as epochal as the steam engine or the telescope had been previously. It was also the first portable energy source in the history of mankind.

Volta refined his battery, using small round plates of copper and zinc, interleaved with cardboard discs soaked in brine, and then made his discovery public on March 20 1800, reporting it in the *Philosophical Transactions of the Royal Society*. Like many inventions, it was the culmination of an evolutionary process, in this case one that pulled together threads of discovery that had originated 200 years earlier.

In 1600, Dr William Gilbert, court physician to Elizabeth I, noticed that a pivoting piece of metal would swing towards objects that had been charged with static electricity by rubbing them with fur. It was the first recorded observation of the properties of electrical charge. Then in 1663, Otto von Guericke, mayor of Magdeburg in Prussian Saxony and inventor of the air pump, devized an electricity generator. This primitive device rotated a ball of sulphur. When von Guericke held his hand against it, the ball emitted sparks of static electricity.

The next breakthrough came in 1745, when Ewald Georg von Kleist, a Pomeranian administrator and cleric, invented the Leyden jar simultaneously with Pieter van Musschenbroek, professor of mathematics and physics at the University of Leyden in Holland. This glass vial, partly filled with water and sealed with a cork pierced by a nail, was the first device that could store electrical charge (nowadays it would be called a capacitor, a term coined by Volta in 1795).

The jar accumulated static electricity by touching the nail with a charged rod. This charge was then released in one burst

Alessandro Giuseppe Antonio Anastasio Volta was born in 1745 of an aristocratic family that had hit hard times. His interest in electricity was pricked by a history of the subject written by the English chemist Joseph Priestley. As a professor at the University of Padua, he devoted himself to studying static electricity. In 1801, the year after he invented the battery, Volta was summoned by Napoleon to demonstrate his inventions. The French emperor bestowed on him the Legion of Honour and made him a count.

by touching the nail with a grounded object; when van Musschenbroek did this with his finger, the enormous discharge was almost enough to kill him.

In America, Benjamin Franklin experimented with a Leyden jar and wondered whether the crackling sparks were a form of lightning, and moreover, whether lightning was a form of electricity. In 1752, he decided to test his theory in a frighteningly dramatic fashion by flying a kite during a thunderstorm.

In 1775, Volta became professor of physics at the Royal School of Cuomo and devised his Perpetual Electrophorus, a device that could be used to generate, store and transport electricity. It had two thin metal plates, one covered with vulcanite rubber, the other attached to an insulated handle. A charge was built up by rubbing cat fur across the vulcanite-covered plate. The second disc was used to transfer the charge to a Leyden jar, enabling a very large charge to be built up in stages.

The potential of this newly discovered physical force was demonstrated in April 1746 when, at the grand convent of the Carthusians in Paris, about 200 monks arranged themselves in a long, snaking line. The monks held one end of an eight-metre iron wire in each hand and together, their connecting wires formed a line more than a mile long.

With the line complete, the Abbot, Jean Antoine Nollet, a noted French scientist, took a Leyden jar and, without warning, connected it to the line of monks, giving all of them a powerful electric shock. The simultaneous exclamations and contortions of a mile-long line of monks revealed that electricity could be transmitted

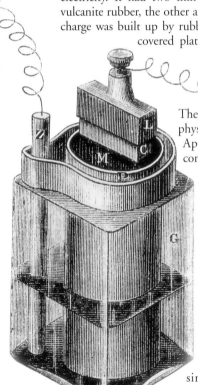

Georges Leclanché's dry cell, with carbon and zinc electrodes in an ammonium chloride solution. It was the final stage in the development of the modern battery

Benjamin Franklin and his son fly a kite in a thunderstorm. The kite rose on a silk thread attached to a piece of wire that would attract lightning. A key tied to the lower end of the thread sparked when Franklin held his hand near it, proving that lightning was a form of electricity. It was a reckless experiment that killed the next two people who tried it.

over a great distance; and as far as Nollet could tell, it covered that distance instantly, suggesting it could be used to send messages over great distances incomparably faster than a human messenger could carry them.

Luigi Galvani, an Italian anatomist and friend of Volta, used an electrophorus to investigate the effect of static electricity on dissected frogs' legs. He noticed that they twitched when touched with a charged rod. He then laid a row of frogs' legs out during a thunderstorm on brass hooks resting on an iron lattice. To Galvani's surprise, the legs continued twitching after the storm. Realizing they would twitch whenever they were touched with two metals, Galvani proposed that it was the result of something he called "animal electricity".

Volta heard of Galvani's experiment and wondered whether the electrical current was generated not by some property of the muscle, but because of the contact between two metals. In 1794, he began experimenting with pairs of metals on their own and found that he could generate a current when a pair was submerged in an acidic solution. It triggered a fierce controversy between the adherents of animal-electricity and metal-electricity, but Volta sealed his argument five years later, in 1800, when he invented his voltaic pile, a series of copper and zinc strips in a salt solution.

Volta's pile had the disadvantage that the current it produced rapidly diminished because of polarization (bubbles of hydrogen built up at the copper electrode, where they acted as an insulator). Several scientists attempted to improve it, notably John Daniel, an English chemist who in 1836 developed what is regarded as the classic form of the voltaic cell.

Daniel's cell avoided polarization by having a copper electrode immersed in copper sulphate solution forming the outer shell of the cell and a zinc electrode immersed in zinc sulphate solution at the centre. A porous cup kept the two electrolyte solutions separate. Instead of hydrogen forming, copper (from the copper sulphate) was deposited at the copper electrode and the current continued flowing.

The final episode in the early development of the battery came in 1866 when Georges Leclanché, a French chemist, invented the dry cell. It had carbon and zinc electrodes in an ammonium chloride solution, and a mixture of carbon grains and manganese dioxide to soak up any hydrogen that was produced. This is the formula used in the dry batteries that power so many of today's electrical gadgets.

Volta's greatest honour came after his death in 1827, when the scientific world used "volt" as the term for electromotive force, the pressure that propels electricity through wires and other conductors.

1803 The spray gun is invented by Dr Alan de Vilbiss.

Punchcard

Joseph-Marie Jacquard

1804

One of the mainstays of electronic computers, and the technology that IBM was built upon, has its origins in a weaving loom that was devised shortly after the French Revolution.

In 1790, Joseph-Marie Jacquard came up with the idea of adapting punchcards (which had already been used by Jacques de Vaucanson, a French inventor, to guide hooks on a silk loom) to weave intricate patterns with different yarns to make textiles such as brocade and tapestry. The punchcard enforced the needle motions needed to produce a pattern and a new pattern could be woven simply by changing the punchcard.

Joseph-Marie Jacquard's

The French Revolution prevented Jacquard from developing his idea. When he eventually produced his punchcard looms, in 1804, they caught on immediately but they also angered weavers, who feared that the new technology would cost them their jobs.

Jacquard was physically attacked and his looms in Lyon were burnt. However, in 1806, his invention was declared public property and he received a pension and royalty on each machine.

The Jacquard Loom is regarded as the first programmable machine. In allowing or preventing a needle from making a particular stitch at various points across the weave, the punchcards were acting as simple "yes/no" mechanisms, the principle that lies behind the electronic digital computer.

In the 1830s, Charles Babbage, the grandfather of the modern computer, realized the potential of Jacquard's punchcards and adopted them as the programming medium for the analytical engine that he never completed.

In 1884, Herman Hollerith devised a tabulator that later used punchcards to analyse the 1890 US census three times faster than previously. He founded the Tabulating Machine Company, which became IBM, for many years the world's largest computer company.

A Jacquard Loom from Spitalfields. This was the first programmable machine. The punchcards it used to produce different textile patterns became the basic principle of the electronic digital computer.

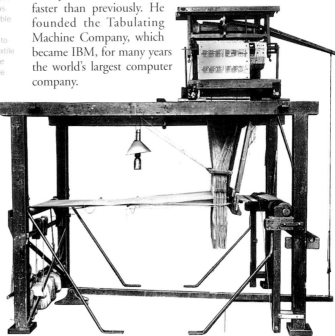

Steam locomotive

Richard Trevithick, George Stephenson

1804

Britain's earliest known railway was at Nottingham, where in 1603 Huntingdon Beaumont, a mining entrepreneur, had leased coal pits on Sir Percival Willoughby's Wollaton estate. By 1604, Beaumont had laid rails for his wagons to carry the coal "with great ease and expedicion" from the pits to the city. But the line did not make a profit and was disbanded within 15 years.

By 1608 Beaumont had introduced rails at his Northumberland pits. His speculations there failed, too, but his ideas were soon being imitated across the Northeast. At about the same time, simple railways were beginning to be seen on the banks of the River Severn at Coalbrookdale; both here and in the Northeast they were to play an increasingly important role in Britain's steady industrialization. Today, Beaumont's name is almost forgotten but perhaps we owe him the credit for sowing the seed of the modern railway.

The germ of the railway goes back even earlier on mainland Europe. Mining often involved negotiating long tunnels dug into a hillside. Railway tracks provide a strong, smooth path which automatically steers a vehicle and stops it getting bogged down. The power needed to move a given load on a railway is a tiny proportion of that on a conventional road or track, and so rails provided a useful method of guiding barrows of metal ore out of mines through narrow, rocky-floored corridors.

Evidence from several sites across central Europe suggests that such rudimentary lines were common by the end of the 15th-century.

Even more unexpected is the use of railways by the ancient Greeks. Stage managers in classical theatre knew they could create spectacular effects by sliding scenery on and off the stage on rails. Remnants can be seen at Greek theatres at Sparta and Megalopolis. Perhaps more significant is the remains of a railway system that carried boats across the isthmus of Corinth. It ran alongside the route taken today by the Corinth Canal and appears to have been used for a thousand years from about 500 BC. This sophisticated system was not developed further by the ancients. The Romans could have had a horse-drawn intercity rail service had they been less conservative in their technical innovation.

By the time George Stephenson was born in 1781, there were more than 100 miles of horse-drawn railways in the Northeast and probably the same amount again elsewhere in Britain. The one by his home at Wylam, west of Newcastle, was 30 years old. Almost all were bringing coal from mines to the nearest navigable waterway.

'Catch me who can'. Richard Trevithick's locomotive which entertained London in 1808

TREVITHICKS,
PORTABLE STEAM ENGINE.

Catch me who can.

Mechanical Power Subduing
Animal Speed.

When Stephenson was 20, working as a colliery mechanic and teaching himself to read, the Cornishman Richard Trevithick was pioneering relatively compact high-pressure steam engines, used mainly to pump water from mines. Trevithick mounted steam engines on road wheels, with mixed results on the poor surfaces of the time. The real opportunity was to try them out on railways, which were now beginning to use iron in place of wooden tracks.

In 1804 Trevithick demonstrated the potential of iron tracks and steam power by building a locomotive which easily hauled more than 20 tons along the Pen-y-Darren tramway near Merthyr. Four years later he showed something similar on a circular track in London's Euston Square, for a while as fashionable an attraction as the London Eye is today.

But over and over again Trevithick failed to convince others of the undoubted worth of his ideas. He gave up on locomotives, emigrated to Peru to install mine pumping engines, lost all his wealth and was bought a ticket home by Stephenson's son, Robert, before dying in penury in 1833.

Trevithick's ideas lived on in the North after his departure for Peru. The Middleton Colliery in Leeds, managed by its Tyneside owners, saw the first successful application of locomotives, built by Matthew Murray and John Blenkinsop, in 1812. These operated regularly for more than 20 years but they depended on a rack-and-pinion arrangement to ensure a good grip on the polished iron rail, a complicated process which proved an evolutionary blind alley.

Back on Tyneside, William Hedley, Timothy Hackworth and Jonathan Forster were soon building locomotives to move coal from Wylam, along that same line past Stephenson's birthplace to the staithes at Lemmington, on the tidal Tyne. Two of them, *Puffing Billy* now in the Science Museum and *Wylam Dilly* in the Museum of Scotland, survive as the world's oldest remaining locomotives.

Stephenson was soon developing his own locomotive ideas at nearby Killingworth. He, the Wylam pioneers and others, some self-taught, others qualified mining engineers, ensured that from 1810 to 1830, Tyneside was the Silicon Valley of the locomotive age. They vied with each other, sparked and stole ideas from rival camps and sometimes even co-operated, to develop better locomotives and track systems. What set

George Stephenson's 'Rocket' stole Tnevithick's glory when it won the 1829 Rainhill Trials.

Stephenson apart, however, was his vision of a national steam railway network, linking cities and carrying passengers as well as freight.

Darlington's Quaker businessmen employed Stephenson to construct a public railway 25 miles long from the West Durham coalfield to the coast at Stockton. By the time it opened in 1825, progress on locomotive design was such that they chose to rely on Stephenson's engines to handle the intensive coal traffic they anticipated. In September, Stephenson's *Locomotion* hauled the opening train, amid great public excitement, down the line to Stockton. The success of the Stockton and Darlington Railway established Stephenson as Britain's leading railway and locomotive builder. It also encouraged others to see the potential.

Soon railways were being promoted all over Britain. But it was the line from Liverpool to Manchester that opened the floodgates. Stephenson was the engineer of this first intercity line. Even so, in 1829 the directors were not convinced that the current locomotives, pottering along at five or six miles an hour, were what they were looking for. So they challenged all-comers to come up with something better.

The subsequent Rainhill Trials were a public relations coup both for the railway and for Stephenson, who was by this time working in partnership with his son, Robert. Their entry, *Rocket*, was the boost the industry needed. The Stephensons brought together the latest thinking (some of it their own and some borrowed) to improve power and efficiency. Thousands of spectators were astounded to see this diminutive machine haul useful-sized trains at a remarkable 30 mph. The future of the railway was assured.

By 1850, after a couple of periods of "railway mania" – dizzy periods of investment-led expansion, not unlike today's internet revolution – Britain was criss-crossed by 6,000 miles of railway. Abroad, the railroad was already feeding the westward movement of the American frontier. Great empires were beginning to be bound together by the iron rail in India and Russia.

The railway brought cheap long-distance travel and it helped workers afford to live away from their workplaces in new suburbs. Big cities were no longer dependent on local food supplies and, crucially for British and, later, world industrialization, cheap fuel meant factories no longer needed to be set beside fast-running streams or coal mines to get power. The railway has never been beaten for mass commuting nor for long-distance bulk freight.

Canned food

Nicholas Appert

1809

Frustrated at the length and vulnerability of the supply lines needed to feed his undernourished troops, in 1795 Napoleon offered a 12,000-Franc prize to anyone who could devise a method of finding a way to preserve food.

Fourteen years later, Nicolas Appert, a French confectioner, unveiled his method of food preservation, although he could not explain why it worked. Appert sealed food tightly inside a glass bottle or jar, which was then heated for some time. Provided the jar was not opened, the food inside remained fresh.

In 1810, Peter Durand, an Englishman, invented the tin can and received a patent from King George III for using pottery, glass, tin or other metals for heat-preserving food. Durand began supplying the Royal Navy with canned food; a can of veal and peas he supplied in 1818 was still fresh when opened in 1938.

However, canning was not widely adopted until 1846, when a method was invented that increased the number of cans that could be produced in one hour from six to 60. Shortly thereafter, in 1855, Robert Yeates, an Englishman, invented the first can opener. Until then, a hammer and chisel had to be used to open cans. Surprisingly, his can

Napoleon started the hunt for a way of keeping food fresh. The process of heating and sealing food, invented by a confectioner, kept micro-organisms at bay. All that was needed was for someone to invent the can opener.

opener did not catch on until about 10 years later, when makers of canned beef started to give it away free with their product.

Fifty years after Appert invented heat preservation of food, Louis Pasteur was able to explain how it worked: the heat killed any micro-organisms in the food, and sealing the can prevented any other micro-organisms from attacking the food and spoiling it.

Precision lathe

Henry Maudslay

1810

The lathe is an invention of deceptive simplicity: ultimately, all it does is turn components steadily while tools go to work on them. But few inventions have played so big a role in so many others: everything from self-tapping screws to jet engine shafts are made on them. They are truly an engineer's right arm.

While foot-operated wood lathes have been around in Europe since at least the mid-16th century, it was the emergence of precision lathes that unleashed their potential. John Harrison famously used precision lathes to make the components for his exquisitely accurate chronometers. The invention of the precision screw-cutting lathe by Henry Maudslay in 1810 revolutionized engineering by turning out nuts and bolts with perfect threads. (One of

Maudslay's other crucial "products" was the family of engineers he trained, including Joseph Whitworth, of screw standard fame, and Joseph Clement, who became chief engineer to Charles Babbage, grandfather of the computer.)

The simplicity of lathes also masks another of their traits: they are among the most potentially dangerous bits of kit in the engineering machine-shop. Anything that gets caught up in them – the silk ties of visiting dignitaries, for example – will be dragged in towards the rotating metal with appalling relentlessness. Components being turned can also fly off at lethal speed.

Even the stuff that lathes produce has its dangers. Metal-turning produces razor-sharp swarf. In Bath, a machinist at a local firm unwittingly became surrounded by a coil of swarf. It pulled tight, and sliced through his legs.

Spectroscope

Joseph von Fraunhofer

1814

In 1835, the French philosopher Auguste Comte declared that there was no point trying to fathom the mysteries of the universe, for its nature would remain forever inaccessible. Yet some 20 years earlier, a German physicist named Joseph von Fraunhofer had invented an instrument that would ultimately prove Comte wrong on a truly cosmic scale.

An expert optical instrument maker, Fraunhofer took great pains over the glass components used in his devices. It was while conducting tests on prisms using a beam of sunlight in 1814 that Fraunhofer noticed that the spectrum of colours they produced was crossed by hundreds of dark lines.

Measuring the wavelengths at which they appeared, Fraunhofer discovered that the lines showed up in the same positions in moonlight – itself made up of reflected sunlight.

One obvious possibility was that the lines were created by the glass prisms – a possibility Fraunhofer refuted by swapping the prisms for a diffraction grating, a device which could create spectra without using glass. The resulting spectrum still showed the lines in the same places.

The biggest surprise came when Fraunhofer used a telescope to shine starlight through a prism, thus creating the first-ever "spectroscope". Once again the lines appeared but, curiously, not always in exactly the positions of those in sunlight.

Fraunhofer died at 39, never understanding the cause of the lines, which still bear his name. However, his headstone carried what proved to be a prophetic judgement on his discovery: *Approximavit sidera* – "He approached the stars". During the 1850s, the German physicist Gustav Kirchhoff showed that Fraunhofer lines are the optical "fingerprints" of chemical elements.

Fraunhofer's spectroscope could reveal the different chemical make-up of the stars, and would ultimately do the same for galaxies on the edge of the universe, precisely the feat which Comte had declared forever impossible.

Rarely has an invention proved so potent, or a philosopher more comprehensively wrong.

Fraunhofer's prism spectroscope, which defied the philosophers and showed the chemical make-up of galaxies and stars.

Miner's lamp

Sir Humphry Davy

1815

The huge demand for coal in the early 19th-century brought with it a terrible toll in lives, many the victims of underground explosions. The principal cause was "firedamp", methane gas which seeped out of the coal seams. Miners carried caged canaries with them to detect the gas, but it was often too late by the time the canary fell off its perch.

In 1815 the Society for Preventing Accidents in Coal Mines asked Humphry Davy, Britain's leading chemist of the time, to find a solution. After a few weeks' research, Davy came up with a new form of miner's lamp fitted with a two-layer metal-gauze chimney which surrounded the flame. By dissipating the heat of the flame, the metal chimney prevented firedamp from reaching its detonation temperature.

Davy refused to make money from his invention, but in 1818 he was made a baronet in recognition of his life-saving services to miners.

Sir Humphry Davy.

Mineworkers in Tilmanston Colliery, Kent, in 1930. Numerous miners' lives were saved by the invention of the Davy lamp which detected methane gas.

Metronome
Johann Nepomuk Maelzel

1816

1816 David
Brewster, a
Scottish physicist,
invents the kaleido-
scope.

Conceived in 1812 by a Dutch inventor, Dietrich Nikolaus Winkel, but patented in 1816 by a German musician, Johann Nepomuk Maelzel, the early mechanical metronome consisted of a pendulum swung on a pivot activated by a hand-wound clockwork movement.

The audible ticking is the sound of a wheel passing a striker in its escapement. The speed of the ticking can be adjusted on modern models by moving a sliding counterweight up or down the arm. As well as its customary use for marking musical tempo, the metronome featured as an instrument in György Ligeti's *Poeme symphonique* and as a stimulus in Ivan Petrovich Pavlov's animal learning experiments.

Photography
Niepce, Daguerre, Talbot, Eastman

1816

Frustrated by an inability to draw, William Henry Fox Talbot (below and opposite outside his studio) invented a photographic process that was easy to use.

Photography (literally, drawing with light) evolved from a combination of two strands of natural philosophy that today we call optics and chemistry, but which first came to fruition with the daguerreotype and William Henry Fox Talbot's photogenic drawings in 1839. All of the relevant pieces of the jigsaw were in place well before 1800 and the wonder is not so much that photography first appeared in the opening decades of the 19th-century but that it was not discovered much earlier.

Since ancient times it had been observed that light passing through a small hole made in the wall of a darkened room (*camera obscura*, in Latin) casts an inverted image of the outside scene on to the wall opposite the hole. By the 17th-century, portable versions of the *camera obscura* (small wooden boxes fitted with lenses to give bright, clear images) were in common use as artists' sketching aids. It was from these that the photographic camera evolved.

During the 18th-century, a number of scientists had noticed that some chemical compounds, in particular silver salts, darken when exposed to sunlight. Johann Heinrich Schulze, a German professor of anatomy, demonstrated the fact in 1727 by using sunlight to record words on the salts, but he was unable to preserve the images permanently.

In January 1839, Louis Jacques Mandé Daguerre, a successful artist and showman, announced the invention of the first practicable photographic process at a meeting of the Académie des Sciences in Paris. Ten years earlier, Daguerre had entered into partnership with another Frenchman, Nicéphore Niepce, an amateur scientist who had succeeded in capturing and fixing his first photographic images in 1816. However, Niepce's process was impractical, requiring exposures of at least eight hours. As a businessman, Daguerre was aware of the commercial possibilities of capturing images formed by the *camera obscura*. After Niepce's sudden death in 1833, Daguerre succeeded in producing images on silver-plated sheets of

copper, made light-sensitive by iodine vapour, which were then developed by exposure to mercury. News of Daguerre's discovery was received with shock and dismay by an Englishman who had for the previous few years been working, totally independently, on his own system of photography. William Henry Fox Talbot was a gentleman, scholar and amateur scientist with an ancestral home at Lacock Abbey in Wiltshire. A gifted polymath, Talbot was interested in many aspects of natural philosophy and mathematics. He was also, for a short time, the Member of Parliament for Chippenham.

Talbot's interest in photography stemmed from his frustration at his inability to draw. In 1833, while on honeymoon on Lake Como, he attempted to sketch the scenery, later describing the results as "melancholy to behold". Reflecting, by comparison, on the images that he had been able to view using his camera obscura ("fairy pictures, creations of a moment and destined as rapidly to fade away"), an idea

Louis Daguerre.

occurred to him: "How charming it would be if it were possible to cause these natural images to imprint themselves durably, and remain fixed upon the paper!"

On his return, Talbot began experimenting with paper coated with solutions of light-sensitive silver salts. He was soon able to make silhouette negative images of leaves and lace placed on sensitized paper that was then exposed to sunlight. These images, which he called photogenic drawings, were fixed by treating them with a strong salt solution.

In 1835 Talbot made a momentous discovery. If another sheet of sensitized paper was exposed in contact with a negative image, the result was a positive photograph in which light and shade were correctly represented. He had discovered the fundamental principle on which modern photography is based.

Talbot's active intellect soon became distracted with other projects and he largely forgot about his photographic work until, early in 1839, news reached him of Daguerre's announcement in Paris. Anxious to establish his prior claim to the invention of photography, Talbot quickly made his discoveries public. As more details gradually emerged, however, it soon became clear that the discoveries of Talbot and Daguerre were completely different. Daguerreotypes were unique, one-off images, made on silver-plated sheets of copper, whilst Talbot's paper negatives could be used to make an unlimited number of positive photographs. The daguerreotype process enjoyed some short-term success but it was effectively a dead-end and was extinct within two decades. The future lay with Talbot's negative/positive process, called the calotype.

Talbot's photogenic drawing process was far from perfect, requiring exposures of up to one hour. In September 1840, he made another breakthrough. He discovered that even a brief exposure in the camera was enough to produce an invisible, latent, image which could then be revealed, or developed, by further chemical treatment. Exposure times were reduced to less than a minute, opening up the possibility of a range of new applications, including portraiture.

Frederick Scott Archer's discovery, in 1851, of the wet collodion process, which used glass plates as a negative support, led to a dramatic growth in commercial photography. The wet collodion process combined the clarity and detail of the

A daguerreotype (left) and a Chevalier folding camera.

daguerreotype with the calotype's ability to make multiple copies and it quickly superseded both earlier processes. Soon studios in every town in Britain were turning out *cartes-de-visite* portraits and by the late 1850s, photography had ceased to be a novelty.

Despite its ubiquity, photography remained largely the preserve of the professional operator. It was an expensive and time-consuming activity that required considerable chemical knowledge and technical skill.

In the early 1880s, the commercial introduction of the gelatin dry plate transformed photography. Unlike wet collodion plates, dry plates could be bought ready coated, saving the photographer a great deal of work and inconvenience. They were also much more sensitive. Exposure times were reduced to a fraction of a second. For the first time, cameras could be held in the hand instead of having to be put on a tripod, and candid photography became a possibility. Movement could be arrested and captured, paving the way for the invention of cinematography.

In 1888, George Eastman, an American dry-plate manufacturer, introduced a camera that was to revolutionize photography and begin the process of turning it into a popular medium. Eastman called his new camera the Kodak, a word that he had made up.

Instead of glass plates, the Kodak, which cost five guineas, was loaded with enough film (originally paper but later celluloid) to take 100 photographs. After the film was used up,

The first Kodak (left) and a Prototype made by Leica for the first Camera to use 35mm film.

the entire camera was returned to the factory where the film was processed. The camera was loaded with fresh film, then returned to its owner, together with a set of prints. For the first time, photography did not require any chemical knowledge or technical skill.

In 1900, Eastman introduced the camera that was to complete photography's transformation. At five shillings, the Brownie was the first camera that almost everyone could afford. Thirty years later cameras became truly portable when 35mm film was used.

Stethoscope
René Théophile Hyacinthe Laënnec

1819

I n 1819, a young Frenchman devised a way for physicians to listen inside the human body with ease and to scrutinize heartbeat, lung and stomach functions.

René Théophile Hyacinthe Laënnec called his invention the stethoscope, after the Greek stethos, for chest. The early stethoscopes built by Laënnec were simply hollow wooden cylinders which acted as sounding boxes or resonating chambers. They were about 30 cm long and as they were held against the chest of the patient, the physician would press his ear to the lid. It was not until the invention was refined by, among others, the American physician George Philip Cammann, that two earpieces were used,

The more familiar binaural stethoscope consisted of a bell-shaped chamber which was pressed up against the patient's skin, connected to a pair of listening earpieces via a "Y" connector and rubber tubing. Within the bell-shaped receiver head, which amplified low-frequency sound picked up on contact with the skin, was a diaphragm to amplify high-frequency sound discernable to the human ear.

The stethoscope, which can check heart, lung and stomach functions.

Waterproof clothing

Charles Macintosh

Charles Macintosh was famous not for his pioneering work as an industrial chemist, nor his patenting of a process to convert iron into steel, nor his development of a series of important fabric dyes. It was something he invented that was rather more useful to the man on the rainy streets of his native Glasgow that made his name: the waterproof coat, the Macintosh.

Macintosh had been experimenting with the waste products of gasworks. He tried dissolving rubber in heated naphtha, a by-product from the production of coal tar or town gas. In 1823 he found he could make fabric waterproof by coating it with the new dissolved rubber solution. A further refinement of the new fabric was to make a "sandwich" of two layers of wool cloth which were bound together with the rubber preparation.

The garments soon became popular even though the first ones were stiff in cold weather and sticky in hot weather. They sold even more when Macintosh improved the coat in 1839 with a more durable "vulcanized" rubber coating, which made it more comfortable in all climates.

Equipement de Guerre. Weatherproof BURBERRY
POUR OFFICIERS FRANÇAIS ET ANGLAIS

" Sans égal pour la campagne d'hiver, car il assure la chaleur, le confort et la protection, et diminue les risques auxquels les variations de température exposent la santé."

LE BURBERRY
Modèles de cavalerie ou d'infanterie, avec ou sans doublure mobile en poil de chameau. La pluie, le vent, le froid, ne peuvent pénétrer ce grand pardessus qui, quoique bien imperméabilisé, est léger et perméable à l'air.

Les véritables vêtements Burberry portent l'étiquette BURBERRYS

Catalogue militaire illustré envoyé franco.

" Le BURBERRY m'a rendu les meilleurs services pendant la campagne. "
— Général d'Urbal.

LE " TIELOCKEN "
Un manteau, avec ceinture, breveté, qui assure une double protection de la tête aux genoux et s'attache sans boutons. Porté par lord Kitchener.

BURBERRY TRENCH-WARM
Assure le confort et les services de trois manteaux, chacun pouvant être porté séparément. Un WEATHERPROOF qui supportera des heures de pluie. Un SHORTWARM en poil de chameau, et UN EPAIS PARDESSUS, pour les temps rigoureux.

Uniformes français et pardessus en tissus réglementaires

GREAT COATS
British Warms, Manteaux d'aviation et tout ce qui compose l'équipement réglementaire.
Tout faits, de suite ; ou sur mesures, de 8 à 10 jours.

BURBERRYS   8 et 10, BOULEVARD MALESHERBES   PARIS
Haymarket, LONDRES

Suspension bridge

Telford, Brunel, Finney

Only the most philistine can fail to be impressed by suspension bridges; their elegance, economy and sweep. They are truly the beauty of the laws of Nature made visible. The shape formed by their main cables (a curve known as a catenary, after the Latin for chain) is a manifestation of the Principle of Least Action, a mathematical expression of Nature's penchant for elegance.

Exactly who should be credited with constructing the first suspension bridge is a matter of some dispute. Many ascribe it to the Scottish engineer Thomas Telford, who built the 176-metre Menai Suspension Bridge between north Wales and Anglesey in 1826. Some give it to Isambard Kingdom Brunel for the Clifton Suspension Bridge near Bristol (its main claim to fame is that it ever got finished at all, after all the wrangling), others to an American engineer, James Finney, who completed a form of suspension bridge in 1801.

Ultimately, of course, the credit should go to the unknown genius who many centuries ago strung together lianas between trees in some tropical jungle, Indiana Jones-style. Among the cognoscenti, however, it is John Roebling, a German-born American engineer, who is most often mentioned as the father of the modern suspension bridge, for his work on the Brooklyn Bridge, opened in 1883.

Roebling was the first to use steel hawsers for the four supporting cables, woven and erected according to a process he himself devised and which is still used today. He also developed pioneering methods to prevent excessive flexing of the deck.

Roebling never lived to see the product of his ingenuity. He died from tetanus after a tragic on-site accident just after work began on the project. His son, Washington, completed the bridge, with its then world-beating span of 486 metres.

Suspension bridge engineers have since became more daring, though not without learning some bitter lessons. The most famous was in 1940, when the Tacoma Narrows Bridge in Washington state tore itself apart in strong winds. The film footage of the disaster, with the bridge twisting in the breeze,

is still shown to civil engineering students as a warning of building bridges so light, sleek and elegant that they lack torsional strength.

Today's suspension bridges, such as our own 1,410-metre Humber Bridge, are sophisticated constructions, each component the result of meticulous calculation. Decks are shaped by the laws of aerodynamics, while the tops of the towers of the Humber Bridge are 3.606 cm further apart than their feet, to take into account the curvature of the Earth.

Oppoaite: Brooklyn Bridge in New York, the largest in the world when it was completed in 1883, and first to use steel hawsers. Its innovative architect, John Roebling, is often regarded as the father of the modern suspension bridge.

Friction match

John Walker

1826

The friction match is an invention which truly hides its light under a bushel. For until John Walker came up with it in 1826, mankind had no simple and reliable means of getting access to fire.

The controlled use of fire is rightly regarded as one of the key discoveries in the history of mankind. But for countless millennia our ability to exploit it was stymied by the problem of making fire on demand. Rubbing sticks together or striking flints into tinder boxes was all very well provided everything was bone dry and you weren't in a hurry.

Solving this problem was one of the first challenges tackled by early modern chemists. White phosphorus, which ignites spontaneously in air, was the first chemical element to be discovered by scientific means, by the German chemist Hennig Brand around 1669. Eleven years later, the English chemist Robert Boyle exploited its properties to make the first crude matches in the form of wooden sticks dipped in sulphur which were lit by passing them through paper coated with phosphorus. They failed to set the market alight, however, not least because white phosphorus is all too keen to burst into flames, and had to be handled with great care.

According to inventing folklore, the breakthrough came in 1826 when Walker, a Stockton-

based chemist, finished mixing some compounds together with a stick, and brushed it across some stones. The head of the stick burst into flames.

Walker worked out that the mixture – potassium chlorate and antimony sulphide – was perfectly stable, but had a low ignition temperature: so low, in fact, that the feeble heat generated by the friction of the stick scraping across the stones was enough to ignite it.

By April 1827, Walker had sold his first box of matches to a Mr. Hixon, a local solicitor. But the matches were still tricky to light and gave off appalling fumes. Continental chemists found the answer, adding a judicious amount of white phosphorus to the mix.

The cleverest innovation followed the discovery of another, more stable form of phosphorus, red in colour, in 1845. Ten years later, J. Lundstrom of Sweden made this the basis of the safety match, in which the chemicals needed to make fire, including red phosphorus, were divided between the match head and the striking surface. Unless the two came together, providing all the necessary ingredients, nothing happened.

Braille

Louis Braille

1827

I t was partly by chance, and partly due to his thoughtful nature, that Louis Braille, a young Frenchman, invented the first practical printed script for the blind.

Braille was blinded by an accident at his father's shoemaking workbench at the age of three, and at 11 he was sent to the National Institute for the Young Blind in Paris, where pupils learnt to read from primitive embossed paper texts. Production of the texts was so elaborate and time-consuming that it was

1827 Charles Wheatstone invents the microphone.

impossible for Braille and his classmates to learn to write and type as well as read.

One day, an artillery officer, Charles Barbier, visited the school to demonstrate a military code he had devised called "night writing". It was a crude alphabet made from a matrix of 12 embossed dots. He had proposed that it should be used in the battlefield at night by troops. But the average soldier had found it too complicated.

Braille at once saw the usefulness of the code. Although the system was still far too elaborate, he realized that it could be read with just one hand.

In 1827, after spending six years simplifying the code, Braille published his first Braille book. It was not an overnight success and took many years to catch on, but by the time of his death in 1852, Braille had liberated many blind people. For the first time they had a chance not just to read books, music and even mathematical tables; they could also write for themselves by using an embossing stylus.

Braille, a language of touch, was developed from a French military code.

Sewing machine

Barthelemy Thimonnier

1830

To Charles Frederick Wiesenthal in 1755, it seemed obvious that if you were carrying out embroidery work it could be done quicker if the needle did not have to be turned to complete the stitch. Why not have a needle with a point at both ends and an eye in the middle? This simple invention was taken up 100 years later when it became a major component of the sewing machine, an invention that made a far greater impact on society.

The first patent for a sewing machine had been submitted by Thomas Saint, a London cabinet maker, in 1790. Although it would certainly have worked, it is unclear if any were made. Several attempts to produce a viable machine were made over the next 50 years. The best was by a Frenchman, Barthelemy Thimonnier, who, in about 1830, developed and put into

The first practical sewing machine made by Elias Howe, in 1845.

industrial operation chain-stitch machines. Most of them, however, were destroyed by rioting mobs in 1841, and the revolution of 1848 caused production to stop.

It required the obsession of a poor New England farmer named Elias Howe to produce the first practical lock-stitch machine and in 1845 to patent the concept of using a Wiesenthal needle and a double-pointed shuttle to work in combination to produce a secure stitch. Though the machine was not successful, the operating principle was right. The proposed selling price of Howe's machine was $300, but a seamstress earned about $7 a week and could be laid off instantly, so the machine was not viable. With no sales in America, Howe came to England where he sold the British manufacturing rights cheaply to William Thomas, a corset manufacturer.

By 1850, Howe returned to America to find others selling sewing machines that were simpler, faster and cheaper to make. Howe made few machines himself, yet by suing those who infringed his patents and by collecting royalties of $5 for every machine made in America, he became the fourth richest man in the country.

One of the first to be successfully sued by Howe was a former itinerant actor, salesman and inventor, Isaac Merritt Singer. He had acquired patents and combined them to produce a rugged and reliable industrial machine. He later became America's largest producer of domestic machines.

Lawnmower

Edwin Budding and John Ferrabee

1830

The lawnmower is a classic case of an invention that brings within reach of everyone something only the super-rich could once afford. Until the 1830s, you had to be wealthy to justify the expense of the small army of scythe-wielding labourers needed for well-kept lawns. Today, millions of people have lawns kept to a standard higher than the aristocrats could manage. It is a privilege granted to us by the lateral thinking of a farmer's illegitimate son, Edwin Budding.

The story goes that in 1830 Budding was working in a textile mill when it dawned on him that the rotary machines used to trim the nap on velvet could just as easily trim grass. Cannily recognizing that he had hit on a great idea but lacking all the skills to exploit it, Budding went into partnership with a businessman, John Ferrabee. Even more cannily, the two men allowed other companies to make their patented mower under licence, bringing in a nice regular revenue stream.

Their first lawn mower had 19-inch (46 cm) cutting blades arranged in a cylinder, propelled via gears by the roller at the back, with the grass thrown into a tray at the front. In short, it was essentially identical to the cylinder mowers on sale in garden centres today.

The source of power for mowers evolved over the coming decades in a kind of fast-forwarded history of propulsion down the ages. Manpower was first replaced by horses (fitted with padded leather boots so as not to damage the sward) and then by steam. The first steam-powered mower weighed two tons and cut a metre-wide swathe in one pass. Petrol-engine models started to appear during the 1890s, and the first electrically driven lawnmowers went on sale in 1926.

It must say something about our passion for gardens that a new type of lawnmower, the hover-mower, became one of the most controversial inventions of recent times. Based on the relatively simple rotary cutting blade design (which emerged long after the original cylinder design with its helical "guillotining" action), the hover-mower made its debut in the early 1960s, but commercial rivalry with traditional mower-makers led to the Great Cuttings Debate. Experts agreed that cuttings provided a useful mulch, but the "collectors" won the day, and the hover-mower makers brought out models that collected clippings. The capitulation had little to do with botanical science but a lot to do with lawn aesthetics.

Sunday relaxation for a Nottingham vicar: the first mowers could cut grass up to seven times faster than a scythe, and there was no skill required. Everyone's lawn could look as if it belonged to the aristocracy.

Chapter 4
Electricity on the move
1830 to 1899

By 1830 the Industrial Revolution was in full swing. Steam power had transformed the landscape, mechanizing textile production, mining and manufacturing, leaving the world poised for the transformation that dominated the second half of the 19th-century: the communications revolution.

Communication was so completely changed between 1830 and 1899 that it can be regarded as a revolution within the Industrial Revolution. In the 18th-century, roads and canals were improved in order to move the products and raw materials around Britain. Then, following the harnessing of steam, the first public railway line between two cities (Liverpool and Manchester) opened in 1830 and the locomotive, in the form of the traction engine, took to the road and field. By 1837, Isambard Kingdom Brunel's *Great Western* was crossing the Atlantic.

But the biggest changes derived from electricity, which came of age with Michael Faraday. He was the first person to fully understand and exploit the relationship between electricity and magnetism, which could be used to instigate movement in the motor or generate electricity in the dynamo.

A series of innovations followed in the wake of electricity's mechanical liberation from the shackles of the chemical battery. The electric telegraph, fax machine, telephone and wireless radio replaced the horseback messenger, semaphore and drum beats as the means of conveying information over long distances. Made-up languages, such as Esperanto followed the spread of telecommunications technology, their inventors believing that all people would be able to converse in a common tongue.

Electricity also enabled the invention of the refrigerator, which brought meat from Argentina and America, butter from New Zealand

and vegetables from Europe, changing eating and retail habits and eventually liberating the housewife from daily shopping. It also made possible X-rays, escalators and the electric chair.

Perhaps surprisingly, electricity took longer to take off in Britain than in America and Europe. The obvious benefit of electric light, even after Joseph Swan and Thomas Edison had invented the incandescent bulb, was not sufficient to usurp steam, coal and coal gas as the primary power source. In fact, the gas mantle was invented more than 10 years after the light bulb, but it took many years before the bulb lit more homes than the mantle. From 1792, when William Murdock, an agent for Watt and Boulton's steam-engine company, became the first to install gas lighting, gas was widely used and it was not entirely replaced by electricity until the mid-20th century. Instead, it was the arrival of electric trams and underground trains, and the demands of industrial motors that sparked electricity's emergence.

Demand for electricity prompted research into efficient and powerful engines, resulting in the invention of the steam turbine by Sir Charles Parsons, and the internal combustion engine, fuelled by derivatives of oil; first kerosene, then petrol. The internal combustion engine became the prime mover of technological development in the late 19th and early 20th-centuries, and carried the focus of innovation away from Britain to America and Europe.

The success of petrol engines was assured by the first striking of oil at Titusville in Pennsylvania in 1859. "Black gold" went on to exert an influence beyond automotive power as its fractions were used by the emerging organic chemistry industry to produce dyes, explosives, plastics, artificial fibres and pharmaceuticals, which with immunisation, antiseptics, X-rays and synthetic drugs such as aspirin ensured the rapid increase in life expectancy in the latter half of the 1800s.

The benefits of medical science, improvements in agricultural productivity and the raising in living standards created by inventions stemming from electricity and steam power, generated a strong belief in Western society in the desirability of technological change. Monuments to engineering, such as the laying of the first Atlantic telegraph cable, the construction of the Eiffel Tower and the Brooklyn Bridge, the building of the Suez and Panama canals, the connection of major cities by rail and the appearance of motor cars on roads impressed upon society the acceptance of technological innovation not just as a fact of life but as a way of life itself.

Dynamo & Motor

Michael Faraday and Thomas Edison

Michael Faraday, a blacksmith's son, unshackled electricity from the constraints of the chemical battery.

The hero of the electric age is Michael Faraday. A largely self-taught scientist, he relied more on practical exploration, observation and his talent of detailed pictorial imagination than on theoretical calculations to make a number of fundamental discoveries and inventions. Faraday, a blacksmith's son, was born in Newington, Surrey, in 1791 and with no formal education he went on to unshackle electricity from the constraints of the battery.

He became fascinated with science while apprenticed to a bookmaker from the age of 14. When he was 21, he attended a lecture by Sir Humphry Davy at the Royal Institution, after which he pestered Davy for a job. Davy appointed him his assistant and took him on a grand tour of Europe, during which he met Count Alessandro Volta, inventor of the battery.

In 1820, Hans Christian Oersted, a Danish physics professor, revealed the link between electricity and magnetism when he showed that an electric current could deflect a compass needle. The following year, Faraday applied Oersted's discovery of electromagnetism to invent a device he called "the rotator", which had a wire suspended over a strong permanent magnet. When the wire was connected to a battery, the electric current generated an opposite magnetic field to that in the magnet, causing the wire to be repelled and rotate around the magnet. It was the first electric motor. Although no more use than a toy, it led Faraday to his most important breakthrough, when in 1831 he showed the reverse effect of the rotator, namely that electricity was induced in a wire when it was moved through a magnetic field, a process called electromagnetic induction.

He used two coils of wire wrapped around either side of an iron ring. A current passing through the first coil produced a magnetic field which ran round the ring, inducing electricity in the second coil. This is the basis of the transformer, which can step up or step down the amount of current according to the ratio of the number of turns in each of the coils.

Faraday's invention produced only a single flash of current in the induction coil when the current-carrying coil was switched on. To produce a continuous current, he used a copper disc that spanned the opposing poles of a horseshoe magnet. One wire touched the edge of the turning disc, the other was fixed to its axis. As the disc was turned in the magnetic field, a meter between the wires revealed that a steady current had been produced. This was the first electric generator or disc dynamo.

Once the principles of dynamos and motors had been demonstrated, progress was rapid. In France Hyppolyte Pixii improved Faraday's design to create the first effective generator in 1832. By 1837, the electric motor had been improved by Charles Grafton Page, a medical physician who is regarded as America's first professional inventor.

By the 1850s, private companies used steam-powered generators to power electric arc lights in the street and in theatres, often selling surplus power to neighbours. Despite its obvious appeal, many argued at the time against the use of electricity, pointing out that employing a steam engine to turn a dynamo to generate electricity was an inefficient use of coal, or gas. But the sheer convenience of consumers being able to

Artificial lightning in Nikola Tesla's laboratory. Tesla invented generators and electric motors for alternating current, which was promoted by George Westinghouse. It proved more efficient than Thomas Edison's direct current for distributing electricity over long distances.

1831 Charles Darwin sets sail aboard *The Beagle* to explore South America and the Pacific, leading him to develop his theory of evolution by natural selection.

Safety fuse invented by William Bickford, a leather merchant, making detonations safer.

switch on a source of power without having to stoke a fire soon won the argument.

The great American inventor Thomas Edison determined to bring electric light to homes, and he created much of the infrastructure needed by the fledgling electricity industry. When he began working on the dynamo, it was at best 40 per cent efficient; he improved it to 82 per cent. But despite his inventive flair, Edison made a major misjudgment. His generators produced direct current (DC), which was unsuited to supplying electricity over long distances from his power station in New York.

Alternating current (AC), which can be transformed from low voltage and high current to high voltage and low current, proved a far better system for distributing electricity. Low current generates less heat and thus loses less power when sent over long distances.

George Westinghouse, an American industrialist, believed in alternating current and promoted its system of generators and motors devised by Nikola Tesla, a Croatian electrical engineer who emigrated to America. But the first efficient AC transformer, taken up by Westinghouse, was created by William Stanley, a New York electrical engineer.

Edison did not give up and direct current continued for a while. In 1881, Godalming in Surrey became the first town in Britain to combine public and private use of direct current electricity with a water-wheel power station. That autumn, Edison opened a DC power station at Holborn Viaduct in London, while in 1887 the Hammond Electric Light Company opened a DC power station to supply an entire district in Brighton.

The ease of AC transportation, however, meant that direct current's days were numbered. The London Electricity Supply Corporation opened an AC station designed by Sebastian de Ferranti in Deptford, southeast London, shortly after the first practical AC power station was built at Lauffen in Germany in the early 1890s. It generated power at 10,000 volts, which was stepped down through transformers for local use, a procedure that required Ferranti to design a new type of cable to carry the high voltages. Today the AC system is standard across the world.

Electric Telegraph

Gauss and Weber, Henry, Wheatstone and Cooke

1833

The invention of a workable electric telegraph system revolutionized communications, but no single inventor can claim total credit. Partly this is because it built upon earlier inventions, such as Claude Chappe's mechanical telegraph, but mainly because the electric telegraph as it became widely used was the fruit of several inventors' efforts.

The first known proposal for passing a current down a wire in order to make something intelligible occur at the other end dates from 1753, when *Scots' Magazine* published a letter from a mysterious "C. M." who suggested a 26-wire system (one for each letter of the alphabet) powered by a frictional generator.

Dr Samuel Thomas von Soemmering's telegraph of 1810 employed a similar system. It had a voltaic cell, invented in 1800 by Alessandro Volta, linked by wires to an array of electrodes immersed in water, each of which corresponded to a letter of the alphabet. Passing a current caused electrolysis of the water, indicated by a stream of bubbles emerging from the activated electrode.

The Cooke and Wheatstone telegraph

In 1816, Francis Ronalds strung more than eight miles of wire around his Hammersmith garden and succeeded in making an electrically-charged pair of pith balls twitch when a current was passed down the wire. Ronalds was keen to demonstrate his system to the Admiralty, but they were unmoved and rebuffed him in a letter stating that, since the peace with France, "telegraphs of any kind are now wholly unnecessary".

Most early experiments were hampered by the lack of a simple means of detecting the presence of electricity in a wire. After the Danish physicist Hans Christian Oerstadt discovered electromagnetism, with his observation that an electric current flowing in a wire gave rise to a magnetic field, instantaneous, long-distance communication was made possible.

Attempts at electric telegraphy still hit snags. The longer the wires grew, the weaker became the signal at the receiving end. This inhibited Samuel Finley Breese Morse, a portrait painter and dilettante inventor from Massachusetts, who first had the idea of building an electric telegraph when returning from Europe in 1832. (See Morse code, page 259.)

In 1833, two Germans, Gauss and Weber, produced a battery-powered telegraph system linking Göttingen's Physical Laboratory with its Observatory, two-thirds of a mile away. By 1835 they had adopted magnetic induction (electricity generated by moving a coil of wire in the field of a vertical magnet) to send messages to a receiver.

The following year an Englishman, William Fothergill Cooke, was inspired to build an electric telegraph based on his observations of apparatus designed by the Russian telegraph pioneer Baron Pavel Schilling. Cooke's first system employed three moving needles whose various positions signified different letters. However, both Cooke and Morse, who were

Telegraph operators on July 3 1897 send Queen Victoria's Jubilee message to the Empire. The electric telegraph linked the world for the first time and had a dramatic effect on business and government, as well as ordinary people, who could now conduct, among other things, long-distant romantic affairs.

unaware of each other's experiments, were unable to send messages over wires of any great length. Lacking scientific training, neither of them was aware that a solution to their common problem had already been developed.

In 1830, Joseph Henry, an American physicist, had realized that a series of small, connected batteries, rather than one large one, would enable signals to travel much further. Cooke was introduced to Professor Charles Wheatstone, who also knew of Henry's work, and the pair built, in 1837, a successful experimental telegraph between Euston station and Camden Town. Wheatstone was disdainful of Cooke, and the two were constantly falling out. However, by 1843 they had a link between Paddington and Slough, while Morse, in America, had also become aware of Henry's work and connected Washington DC to Baltimore, 40 miles away, the following year.

The new technology entered popular consciousness as the wonder of the age. The Slough-Paddington telegraph was promoted as a spectator attraction at a charge of one shilling a head, a contemporary poster declaring, "The Electric Fluid travels at 280,000 miles per second".

International communications soon followed. Wheatstone's vision of a submarine cable link to France in 1840 had been realized by 1851. The design of the wire-armoured cable (an unarmoured version laid in 1850 had been cut through by a French fisherman the following day) set the pattern for future ventures. The first successful transatlantic cable was completed in 1866, an earlier, thinner cable of 1858 having failed after only one week.

As telegraphic traffic increased, ways to maximize the potential carrying power of the lines were sought. In 1853, Wilhelm Grindl, an Austrian, proposed a method of doubling a wire's capacity. This "duplex" system was perfected in 1872 by Joseph B. Stearns of Boston, allowing remote operators to telegraph over the same wire at the same time. Two years later, the American inventor Thomas Edison doubled the duplex's capacity with the introduction of his "quadruplex". This enabled four messages to be sent over the same wire simultaneously, thereby adding $15 million (£110 million in today's money) to the value of Western Union.

The electric telegraph revolutionized communications between individuals and nations and had far-reaching effects on society at all levels. Romances blossomed over its cables, there were weddings by telegraph and it spawned new crimes, prompting governments to attempt with little success to regulate the medium. In the same way as e-commerce is currently changing business practices, the telegraph made dozens of new enterprises possible and made others redundant. The electric telegraph network laid the foundations for the mass communications technologies that followed.

Refrigeration

1834

Jacob Perkins

The challenge of producing ice-cold temperatures on demand has parallels with the problem of doing the same with heat, and both were solved within a few years of each other.

For most of history, mankind had to rely on nature to provide a source of cold material, in the form of snow and ice, just as lightning had been relied on to provide a source of fire. Storage of the snow and ice in an underground cellar lined with natural insulation such as straw was invented by the Greeks, and you can find such "ice houses" in the grounds of country homes to this day.

The secret of creating cold on demand was found many centuries ago in the Middle East, perhaps after someone realized how the body does it: by sweating or, as scientists call it, evaporation.

When a fluid evaporates from a surface, the escaping molecules take heat away with them, which is why a damp cloth draped over a milk bottle keeps it cool. The Egyptians used nighttime evaporation of water in trays to make ice cubes in around 500 BC.

But water evaporation is an unreliable way of chilling things and when the water runs out it does not work at all. In 1834, Jacob Perkins, an American living in London, solved both these problems. First he swapped the damp cloth for pipes filled with highly volatile liquids whose molecules evaporated very easily. Then, to make sure they kept performing their cooling job, he fed the evaporated molecules into a

1834 Samuel Hall, an English engineer, invents a steam condenser, reducing the amount of water steam ships carried by turning waste steam back into water.

compressor, which forced them back into a liquid state, ready to evaporate all over again. Perkins had thus created the first refrigerator.

For some reason he never pushed his invention, leaving it to others to exploit its huge potential. By 1856 Alexander Twining was making ice in commercial quantities, and in 1880 the first cargo of refrigerated meat arrived in England from Australia, opening up a worldwide trade based on refrigeration.

Refrigerators are widely thought of as a Good Thing, but their success is hardly unalloyed. Until recently, the volatile liquids they used were either lethal or environmentally nasty, and by allowing supermarkets and customers to keep food for months on end, they kill off local grocers' shops. The year-round availability of fruit and vegetables that refrigeration has given us has also robbed us of the seasonal thrill of being able to get fresh asparagus and strawberries. And a larder is still a better place to keep cheese.

An early 'ice safe' for storing food.

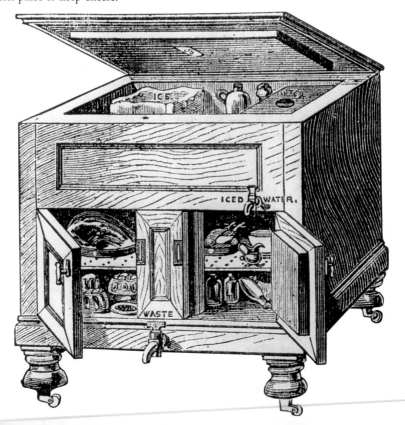

Propeller
Francis Pettit Smith

1835

The propeller is a perfect example of a great idea that has been staring people in the face for centuries. It is simply the reverse of the water-raising screw invented by Archimedes more than 2,000 years ago. Instead of the screw remaining stationary relative to its surroundings, while the water within it moves as the screw rotates, the propeller is a rotating screw moving through stationary water. Its propulsive force comes from the steady flow of the mass of water pushed rearwards by the propeller.

A number of people toyed with the idea of using a screw-like device to propel objects through the air. James Watt seems to have been the first to think about using it for ship propulsion, in 1770. In the early 19th century, a slew of patents were granted to various people for designs of screw propeller. In 1835 Francis Pettit Smith carried out experiments with a small boat driven by a wooden propeller, and went on to launch the first purpose-built propeller-driven ship, the *Archimedes*, in 1838.

Isambard Kingdom Brunel built the *Great Britain*, the first propeller-driven ship to cross the Atlantic (opposite).

Among those to be impressed by the ship's performance was Isambard Kingdom Brunel, then in the process of revolutionizing transatlantic travel. His paddle-powered steamship the *Great Western* had already crossed the Atlantic in just 15 days, but Brunel aimed for even greater speeds. After seeing the *Archimedes*, he decided that his next ship, the *Great Britain*, would also be equipped with a propeller, and in 1845 it became the first propeller-driven ship to cross the Atlantic.

Paddlewheels can be almost as effective as propellers in terms of propulsion. But paddles are easily damaged by storms (or enemy ships) and are given to lifting clear of the water in rough seas. That's why you now tend to see them only on the placid rivers of the Deep South of America.

There's no escaping the fact that the propeller is a very efficient form of propulsion. The propeller of the *Great Britain* had about 90 per cent of the efficiency of its modern counterpart. That's not to say they're now perfect. Nuclear submarine designers are constantly trying to minimize the

1835 William Henry Fox Talbot invents negative photography, the principle on which modern photography is built.

noise of the churning propeller, which can give away the submarine's position. One solution, invented in 1853 by the great engineer Henry Maudslay, is the variable-pitch propeller, which can be trimmed at high speed to minimize "cavitization" (bubbles on the propeller), a primary source of noise.

Mechanical Computing

Charles Babbage

1835

The demand for computational power began after the death of Isaac Newton in 1727, when many of the world's leading minds became focused on exploiting his scientific legacy. From the calculation of comet orbits to the prediction of tides, Newton's laws of motion and gravity allowed a host of phenomena to be understood in exquisite detail. Yet a huge hurdle faced anyone wanting to exploit the full power of Newton's insights: hideously complex calculations.

By the middle of the 18th century, the Industrial Revolution produced the same problem outside the academic world. Bankers, surveyors, insurers and navigators all found themselves routinely involved in tricky calculations of everything from compound interest owed to distances travelled.

Ada, Countess of Lovelace, daughter of Lord Byron, was the first computer programmer.

And all had no choice but to put their trust in vast tables of logarithms and other mathematical quantities. Compiled under the direction of mathematicians, these tables were calculated by armies of "computers". These human adding machines, often teenagers, each worked on a small part of a formula for the quantity in question. Their individual efforts would then be pooled to give the final figure for entry in the mathematical tables.

It was a recipe for errors and they duly appeared in their thousands. By 1784, fears about the reliability of mathematical tables reached a level where the French government decided to start from scratch, commissioning Baron Riche de

Prony to prepare a definitive set, calculated to 19 places of decimals. Compiled by a team of around 100 mathematicians and "computers", the tables filled 17 hand-written volumes. Yet they, too, were shot through with errors.

It was while idly glancing through a set of such tables one evening in 1812 that a 21-year-old Cambridge student hit upon a solution: why not get a machine to calculate the tables automatically? This idea led Charles Babbage to become the grandfather of the modern computer.

Babbage set about turning his vision into reality around 1820, with a device he called the Difference Engine. Using a series of gears and axles, the device was designed to exploit a

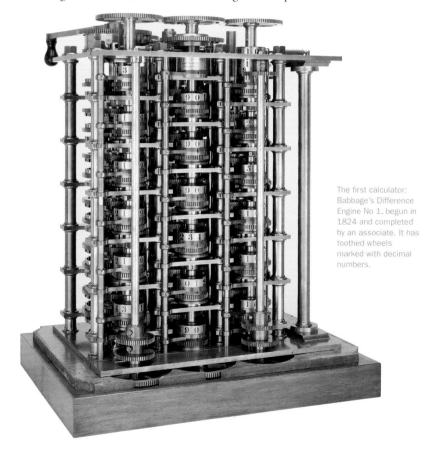

The first calculator: Babbage's Difference Engine No 1, begun in 1824 and completed by an associate. It has toothed wheels marked with decimal numbers.

mathematical trick for calculating complex quantities from simple components known as the Method of Differences. Crucially, Babbage also designed the machine to print out its own results, rather than allow a human to write down the answers and introduce mistakes.

By 1822, Babbage had a small prototype working. It was clear, however, that the full-sized Difference Engine, with an estimated 25,000 parts, would require far more effort and expenditure. After winning government backing, he set to work again, only to become embroiled in a long dispute with his chief engineer, Joseph Clement.

By 1832, almost a decade after starting, Babbage had only a fraction of the Difference Engine completed. Despite working perfectly, it failed to impress the government, as did Babbage's revelation that he was working on plans for an entirely new device that would make the Difference Engine obsolete. Known as the Analytical Engine, this locomotive-sized machine would act like a mechanical mathematician, actually solving equations. It featured components now found in every personal computer, such as a mechanical memory and a central processor. It could be programmed using punched cards, an idea Babbage borrowed from the Jacquard Loom. His friend the amateur

Charles Babbage.

mathematician Ada, Countess of Lovelace, even devised a program for calculating so-called Bernoulli Numbers using the device, and thereby became the first computer programmer.

Yet the staggering complexity of the Analytical Engine (plus Babbage's constant revisions of its design, and the understandable scepticism of the government) ensured that it was never built. Nothing approaching its power or versatility would be seen for over a century.

Even so, Babbage's vision of performing calculations using machines came to fruition within two years of his death. In 1873 the Scottish physicist Sir William

Thomson unveiled an arrangement of pulleys, gears and axles whose movements sketched the rise and fall of tides on a rotating drum. Over the next decade, a variety of these tide-prediction machines were devised, the accuracy of which was superseded only by electronic computers in the mid-1960s.

Born in London in 1791, Charles Babbage was the son of a wealthy banker, inheriting in his mid-30s a fortune today worth the equivalent of £5 million.

It was always clear that Babbage would not follow his father's career. As a maths student at Cambridge, his non-conformist attitudes led him to become a founder of the Analytical Society, dedicated to introducing continental mathematics into the "moribund" university syllabus. Later he would turn his attentions to the Royal Society, supposedly the world's leading scientific body, by launching an attack on its indolence.

While much of his life was embroiled in his visionary quest for a mechanical computer, Babbage applied his creative mind to other questions. A spate of rail accidents prompted him to recommend that trains be fitted with "black box" recorders, while a trip in a diving bell led him to advocate submarine warfare. He is also credited with inventing the idea of skeleton keys and the railway cow-catcher.

A great socialite, Babbage regularly invited influential people to his dinner parties. However, the failure of his Difference Engine project, and the indifference of the government to his ideas, was an enormous disappointment, and he died, lonely and embittered, in 1871.

Morse Code

Samuel Morse

1837

On May 24 1844, the auspicious words "What hath God wrought!" transmitted from Washington DC to Baltimore, heralded the inaugural public demonstration of Morse code, seven years after Samuel Finley Breese Morse, an American painter, invented his telegraphic system of dots and dashes.

While returning by ship from studying art in Europe, Morse overhead a conversation on the new science of

1838   Professor M.H. von Jacobi, a German, invents electrotyping, enabling large intricate print runs.

electromagnetism. Realising its potential, he entered into a partnership with Alfred Lewis Vail, an engineer, to develop a machine to transmit a code he had devised.

In 1837, Morse was granted a patent for his first electromagnetic telegraph, which used moulded metal type to send the Morse signal. The transmitter was soon deemed too complex and was abandoned for the familiar Morse key transmitter.

Morse's bullish powers of coercion surpassed his skills as an inventor and he never acknowledged the extent to which Joseph Henry, the American pioneer of electricity, helped in designing his transmitter. After aggressively lobbying Congress, Morse obtained the $30,000 to build his 40-mile telegraph line from Washington to Baltimore.

In 1851, International Morse Code, a simplified and extended Morse system, was devised by a conference of European nations. Its celebrated international distress signal – three dots, three dashes, three dots – survived as the maritime Mayday code, until it was abandoned in favour of satellite communications on February 1 1999.

Fuel Cell

Sir William Robert Grove

1839

Fuel cells have offered an efficient and green source of energy for more than 160 years by converting the chemical energy of fuels into electricity without combustion. Now they are on the brink of making their biggest leap forward, promising to replace the combustion engine to power the motor car.

The technology has come a long way since Sir William Robert Grove, a Welsh judge, inventor and physicist, conceived the first fuel cell in 1839. It catalyzed hydrogen and oxygen on platinum electrodes to make electricity and water. The principle was the same as a battery but with the crucial difference of an unending supply of the raw materials.

The need for high-efficiency power supplies for satellites and manned spacecraft created a new impetus for fuel cell development in the early 1960s. General Electric devised an electrical power system based on fuel cells for Nasa's Gemini and Apollo space capsules. Nowadays, the space shuttle's electricity is provided by fuel cells that generate a convenient by-product: drinking water.

Clean-air legislation has been the main impetus behind a recent drive to make cheaper fuel cells, with possible application for the motor car. Research has focused on hydrogen-bearing methanol, petrol and methane, which are safer than hydrogen but they use a "reformer" to extract the hydrogen from these hydrogen-rich fuels, a process that causes some pollution.

Vulcanization of rubber
Charles Goodyear

1839

Rubber is the most extraordinary natural polymer, with probably more varied uses than any other. From wet suits to car tyres, gaskets to surgical gloves, it's the perfect material for all sorts of applications, as long as it's used with care.

Discovered in the jungles of South America in 1736, rubber was long regarded as a tricky material to use, becoming brittle in cold weather and squidgy in summer. Then in 1839 an American named Charles Goodyear accidentally dropped some rubber mixed with sulphur on a hot stove and invented vulcanization. His invention was not entirely a fluke: he had already spent some years looking at ways of making rubber more robust. Even so, his process was all too easily copied, and he spent the rest of his life trying to prevent others infringing his patents.

Seeking new markets, he took out patents in Europe, only to have them rejected on technicalities. The process did make money, but not before Goodyear died with debts totalling £4 million in today's money.

Vulcanization boosts the resilience of rubber, but some applications call for untreated rubber's ability to "remember" and return to its shape. That's when it has to be used with particular care, as NASA discovered to its cost in January 1986. The space shuttle Challenger took off in weather so bitterly cold that a 12-metre rubber gasket on one of its boosters lost its flexibility and allowed hot gas to escape. The resulting explosion led to the deaths of all seven astronauts.

Postage Stamp

Rowland Hill

1840

1840 First electric clock invented.

1841 Standardized screw threads developed by Joseph Whitworth, an English engineer.

German publisher Christian Bernhard Tauchnitz invented the paperback book but it does not become mass-market until 1935, when Penguin Books are launched in England.

1842 The first underwater tunnel is opened in London. Built by Marc Isambard Brunel, who invented the tunnelling shield to bore it, the tunnel now houses the East London Line of the London Underground system.

Ancient Egyptians had a postal network in place by 2000 BC but it took a brilliant display of lateral thinking by a Victorian to turn a potentially complex and expensive system into a cost efficient service accessible to all.

In 1837, a pamphlet prepared by Rowland Hill, an English administrator and tax reformer, pointed out that the existing system, in which the cost of every letter was worked out according to its destination and weight, was unnecessarily cumbersome and expensive.

He demonstrated that, contrary to common assumption, the principal cost involved in sending a letter was not transporting it to its destination, but the handling and sorting needed to work out the fee for each one.

Hill thus proposed switching to a simple, flat fee paid in advance of one penny per half-ounce for delivery of a letter anywhere in Britain.

Backed by the government, Hill came up with two different methods: pre-paid envelopes and adhesive stamps. The latter, which launched as the Penny Black and Two penny Blue on May 1 1840, were an immediate success and spread worldwide.

Fax machine
Alexander Bain

We owe the development of the fax to a Scottish clockmaker, Alexander Bain, who, surprisingly, was granted a patent for his creation more than 150 years ago. Equally remarkable is that even now, after the invention of electronics, Bain's original concept is the basis for modern facsimile machines.

In Bain's invention, a wire linked two electric pendulums swinging synchronously. The sender used an electrically conductive material to write a message or to draw a picture on a piece of paper, and wrapped the piece of paper around a drum that rotated under a needle attached to one of the swinging electric pendulums.

As the needle passed over the text or drawing, electrical impulses were generated, which were transmitted to the pendulum swinging at the other end of the wire. The impulses caused the receiving pendulum to make marks on a piece of paper, thus reproducing the original message or picture.

The way that fax machines work today is remarkably similar to Bain's invention. Fax machines digitize an image, dividing it into a grid of dots. Each dot is given a value of one or zero depending on whether it is black or white. The sequence of zeros and ones, known as a bitmap, is transmitted as computer data, and translated at the receiving end back into dots to reproduce the picture.

Bain never performed a fax transmission, but it is clear from his patent application for "improvements in producing and regulating electric currents and improvements in timepieces and in electric printing and signal telegraphs" that his invention made facsimile transmission entirely feasible. The principle was proved sound by Frederick Blakewell, an English physicist who demonstrated a working facsimile machine, also with rotating cylinders and styluses, at the Great Exhibition in 1851.

In 1865, Giovanni Caselli, an Italian abbot born in Siena in 1815, introduced the first commercial fax service. It had the enthusiastic support of Napoleon III, who made the telegraph wires needed available after being shown Caselli's "pantelegraph". A fully legible archive of letters sent between Paris and five other French cities during this period are displayed in French museums.

The pantelegraph was later tried in England between London and Manchester. Meanwhile, in China, the Emperor realized the advantages of the fax principle – Chinese text being almost impossible to send via the conventional telegraph because of its thousands of ideograms. However, negotiations between Peking and Caselli petered out.

To Caselli's misfortune, the world had started to invest heavily in conventional telegraph services and several national telegraph authorities promoted development of the Morse system over Caselli's fax principle. The pantelegraph was associated with the transmission of images. The advantages of also using it to send text were not fully perceived in the 1860s. It was an early example of a new, superior technology being squeezed out by an existing system. Caselli's invention fell into disuse and he died a disappointed man in Florence in 1891.

In 1902 Dr Arthur Korn, a German scientist, invented the principle of photoelectric reading, the principle at the heart of today's fax machines. By 1910 newspapers were regularly sending and receiving pictures and in 1922, Dr Korn transmitted images between Europe and America by radio.

The fax was expected to become a common household appliance and millions of dollars were spent on developing it in the 1920s, but it was another 40 years before the fax machine broke out of the newspaper offices to become a familiar item of business equipment.

Anaesthesia

William Morton

1845

Surgery without anaesthesia is regarded as one of the horrors of the past, yet it is a rather misleading image: acupuncture was used in the Orient and anaesthesia in the form of large doses of alcohol has been practised since ancient times. But alcohol is far from ideal; a dose large enough to induce unconsciousness is perilously close to the amount capable of inducing death.

In the 1840s, three new means of inducing unconsciousness emerged: nitrous oxide, ether and chloroform. All were quickly seized on by doctors hoping to improve the lot of their patients. The intoxicating effects of nitrous oxide had been recognized as early as 1800 by the English chemist Humphry Davy. Trying it out on himself he found that he felt no pain under what was then known as "laughing gas", and he suggested it could be used as an anaesthetic. Laughing gas parties where people would sit around doing nothing else but inhaling the gas soon became a craze.

The American chemist Charles Jackson had by 1845 found a more effective anaesthetic: ether, a volatile compound that rapidly produced a loss of both consciousness and sensation. In 1846, William Morton of the Massachusetts General Hospital used ether while removing a patient's tooth. Morton kept his methods secret but his patient told his story to a newspaper and Morton was forced to demonstrate the use of ether during an operation at the hospital. The demonstration turned ether anaesthesia into a standard surgical procedure.

Doctors in Europe, inspired by the American experience, began to look for new anaesthetic compounds. Tests on animals and then humans prompted the Scottish obstetrician Sir James Young Simpson to adopt chloroform for use in childbirth. It immediately brought him into conflict with Calvinists, who opposed such use of pain relief. In 1853, Simpson – then obstetrician to Queen Victoria – persuaded the monarch to use chloroform for the birth of Prince Leopold. It silenced his critics.

Anaesthesia today is sophisticated. Various mixes of agent are used according to the type of operation, and patients are constantly monitored. Even so, the precise way in which anaesthetic compounds induce a loss of consciousness remains unclear.

A 19th-century inhaler for an ether anaesthetic.

Pneumatic tyre

Robert Thomson

1845

Civil engineer Robert Thomson wanted to smooth out the bumps of 19th-century roads. For thousands of years, wheels had been shod with nothing but wooden rims, sometimes with metal tyres for extra protection. Carriages, carts and the emerging bicycle all needed a more comfortable ride.

In 1845, Thomson patented the idea of using air-filled hollow leather tyres. These "Aerial Wheels" proved a reasonable success, but even the toughest leather did not last long on the primitive roads of the day. Thompson's patent proved to be a blind alley, and he made far more money from his next product: solid rubber tyres for carriages and carts using refined "vulcanized" rubber.

The tyre company, Dunlop, was started by a racing cyclist who saw the benefits of pneumatic tyres.

In 1888, John Dunlop, a Scottish vet, invented a more effective design with a rubber inner tube nestling within a tough outer cover. This was the first modern rubber pneumatic tyre, with air held in under pressure by a valve.

Some thought the new tyres would never catch on, but racing cyclists found the pneumatic tyres worth extra seconds in a race. One of them, Harvey du Cross, started a company to make the Dunlop-type tyre that grew into the giant company of the same name we are familiar with today.

In 1890, Charles Welch of Coventry found that sturdy wire reinforcements within the tyre helped keep it attached to the wheel, especially when combined with a rim sized to match the tyre.

Nitroglycerine

Ascanio Sobrero

The first new explosive since the invention of black powder in China a thousand years earlier appeared with a bang – quite literally – in a laboratory in the University of Turin in 1846.

Slowly adding glycerine to a mixture of nitric and sulphuric acids, Ascanio Sobrero found that he had made a liquid of nightmarish instability: nitroglycerine.

Even the slightest shock caused the chemical to explode, releasing a shock-wave of 20,000 atmospheres pressure and 5,000°C temperature, that expanded at more than 20 times the speed of sound.

Horrified by his creation, Sobrero did not pursue the commercial use of nitroglycerine. Not until 1859 did anyone have the courage to attempt large-scale manufacture, and even then at the price of personal tragedy. No sooner had the young Swedish chemist Alfred Nobel begun mass production of nitroglycerine than an explosion ripped through the plant, killing five people, including his youngest brother.

Nobel resolved to tame this chemical beast, and began experiments on board a barge in the middle of a lake. In 1866, he noticed that a chalky packing material called *kieselguhr* had safely mopped up a potentially disastrous leak of nitroglycerine. Once absorbed in the *kieselguhr*, the nitroglycerine lost its terrifying instability.

Nobel realized that this combination held the key to the commercial exploitation of nitroglycerine, and in 1867 he patented it under the name Dynamite. He went on to find other combinations that boosted the explosive performance of Dynamite, and developed the so-called mercury cap, based on fulminate of mercury, to act as a detonator.

Dynamite became a huge success, and despite endless patent battles Nobel amassed a fortune today worth around £150 million, which after his death in 1896 was used to endow the annual Nobel Prizes.

Antiseptics

Ignaz Semmelweis

1847

1847 Charles Babbage invents the ophthalmoscope but fails to exploit it. In 1851 it is reinvented by Herman Helmholtz, a physiologist.

Perhaps the single most important of all medical inventions, antiseptics have saved the lives of countless millions. Yet their use was initially bitterly resisted by the medical profession.

In 1847, Ignaz Semmelweis, a Hungarian obstetrician at the Vienna General Hospital, was trying to find out why one in 30 young mothers was dying of puerperal fever on a ward staffed only by midwives. But, Semmelweis realized, the figure jumped to almost one in five on the ward where medical students examined mothers.

He then discovered that the students were arriving at the ward after performing dissections on corpses, still carrying dead tissue on their hands. His suspicion that the students were causing the deaths hardened after attending the autopsy of a medical friend, a pathologist who

A carbolic acid spray, 1875.

had pricked himself while dissecting a corpse. Semmelweis was struck by the similarity between the lesions on his friend's body, and those of women with puerperal fever.

He decided the students were carrying some lethal agent on their hands and forced them to wash in dilute bleaching powder before examining his patients. Within eight weeks, the death rate plummeted almost 20-fold. This failed to impress Semmelweis's superiors, who resented being told they had caused so many deaths, and they soon got rid of him.

Among those who read of Semmelweis's work was Joseph Lister, an English surgeon who was attempting to reduce death rates in his operations.

In 1865, Lister began using carbolic acid to clean his hands and his equipment before surgery, and produced a three-fold drop in mortality rates. But only after Lister showed repeated success with formerly high-risk operations did other surgeons finally begin to use antiseptics routinely.

Light Bulb

Joseph Swan and Thomas Edison

1848

The light bulb has come to symbolize the Eureka moment of invention, that split second when a bright idea pops into the mind. However, the light bulb itself was a far from instant invention. It took 80 years from the first demonstration of electric light for the first light bulbs to go on sale, accompanied by a warning: "This room is equipped with Edison Electric Light. Do not attempt to light with match. Simply turn key on wall by the door. The use of electricity for lighting is in no way detrimental to health, nor does it affect the soundness of sleep."

Even once the light bulb was common place, two men, Joseph Swan, an English chemist, and Thomas Edison, the American inventor, continued to dispute their claims to have been the first.

Electric incandescent light was first demonstrated in 1801 when the British chemist Humphry Davy passed a current through strips of platinum, making them glow brightly. Davy's experiments helped fuel the interest in electric light, but they also highlighted the biggest obstacle: exposed to air, these early filaments burnt away. Even when the English scientist Frederick de Moleyns obtained the first patent for incandescent light 40 years later, using powdered charcoal between two wires, the lamp still burned only for a short time.

Davy next turned his attention to the arc lamp, using an electric arc struck between two carbon electrodes, which he first demonstrated in 1807. Arc lamps were reliable, but they were too bulky and fierce to be used in anything other than public lighting – the first practical example was installed in a Dungeness lighthouse in 1862.

The solution for domestic lighting was to find a sustainable form of incandescence. The problem attracted the Sunderland-born physicist and chemist Joseph Swan who thought that there might be a better filament than the platinum used by Davy. By 1848 Swan was experimenting with strips of paper which he carbonized by saturating them with tar and treacle, baked in a kiln and enclosed within a glass bulb vacuum. In 1860, he patented the first experimental light bulb using a carbon filament. But vacuum-pump technology and battery-cell power supply was still not good enough and the bulb was dim and short lived. It took 15 years before Swan could complete his research. In January 1879, he demonstrated the light bulb in Newcastle.

Joseph Swan's electric light bulb, lit up the Houses of Parliament in 1881. He demonstrated the first light bulb using a carbon filament, but 10 months later Thomas Edison (opposite) had a longer lasting bulb. Their patent disputes were settled out of court.

Swan's design has changed little in 120 years. The filament was fed electricity through the base of an airless glass bulb. When a current passed through the filament, it glowed. Without oxygen, the filament did not burn away quickly. Swan filed a patent in Britain, selling American rights to the Brush Electric Company. But by then, Swan faced a rival.

In 1878, the inventor Thomas Edison, aged 31, announced that he would devise a safe, mild and inexpensive household lamp. Such was the confidence in Edison's new Electric Light Company that illuminated gas stocks immediately fell in

London and New York.

After 1,200 experiments Edison's research team came up with a filament made from a carbonized cotton thread. His prototype was lit on October 21 1879 and lasted 150 hours. On New Year's Eve, 10 months after Swan's bulb had gone on show, Edison publicly demonstrated his lights at his laboratory.

Swan may have got there first, but Edison drove the technology onwards. His laboratory created an infrastructure for electric light, with efficient high-voltage generators, transformers, copper-wire power lines and an electric meter. In 1880, Edison used carbonized bamboo fibre to give his light bulb a life of 1,100 hours. His first lighting system was installed in April 1880 on the steamship *Columbia*, and the following January, light bulbs went up at the New York printing firm of Hinds and Ketcham. Finally, on September 4 1882, a central power station opened in Pearl Street, Manhattan.

Swan continued to promote his invention in Britain and illuminated Parliament with electric light in 1881, the British Museum in 1882, and his own home shortly afterwards. He and Edison began legal action over patent rights, but settled out of court, forming a joint company, Edison and Swan United Electric Company, in 1883.

Knighted in 1904, Swan continued to work on improvements. In 1907 tungsten, which burns at high temperatures without melting, replaced carbon. In 1913 filaments were coiled and the bulbs filled with the inert gases nitrogen and argon. In the 1930s the fluorescent light arrived, a long tube of mercury vapour that lights up as an electric charge passes through it.

The implications of Swan and Edison's work were far reaching. Electric light was cleaner, safer and brighter than gas, and it made domestic and business use of electricity practical and popular.

1848 John Curtis invents chewing gum made from the sap of spruce trees, a native American recipe.

1849 Walter Hunt, an American mechanic, invents the safety pin, when twisting a wire while thinking of a way to pay off a $15 bet. He settled the bet by selling the right.

Skyscraper

1848

James Bogardus

Two developments enabled the development of buildings over 10 storeys high in mid-19th century American cities, where the growth of urban commerce brought pressure on limited real estate. The invention of the safety lift in 1852 by Elisha Otis made buildings above five storeys workable, but the key invention was the use of cast-iron columns and girders in construction instead of masonry.

In 1848, James Bogardus, an inventor and builder regarded as the father of the modern skyscraper, built the Cast Iron building, his own five-storey factory, in New York. It pioneered the use of a cast-iron framework to support the

The skyline on crowded Manhattan island arose after the introduction of cast-iron building techniques.

weight of the upper floors, without the loss of floor space on lower floors created by the use of masonry supports.

When fire ravaged the centre of Chicago in 1871, the city leaders took the opportunity to build bigger, better and taller. The 10-storey Home Insurance Company Building, completed in Chicago in 1885, is regarded as the first modern skyscraper. It was the first to use steel girders, which were stronger than iron, and to hang an outer masonry curtain wall on to the load-bearing steel skeleton.

Giant buildings needed giant foundations. Dankmar Adler, a German-born architect and engineer, sank what he called a "caisson" foundation down to bedrock. On top of this he built a modest 10-storey building in 1892. Essentially a cylinder-shaped excavation filled with concrete, this was the principle used on many subsequent high-rise blocks.

Safety lift

Elisha Otis

1852

What would happen if those simple steel cables hoisting the cart up a skyscraper gave way? The comforting answer is: nothing at all. This gravity-defying miracle was made possible by the ingenuity of Elisha Otis, an American inventor who in 1852 devised the safety elevator.

As an engineer with a bedstead-maker in Yonkers, New York, Otis was asked to design a hoist for lifting heavy equipment about the factory. Concerned about the risk of the hoist's rope breaking, he set about designing ratchets fitted to the sides of the hoist which allowed it to move up and down smoothly, but which snapped into action with any sudden downward motion, preventing a lethal plunge.

Leaving the company in order to market his invention, Otis arranged an impressive public display of its abilities at New York's Crystal Palace in 1854. Ascending to the height of a house aboard a hoist, Otis ordered the rope cut with an axe. The ratchets sprung into action. Otis remained suspended in mid-air.

Within three years, his safety elevator had been installed for passenger use in the New York store of Haughwout & Co.

1853 George Crum, chef at Moon Lake House Hotel at Saratoga Springs, New York, invents the potato crisp after railroad magnate Cornelius Vanderbilt repeatedly sends back his chips, asking for them to be thinner. In a fit of pique, Crum slices potatoes so thin that they cannot be picked up with a fork. They are an instant hit.

Gyroscope

1852

Jean-Bernard-Léon Foucault

In 1851, the French physicist Jean-Bernard-Léon Foucault made crucial breakthroughs that led to an important navigational aid. While experimenting with a giant pendulum, Foucault realized the Earth spins on a north–south axis. He constructed a free spinning rotor, mounted it in a gimbal, and christened it the gyroscope. He observed and recorded for the first time the odd effects of inertia on a gyro. Once set in motion, the rotor would spin around the same axis, fixed in space, regardless of the movement of the surface beneath it. If the gyro was moved, the orientation of the rotor stayed in the same plane.

By making use of these qualities, the gyro was adapted to a variety of uses, of which the most important is probably the gyro-stabilized compass, needed because the iron-and-steel hulls of ships interfered with their magnetic compasses. It was later used to direct rockets to the moon.

Airship

1852

Henri Giffard

Roger Bacon, the 13th-century Franciscan friar with a predilection for experimenting with gunpowder, was the first to think of buoyant flight. He suggested that it could be achieved by filling a thin-walled metal sphere with rarefied air or liquid fire. Francesco Lana di Terzi, an Italian, developed Bacon's idea further in 1670 when he calculated that four such spheres would be needed to lift a boat.

The first practical airship concept came from Jean-Baptiste-Marie Meusnier, a French Engineer Corps officer. In 1784 he devised an elongated balloon driven by airscrews. It never flew, but it served as the impetus for Sir George Cayley, Britain's first aeronautical scientist. In 1816, Sir George improved the concept with an egg-shaped balloon with steam-powered propellers.

Opposite: Crowds gather in Tokyo to admire the Graf Zeppelin airship on its 1929–30 journey round the world.

The first steam-powered airship to fly was the three-horsepower Aerial Steamer, filled with hydrogen. Designed by Henri Giffard, it flew in Paris on September 24 1852 at 7 mph.

After the appearance of the gas engine in 1860 the first gas-powered airship flew at Brno, Moravia, at 9 mph.

The next step was electric. In 1884 Charles Renard and Arthur Krebs built *La France*, an electric airship nearly 60 metres long. With a huge propeller at the front, it was the first that could be steered accurately, provided the weather was calm.

After the invention of the petrol engine, David Schwartz built an aluminium airship with a 12 hp Daimler engine. He died before it flew, but his widow supervized its completion.

Then came the master. In 1893 Ferdinand Graf von Zeppelin, a German aristocrat and soldier, had submitted to the German military a design for a large airship, with the gas bags confined within a rigid cigar-shaped aluminium structure. It was turned down. Zeppelin resigned and set up the Zeppelin airship corporation in 1898 to build his first airship. The LZ-1 was launched from its floating hangar on Lake Constance on July 2 1900. Other models followed, but the loss of LZ-4 in 1908 plunged Zeppelin into bankruptcy.

A wave of national sympathy brought forth donations exceeding six million Deutschmarks which funded the Zeppelin Foundation, and the age of stately, silent and awesome airship travel began. The Zeppelin was a mixed success in the First World War; it undertook several successful bombing raids on Britain, but it made a large and fat target filled with highly explosive hydrogen. Around 40 were destroyed.

The airship reached its zenith when the Graf Zeppelin circumnavigated the globe in 1929. Seven years later, the age of long-haul, lighter-than-air aviation came to a sudden end with the spectacular accident of the transatlantic LZ-129 *Hindenburg* at Lakehurst Naval Air Station, New Jersey.

Helium, a non-flammable gas, had been available for 10 years, but the rise of Hitler had produced an export embargo to Germany. Had *Hindenburg* been gassed with helium as planned, 36 lives might have been saved and the age of commercial passenger airship travel might not have come to an end.

Bunsen burner

Robert Wilhelm Bunsen

1855

I n 1860, Robert Wilhelm Bunsen, a German chemist, published a scientific paper with Prussian physicist Gustav Kirchoff that established him as one of the world's most eminent chemists. It proposed the new science of spectroscopy, the analysis of substances by studying the wavelengths of light they absorb.

As part of the work towards this advance, Bunsen needed a high-temperature gas flame that burnt with near-invisibility. In 1855 he modified a burner designed by a young technician at the University of Heidelburg, Peter Desaga.

This was a simple metal tube connected to a gas supply sitting on a circular base with adjustable air inlets around the base of the burner tube. This enabled Bunsen's burner to precisely mix air and gas to attain higher flame temperatures.

Early Bunsen burners burned with a near invisible flame, reached 1,500°C, and prompted the discovery of new elements.

The hottest point of a correctly set Bunsen should be just above the flame, where the temperature can exceed 1,500°C. This crucial advance gave Bunsen the flame that he needed to do his pioneering work on spectroscopy, and in 1860, it led him and Kirchoff to discover two new elements, caesium and rubidium.

Pasteurization

Louis Pasteur

1856

A plea for help from a French distiller prompted the chemist Louis Pasteur to devise one of the most famous of all industrial processes: pasteurization, the heat-treatment of natural liquids such as milk to combat souring.

In 1856 Pasteur, aged 34 and head of science at the University of Lille, was approached by a local businessman whose failure to make alcohol from sugar beet without it souring had pushed him to the brink of bankruptcy.

Everyone knew fermentation was nothing but a chemical reaction. Pasteur showed that everyone was wrong. Examining the businessman's failed fermentations under the microscope, Pasteur found that part of the answer lay in the type of yeast involved. From his own research, he knew that if fermentation were a chemical reaction, it would produce equal amounts of left and right-handed versions of alcohol molecules. Tests revealed, however, that only one version of the alcohol molecule was formed during fermentation. This lopsided outcome was a classic symptom of a living organism at work. Pasteur thus concluded that yeast must be a living organism, and one that, if left to continue its action, would cause souring.

Louis Pasteur in his laboratory. Until his intervention, fermentation was always thought of as a purely chemical reaction. He proved otherwise.

The clinching evidence came in 1858, when Pasteur showed he could kill yeast by warming it to around 57°C. Doing the same to wine and beer prevented it souring. Taking a batch of wines and heating some while leaving others, he demonstrated that all those treated tasted fine, while some of those left untreated had gone sour. He went on to show that milk went sour for the same reason: the action of living micro-organisms within the liquid. By killing these organisms – some of which, like the tuberculosis bacterium, are deadly – pasteurization saved not only the Lille distillery, but also countless lives.

Synthetic dyes

William Henry Perkin

1856

I f we could be transported back to the Britain of 150 years ago, we would get a similar impression to that of a washed-out colour print from the 1960s. Before 1856 all dyes and colouring materials were those found in nature, their rather muted colours making the early Victorian world a fairly drab place. But of course no one knew any better, for it had been like that for centuries.

Much of the blue came from woad, a plant that grew in Europe and the colour with which our ancestors would paint their naked bodies on festal occasions. Reds came from several imported woods, from cochineal made from certain North American insects, and above all from madder obtained from roots of a plant that grew in many countries. There were yellows of vegetable origin, including saffron and turmeric, and a few orange dyes from woods that again had to be imported. When used for clothes or curtain materials their overall effect was less than dazzling, and the impression of general gloom would be heightened by wallpapers of similar appearance. It was all a bit hit-and-miss: various chemicals were added to the dyers' vat, the function of which was often made clear only when someone forgot to add them.

No one was looking for any kind of breakthrough in dye manufacture, but when one did happen it burst on society like a bombshell. William Henry Perkin, the

William Henry Perkin stumbled by accident on the first man-made colour.

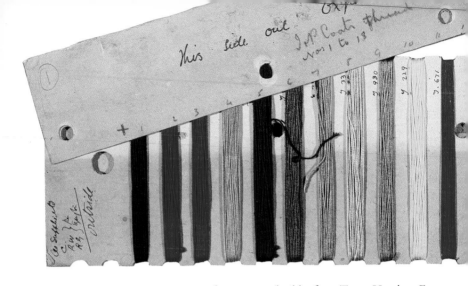

Above: an Indian dye fastness test card from 1904. Before 1856 there were no synthetic dyes and the world was a duller place.

youngest son of a carpenter/builder from Tower Hamlets, East London, was fascinated by chemical experiments and by the time he was 15 he had entered an institution that had been established in London seven years previously. This was the Royal College of Chemistry, which had been set up by the Prince Consort and others to promote the development of chemistry in a country that was already falling well behind others on the continent, above all Germany.

A German chemist, A.W. Hofmann, was called to be its first director. So well did William progress under Professor Hofmann's guidance that within two years he became his assistant. And then came the shock of discovery.

At this time the drug quinine, made from the bark of a rapidly diminishing number of cinchona trees in Peru, was in short supply. Perkin, with all the optimism of an 18-year-old, wondered if he could make some synthetically from laboratory chemicals. During Easter 1856, while all London was celebrating victory in the Crimean War, he set about the task in his little laboratory at home.

He was unsuccessful. Yet within the unpleasant brown-black tar that he produced there lurked a substance that was to revolutionize chemistry and ultimately transform society itself. Out of this disappointing mess Perkin was able to extract a new and beautiful dyestuff, called from its colour just "mauve". Unlike most natural colours, it could dye silk.

Declining Hofmann's advice to remain at his College, Perkin embarked on a series of field tests and eventually set up

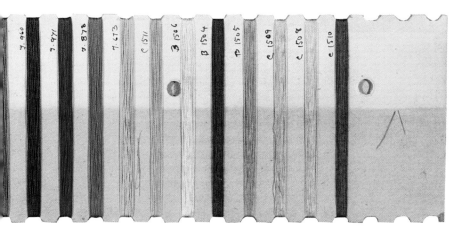

a factory at Greenford in Middlesex to make the material in bulk. Its success was almost immediate and a craze for mauve swept the women's fashion world. Following Prince Albert's death in 1861 it was just the thing to wear after the obligatory Court mourning. Even the humblest citizen would have encountered it in the "penny purple" postage stamps.

Meanwhile others jumped on the synthetic dyestuffs bandwagon, and before long several other wholly new colours were being produced on a large scale in Britain, Germany and France. One of the older natural colours was alizarin, hitherto obtained from the root of the madder plant, and prescribed (by royal decree) as the brilliant red dye for coats in the British army as well as hunting jackets for the aristocracy. This, too, was manufactured at Greenford from 1869. By the time he was 36, Perkin had made so much money that he was able to retire, but he still retained an interest in research and did notable work in his private laboratory.

An article titled 'On Artificial Arlizan', written by William Henry Perkin in the *Journal of the Chemical Society of London* in 1870.

Even more remarkable than the product of Perkin's experiments was their starting point. This lay in the tons of foul-smelling tar accumulating daily at an embarrassing rate in the yards of Britain's gasworks. It was a by-product in the manufacture of town gas, then made by heating coal in special furnaces or "retorts". At first it was thrown into rivers or just burned, then it was used to preserve wooden railway sleepers, then distilled to make volatile liquids such as creosote oil for illumination or timber treatment. This distillation process began to interest some chemists, and from the products there began to emerge a large number of new substances that today we call "organic".

Organic chemicals, once simply the products from living organisms, were becoming recognized more broadly as compounds of the element carbon. The dyestuffs industry became the first large-scale application of organic chemistry. By 1900 its output was thousands of tons, with dozens of new dyes added each year. Meanwhile some of these dyes were shown to have other uses, for staining bacteria in medical research, for example, and later as drugs themselves (as trypan red or trypan blue used in veterinary medicine).

Perkin's 'Original Mauveine', repackaged in 1906.

Several new organic compounds were discovered which were antipyretics, reducing high body temperatures. The most famous of these was aspirin, discovered in 1899. Since then one million billion tablets have been consumed, for all kinds of medical reasons. A different kind of material used hugely in medicine was phenol (carbolic acid), another coal-tar product and advocated by Lister as the first antiseptic to be employed in surgery. Yet other organic compounds, starting with ether and chloroform, have become essential anaesthetics.

Though not strictly medical, synthetic perfumes made life more pleasant, beginning with coumarin that smelt of new-mown hay. This also was first synthesized by Perkin.

When Napoleon anointed himself with several bottles of *eau de Cologne* each day he might not have imagined the choice that organic chemistry would have provided 50 years on. Perhaps he wished to expel the "whiff of grapeshot". But that whiff would have distinctly changed in another half century, for the Swedish inventor Alfred Nobel had discovered that another organic compound, nitroglycerine, known to be a powerful explosive, could be made safe by absorbing on a mineral called *kieselguhr*. The end product was named dynamite and became the first major explosive after gunpowder. A related explosive, also made by Nobel, was nitrocellulose.

Later, similar products from coal-tar found use as military explosives, including picric acid and TNT. The method of making them had been studied in Hofmann's laboratory years before, starting from highly inflammable liquids such as benzene. One day a laboratory assistant, Charles Mansfield, was killed while distilling benzene, but the research did not stop.

Britain was soon overtaken by Germany in its organic chemistry industry, and the First World War was protracted by the slowness of explosive production by Britain. When the British army went into action in France in 1914 the conspicuous red uniforms were now absent. Instead their greatcoats were dyed with khaki. The raw materials for that dye came from factories in Germany.

1857 Slot machine invented.

1858 Charles Moncky's adjustable wrench comes into widespread use. It becomes known as a monkey wrench.

Internal combustion engine

Étienne Lenoir

1859

I t comes in for criticism these days, and its ubiquity in this crowded island is certainly oppressive, but the fact remains that the internal combustion engine is an astonishing invention: a means of turning primordial sunbeams into power.

Petrol derives its energy (10 kilowatt-hours-worth packed in every litre) from the fossilized remains of organisms that mopped up the solar energy striking the Earth hundreds of millions of years ago. The modern internal combustion engine is designed to turn that energy into motive power, which it does efficiently: some engines develop 100 horsepower per litre of capacity (the muscle-power of a horse crammed into every cubic inch).

The basic principles of internal combustion were laid down by the French physicist Sadi Carnot as early as 1824, but it

1859 Jean-Joseph-Étienne Lenoir invents the internal combustion engine. Based on the steam engine, it uses gas explosions to drive the piston.

took another 35 years before they were turned into a working engine. In 1859 the Belgian inventor Étienne Lenoir converted a steam engine so that it took in an explosive mixture of air and coal gas, ignited it with an electric spark, and drove the piston with the hot exhaust gases. The result was the world's first working internal combustion engine.

Delivering one horsepower and costing around £65 each (about £3,500 today), around 1,000 of Lenoir's engines were installed in Britain alone, for jobs such as pumping water.

For all its ingenuity, Lenoir's engine was woefully inefficient. In 1862, the French engineer Alfonse Beau de Rochas laid down practical rules for a much more efficient device – the four-stroke engine. The German engineer Nikolaus Otto turned his rules into a working engine in 1876 and to this day automotive engineers call the four-stroke cycle of intake, compression, power and exhaust the Otto Cycle.

Otto sold around 35,000 gas-powered engines, each developing a few horsepower, over the next 10 years, by which time other German engineers had developed more efficient engines. Carl Benz had a single-cylinder, petrol-driven carriage running in 1885. The next year, two of Otto's former associates, Gottlieb Daimler and Wilhelm Maybach, demonstrated their own petrol car, which featured a "carburettor" for mixing fuel and air.

Rudolf Diesel, another German engineer, came up with the other major internal combustion engine concept in 1892. Compressing air increases its temperature, as anyone who has felt a bike-pump warm up while inflating a tyre knows. Diesel used this effect to do away with the need for a special ignition system, squirting fuel into the high-pressure, 800°C air inside the cylinder of an engine, where it ignited spontaneously.

Diesel's design allowed combustion to take place in a relatively small volume of the cylinder, giving it twice the compression ratio of a petrol engine and thus boosting overall efficiency. Combined with the fact that his engine could use a heavier, less flammable and cheaper kind of crude oil, Diesel's engine was an immediate commercial success, and he quickly made a fortune.

Remarkably, the engine's abilities impressed even the Admiralty, perhaps the most innovation-averse group of people ever to sit round a table. In 1913 they invited Diesel over for talks but he never arrived. The mystery of his disappearance from the deck of the *Dresden* while crossing the Channel has never been solved.

The fundamental design of the internal combustion engine has remained largely unchanged ever since, although many a bell and whistle has been added to boost its performance.

The pinnacle of internal combustion technology is the Formula 1 car. Its compact three-litre engine develops an astonishing 800 horsepower. Such innovations as supercharging, turbo-charging and electronic-engine management have been studied, exploited, then banned by the authorities, who like to keep Formula 1 engineers on their toes. Many of these innovations have ended up in family cars.

A cutaway shows the Otto Cycle (intake, compression, power and exhaust) of a four-stroke engine, which has changed little in more than a century.

Right at the other end of the spectrum is the two-stroke engine. This is the Cinderella of the internal combustion engine story. It's incredibly simple, has no valves, timing gear or oil sump (so it can work upside down), and it is cheap to make. Yes, it's less efficient and kicks out more pollution, but it could be made a whole lot better if only engineers would give it some thought.

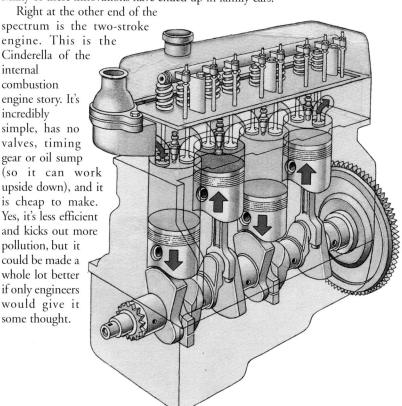

Oil well

Edwin L. Drake

1859

I n 1857, Edwin L. Drake, a tenacious former railway conductor, arrived on the back of a twice-weekly mail wagon at Titusville, an impoverished village in the hills of Pennsylvania.

Backed by James Townsend, a banker, Drake was convinced oil "could be obtained in large quantities by Boreing as for Salt Water." Until then, it came from natural surface seepages.

It took Drake two years to buy farmland that contained an oil spring, to build a drill derrick and to erect a steam engine to power the drill bit. He lost countless workers recruited from the salt drilling industry to the whiskey bottle, until he recruited William Smith, a blacksmith who had built drill bits for the salt borers.

On Saturday, August 27 1859, the drill bit dropped into a crevice at 23 metres and slipped a further 15 cm. Work was abandoned. In the morning, Smith discovered oil floating at the surface and by Monday, every vessel in the vicinity had been filled with oil. Drake attached a hand pump and became the first man to pump oil from a well.

Land prices shot up and Titusville's population tripled in a week. Drake filled every vessel he could get his hands on. With the suddenly plentiful supply of oil, its price dropped so far that the empty whiskey barrels were soon worth more than the black liquid they were bought to hold. Modern drilling methods use a rotary action and the borehole is filled with pressurized liquid to control the oil flow.

Edwin L. Drake, in top hat, struck oil in Titusville, Pennsylvania, and became the first man to pump it from a well.

Cigarettes
Europe

E arly versions of cigarettes were smoked in Central and South America, where natives rolled tobacco in vegetable leaves or corn husks, but for the first 200 years of its use in Europe, tobacco was smoked in pipes or cigars.

In 16th-century Seville, beggars shredded the remains of smoked cigars into paper scraps. These poor man's smokes were called *cigarillos*, Spanish for "little cigars". This practice spread throughout Europe, until in the Crimean War British and French troops began rolling cigarettes from fresh cigar or pipe tobacco.

Soldiers drained cannon fuses of their gunpowder and used the empty paper tubes to roll cigarettes. In the late 1850s the French translated "cigarillo" and coined the word "cigarette", French for "little cigar".

In the 1860s and 1870s, cigarettes could be bought from tobacconists but they were all hand-rolled. Cigarettes became more popular following the blending of Virginia and Turkish tobaccos for a smoother smoke, and James Bonsak's invention in 1880 of a cigarette rolling machine that fed tobacco on to a strip of paper, shaped a cigarette, gummed it shut and cut it to length with a rotary blade.

1860 Frederick Walton ·invents linoleum, the first synthetic floor covering, using linseed oil with resin and cork dust on a flax backing.

The perpetual mouse-trap invented by Colin Pullinger, from Hampshire. It has a trap door to capture and hold dozens of mice.

Eventually, hand-rolled cigarettes were largely replaced by packets of 'tailor-made'.

Bicycle
Pierre Michaux

H ow novel was the bicycle? Transport innovations are frequently secondary applications of power sources first applied in static industrial contexts, such as the steam engines that propelled railway locomotives, and the internal combustion engines that powered motor vehicles and lifted aeroplanes off the ground.

Similarly, the technological principle on which bicycles are based – the conversion of human leg muscle power into rotary motion –

1863 The first public underground railway is opened in London. It runs from Farringdon Street to Paddington.

1864 William Siemens invents the open-hearth process to make steel, which becomes the leading method of production for many years.

The penny farthing increased the size of the bicycle front wheel to address gearing problems.

came from much earlier applications, such as the treadmills first put to use by the Romans to lift heavy loads with cranes.

Like many other notable inventions, the bicycle was not the result of any dramatic breakthrough. Instead, several elements amalgamated over a protracted period of incremental design modifications.

This process explains why a relatively simple mechanical device, which relied only on human muscles to propel its rider at a speed four or five times faster than walking, took until the last third of the 19th century to become popular.

In 1817, Baron von Drais, a German, invented the first known bicycle and introduced it in Paris. In Britain it was called the hobby-horse or dandy-horse, and enjoyed a limited vogue during the 1820s and 1830s, and now looks surprisingly modern.

It owed a lot to cart and carriage design, with its wooden-spoked and iron-tyred wheels, wooden frame and wrought-iron fittings, but its two wheels were of equal size and it was steered by handlebars attached to the front wheel. There was no mechanical connection between the rider and the wheels; the rider simply sat astride, supported by a chest-rest, and pushed off the ground.

Several designs that attempted to link the rider's legs to a drive wheel followed the hobby-horse. The earliest known, by the Scottish blacksmith Kirkpatrick Macmillan, involved a mechanism akin to some early textile spinning wheels.

Reciprocating leg movements were converted into rotary motion through treadles, which were connected by rods to cranks attached to the larger rear drive wheel. The machine was efficient enough to carry its inventor on a hilly 140-mile round trip between Dumfries and Glasgow in 1842.

But Macmillan was no entrepreneur, and the influence of his design was confined to a few copies in the Galloway region. The first machine to diffuse across national borders was the French vélocipède, invented in 1861 by Pierre Michaux and his son Ernest.

By Her Majesty's Royal Letters Patent

THE "ARIEL" BICYCLE.

Fitted with Lever Tension Wheels, India Rubber Tyres, Improved Rudder, Registered Cliptail Sliding Spring, &c.

STARLEY & COMPANY,
PATENTEES & MANUFACTURERS,
ST. AGNES WORKS, COVENTRY.

Known in Britain as the boneshaker, it was the first bicycle with pedals, in this case attached to the hub of a larger front wheel. However, such a configuration required riders to twist their bodies when steering, and yielded relatively low speeds for the pedalling invested because of the inherently low gearing.

The next notable design step addressed the boneshaker's gearing problem by greatly increasing the size of the front drive wheel. In 1870, James Starley of Coventry designed the Ordinary, better known now by its derisive nickname penny farthing. By 1874, it incorporated his enduring innovation of suspension wheels with wire spokes set at a tangent to the hub to resist the winding stresses of pedalling. It was a highly efficient machine, capable of speeds up to 20 mph, but there were obvious difficulties of mounting and dismounting, and of severe injuries resulting from the falls taken by riders on hitting an obstacle.

The position of the rider atop the drive wheel also meant the wheel's radius had to match the inside leg measurement of the rider, rather like a pair of trousers.

The style of dress of early French women cyclists was considered too racy.

The use of the Ordinary only by the relatively long-legged and audacious triggered a search for less hazardous solutions to the gearing problem. The answer was to use a chain to connect the pedals with the drive wheel, a design used in the first "safety" bicycles, such as Harry J. Lawson's Bicyclette of 1879, and the much more successful Kangaroo of 1883.

In 1885, John K. Starley, nephew of James, hit upon the configuration that effectively stabilized bicycle design: the Rover Safety bicycle, with more or less equal-sized wheels and chain drive to the rear wheel. The first frames were curved, but the archetypal diamond-shaped frame was quickly established, as were sprung saddles.

The remaining key design innovation of the 19th-century was the pneumatic tyre, invented in 1888 by John Dunlop, a Scottish veterinary surgeon. When such tyres were applied to bicycles in the early 1890s, they soon ousted the solid rubber tyres that had been fitted to wheels from about 1868.

Not only did they greatly increase the comfort of the bicycle, they also significantly increased its speed, with the result that the Ordinary, the sporting wheelman's preferred "steel steed", soon went out of production. Although there

were developments in gears and braking in the early 20th-century, bicycle design was now fundamentally established.

The bicycle was instrumental in some of the sweeping social and demographic changes that have marked the century and more since its emergence. It brought together courting couples from villages outside walking distance, thus diversifying the gene pool in rural areas.

The bicycle certainly played a significant part in the emancipation of middle- and upper-class women. This was not, as some male Victorian doctors opined, because the exercise could help cure the fairer sex of their myriad nervous disorders; rather, that cycling liberated women from the constricting boned corsets, hooped crinolines and voluminous skirts of the Victorian era.

French women in knickerbockers led the way, rather too fast at first for their staid British sisters and their male chaperons. By the 1890s, the new bicycle craze in America was reinforcing earlier calls for "rational dress" associated with Amelia Jenks Bloomer, the women's rights campaigner. At this time, cycling knickerbockers were also called "rationals" or "bloomers".

In contemporary Britain, bicycles have become the preserve mainly of playful juveniles, sporting types, and those dedicated to challenging the hegemony of the internal combustion engine on the highway.

But these mundane contraptions, along with derivative designs such as the trishaw or pedicab that threads the throngs of many cities of east and south-east Asia, continue to play a substantial role in the work and play of many other nations.

The motoring masses should also recognize the bicycle's historical precedence: it was cycling enthusiasts during the middle-class bicycle craze of the late 19th-century who rediscovered and revived the open road, after the horse-drawn coach had disappeared in the face of the railway juggernaut.

These pioneers in Western Europe and America demanded better road signs and surfaces, literally paving the way for the motorist. And it was often bicycle manufacturers, such as Humber, Morris and Rover in Britain and Peugeot in France, who turned their expertise to the production of motor cars, a relationship that developed into one of the most unequal sibling rivalries in the history of invention.

Opposite: Curiosity gets the better of French comic actor Jacques Tati to his regret.

Torpedo

Robert Whitehead

1866 Thomas Allbutt, an English doctor, invents the clinical thermometer. George Westinghouse devises the air brake after riding a train that narrowly avoids a crash. He bases the first of his 363 patents on the compressed air technology used by tunnelling power drills.

It is rare for breakthrough technology to work straight away, but it happened in 1866 when British engineer Robert Whitehead devised the torpedo.

Whitehead, born and apprenticed in Lancashire, made his name as an engineer in factories throughout Europe. In 1856, while manager of an Austrian maritime engineering firm, he was asked to try to improve the "spar torpedo", a timber pole capped with an explosive charge, which, in battle, was pushed against the hull of an enemy warship and detonated.

Whitehead abandoned this and looked for ways to guide and power a self-propelled torpedo. He was secretive of the prototype that he built with his 12-year-old son and an old man from his works, but it was said to be "dolphin shaped" with four fins.

The two-cylinder engine was powered by compressed air, stored in a steel tank at a pressure of 370 lb per sq in. Fired from an underwater launching tube by compressed air, the tin fish was maintained at the correct depth by elevators controlled by hydrostatic valves. Trim tabs set by guesswork kept the torpedo vaguely on course. With a speed of six knots and a range of up to 300 yards, the weapon was soon in demand.

Whitehead later invented a combination of hydrostatic valve and pendulum that acted on the controls, keeping the torpedo at the right depth. This guidance system, "The Secret", worked so well that it checked the depths of torpedoes until after the Second World War.

Traffic lights

J. P. Knight

The streets of Victorian London were almost as packed with traffic as they are today. The only difference was that the congestion was caused by horse-drawn vehicles, requiring hundreds of police officers to keep the streets clear. In 1868 London was blessed with the world's first traffic lights, set up outside the Houses of Parliament by railway engineer J.P. Knight.

The gas-powered lights were an adaptation of Knight's railway signals, with semaphore arms for daytime and a single gas lamp for night. A constable on the ground pulled a lever to lift the semaphore arms and revolve a red glass bezel in front of the lamp to signal danger. When the arms were lowered, a green glass bezel rotated in front of the lamp. Disaster struck in 1869 when the traffic light exploded, injuring the bobby beneath. The Knight light proved a cul-de-sac in the history of traffic control.

In 1914 in Cleveland, Ohio, the first traffic lights with separate red and green lights were erected, and by 1918 New York had a system controlled by a traffic officer in a well-protected "crow's nest".

In 1920, William Potts, a Detroit police officer, had the idea of adapting for the road the automatic signalling technology coming into use on railways. Within a few years, 15 sets of automatic lights built by Potts were dotted across Detroit.

In 1925, Britain's first electric traffic lights, controlled by a policeman in a hut at the centre of the junction, were installed at the confluence of St James Street and Piccadilly in London. By 1926 traffic lights had reached Wolverhampton, where until 1928, Britain's first automatically switched lights, working on a time interval, kept drivers from colliding.

Modern traffic control systems are switched by a variety of sensors, from induction coils to close-circuit television and even digital recognition systems.

Qwerty typewriter

Christopher Sholes

1868

As a journalist, Christopher Latham Sholes of the *Milwaukee News* was all too aware that handwriting was both slow and sometimes illegible. Around 1860, he began thinking of making a machine that would type words rapidly and legibly on to paper, a goal others had attempted, but without real success: some machines were as big as pianos, and all were much slower than handwriting. Sholes applied for a patent for his first typewriter in 1867.

1868 London painter Waddy Maughan invents the gas water heater.

1869 French chemist Hippolyte Mege-Mouriez invents margarine to win a prize offered by Napoleon III for a butter substitute.

It was hardly an engineering triumph: the keys hit the paper from underneath the machine, preventing the text from being seen while being typed. Worst of all, the keys jammed frequently, yet typists would often be unaware of the problem until they saw the finished text.

The breakthrough came after Sholes saw an article in *Scientific American* that described an idea for a writing machine developed by an Englishman, John Pratt. He took Pratt's design and, in 1868, he filed a patent for a new typewriter that was much faster than a pen.

The secret of its speed lay in the keyboard design, which paradoxically slowed the typist down. Sholes arranged the letters in the now familiar qwerty sequence: this forced typists to move their fingers further than was really necessary to type common letter sequences but it gave the keys time to fall back into place after typing.

The qwerty system acquired dominance chiefly as the result of a race between two Cincinnati typewriting teachers, Louis Taub and a Mrs Longley.

In 1888, Taub, who used a non-qwerty keyboard, challenged Mrs Longley to a speed test. To take the test, Mrs Longley hired Frank McGurrin, a qwerty typist. McGurrin had a crucial advantage over Taub: he knew where all the keys were by feel alone; in other words, he could touch-type. As a result, McGurrin easily won the race, and his victory ensured qwerty became the favoured choice of American typing schools.

Despite its in-built inefficiency, now made redundant by the absence of mechanical keys on electronic word processors, no-one has successfully challenged the dominance of the qwerty keyboard ever since.

The early qwerty keyboard layout remains unchallenged.

Barbed Wire

Joseph Glidden

The Colt and the Winchester are generally thought of as the tamers of the Wild West, but barbed wire has a stronger claim. It opened the American West to settlers, giving them security and preventing cattle barons from driving their herds across their lands.

1874 Henry Parmalee invents the automatic fire sprinkler, using it to protect his piano factory.

The first patent for barbed wire was issued to W.H. Meriwether in 1853 but this did not have any barbs; instead it was a snake wire with a curl in it. In 1867 patents were issued to Alphonso Dabb and Lucien Smith, but it was Joseph Glidden, a 60-year-old rancher from New Hampshire, who invented a method for manufacturing barbed wire and went on to become the Henry Ford of the "thorny fence".

Allen single-line four point, 1876.

Orlando Briggs four point, 1883.

In 1873, Glidden attended a county fair at De Kalb in Illinois with his friend Jacob Haish. There, the pair saw barbed wire exhibited by one Henry Rose. Glidden, saying nothing to Haish, devised a way to thread the barbs on the wire, and patented it. Haish, also remaining silent, devised his own method of manufacture and challenged Glidden's patent, but lost.

Mexican barbed wire.

Ross-type wire with copper bond, 1879.

Soon Glidden's wire was spanning the nation and he became extremely rich, ruthlessly suing anyone who infringed his patent and driving his rivals into the ground. Over the next few decades hundreds of designs were patented, including one with rotary spurs and wheels patented in Britain in 1876 by William Hunt, an American.

Home-made wire barb on two-line horse wire.

Hodge 10-point spur rowel, 1887.

In its heyday there were more than 1,200 types of barbed wire in use, some of which ran to more than a dozen variations. But it was also seen as an icon of oppression, which in the First World War stretched from Switzerland to the English Channel and in the Second World War contained concentration camps.

The thorny fence has now been usurped by the electric fence, a more effective cattle controller and a less hazardous barrier to set up.

Telephone

Alexander Graham Bell

1876

Shrinking the world: by the turn of the century, only 24 years after Alexander Graham Bell invented the telephone, long-distance calls became possible.

The Victorians called it "talking by lightning". The discovery that electricity was capable of both carrying the voice and working an apparatus to reproduce it, enabled man to converse over great distances. Historically, the word telephone (from the Greek *tele*, "far off" and *phone*, "voice" or "sound"), meant any device, such as a foghorn, megaphone or speaking tube, for conveying sound over a distance.

A Frenchman, Charles Bourseul, first propounded the idea for an electrical device capable of transmitting and receiving speech in 1854. His theory involved "speaking close to a moving plate so flexible as to register all voice vibrations". He envisioned the vibrating plate sending signals down a wire by making and breaking a connection to a battery, then turning the signals back into sound by reversing the process. Bourseul never pursued his idea, and it was left to Johann Phillip Reis, a Frankfurt teacher, to attempt to turn the theory into practice.

Reis first demonstrated his "telephone" publicly in 1860. It was able to convey somewhat distorted sounds by means of a cork, a knitting needle, a sausage skin and a piece of platinum. But it was unable to transmit intelligible speech because the vibrating diaphragm made only sporadic contact with the circuit. For reliable speech reproduction, this contact had to be continuous, so that variable acoustic pressure on the diaphragm made smooth fluctuations in the current transmitted. Reis gave up on his recalcitrant sausage skins, and it took a further 16 years for the world's first successful telephone to emerge.

Alexander Graham Bell called his assistant from the next room by phone and made history in March 1876.

The enormous success and growth of the telegraph network inspired ever more efficient ways of maximizing its potential. Thomas Edison's "quadruplex" system of 1874 enabled four messages to be transmitted simultaneously over the same wire. This inspired many others to try methods of increasing wire capacity further. Various individuals had been experimenting with the "harmonic" telegraph, a device which used differently pitched magnetic reeds, each transmitting a unique frequency down a wire. By this means, it was theoretically possible to telegraph multiple messages down one wire at the same time.

Elisha Gray, an inventor from Ohio, and Alexander Graham Bell, a Scot working in Boston, Massachusetts, were both separately engaged on harmonic telegraph experiments during the 1870s. Bell was well qualified in the mechanics of speech and sound, and both he and Gray were equally capable of "inventing" the first workable telephone. However, the glory went to Bell, who managed to pip Gray to the post. He filed his patent application, "Improvements in Telegraphy", on March 3 1876, only hours before Gray applied to patent his

own telephone and a week before Bell actually managed to transmit speech.

Gray did not initially contest Bell's entitlement to the telephone patent. He was discouraged from doing so by his legal advisers who considered the device an insignificant offshoot of his work to improve the harmonic telegraph. History does not record Gray's subsequent words to his lawyers, but the words spoken by Bell to his assistant, Thomas Watson, in the first electrical transmission of the human voice, on March 10 1876, are well known: "Mr Watson, come here, I want you!"

Since Bell's working model used ideas indicated in Gray's Notice of Invention – ideas not proposed in Bell's own patent – the technology Bell used became the subject of some 600 lawsuits, but none was able to prove misappropriation of Gray's ideas.

Bell's device, like that of Reis, was based on a diaphragm connected to a moving wire. Bell is said to have conceived this diaphragm when bellowing into a deaf man's ear, part of an earlier experimental device for teaching the deaf. However, Bell's telephone differed in that the moving wire floated in a metal cup filled with acid, thereby maintaining a continuous contact with the circuit. Sounds acting on the diaphragm caused it and its attached wire to move, thereby changing the resistance in the liquid. The varying current it produced was conveyed to a receiver, and converted into sound by the moving diaphragm of the receiver in a reversal of the process.

The telephone rapidly proved to be a great success. Within 17 months of Bell's first transmitted speech, there were 1,300 in use. The figure multiplied to 30,000 by 1880. Success encouraged a scramble to devise improved systems that would not infringe Bell's patents. Thomas Edison patented an improved transmitter in 1877, but Emile Berliner, inventor of the gramophone, objected, claiming prior Notice of Invention.

Litigation over patents was commonplace in the race to improve the telephone. Bell invented the "photophone" in 1880, a device which transmitted sound on a beam of light, employing a vibrating mirror and selenium cells, an early (and cable-less) precursor of modern fibre-optic transmissions.

Opposite: Alexander Graham Bell on the phone, opening the New York–Chicago link. His arch-rival, Elisha Gray, pursued him with 600 lawsuits.

1876 Gustav de Laval invents the centrifuge for separating cream.

Almon B. Strowger, a Kansas City undertaker, annoyed at having potential customers' calls referred to a rival's telephone, devised the first automatic exchange in 1890. Long-distance calls became possible from 1900, thanks to the electromagnetic overload coil invented by Serbian-American physicist Michael Pupin. The coils, spaced at intervals along a line, decreased loss of signal strength over distance. The telephone had truly come into its own.

Carpet sweeper

1876

Melville Bissell

Keeping carpets dust free has always been a problem for householders, a source of business for salesmen, and a challenge for inventors. In 1811 Joseph Hume patented a mechanical sweeping machine but it never caught on.

In 1859, Sir Joseph Whitworth, an English machine toolmaker, invented a small sweeper containing a fan to blow carpet dust into a box. Householders were unimpressed.

Then in 1876 an American, Melville Bissell, who was suffering dust allergies from the sawdust in his china shop, designed a sweeper, complete with height adjustable brushes that fed into a sealed dust box. It was an instant success and Bissell became a household name.

The English Ewbank followed on Bissell's invention in America.

Moving pictures

Eadweard Muybridge

T he movie industry might be the apotheosis of glitz today, but its origins lie in nothing more glamorous than a simple bet. The story goes that in 1872 the Californian railroad tycoon and racehorse breeder Leland Stanford got into an argument over whether a galloping horse ever has all four hooves off the ground.

Experts and artists alike agreed that horses always kept at least some contact with the ground, but Stanford decided he knew better and bet $25,000 (the equivalent to $500,000 today) that horses sometimes left the ground completely.

As horses' legs move too quickly for anyone to see exactly what happens, Stanford hired an English photographer named Eadweard Muybridge with the intention of capturing the truth on a photographic plate. He had to wait five years for an answer.

Muybridge, born Edward James Muggeridge, was something of an eccentric, which was blamed on serious head injuries sustained in a stage-coach accident after he emigrated to America in the 1850s. In 1875, after being tried and acquitted of the murder of his wife's lover, he lay doggo in Central America.

On his return in 1877, Muybridge set up a battery of 16 cameras parallel to a racetrack in Sacramento, with each camera's shutter connected to a wire stretched across the racetrack. As the horse galloped past, its legs tripped the shutters in sequence, creating a series of photographs showing the position of the horse at each instant.

Muybridge stuck the images on a rotating disc and shone a light through them. The flickering images proved that Stanford was right: the horse did, in fact, sometimes have all four hooves off the ground.

Muybridge went on to carry out similar studies of the movement of animals and humans, his academic interests being shared by a French scientist, Etienne-Jules Marey, who in 1882 invented a single camera capable of taking multiple exposures in rapid succession.

Others, however, were less interested in making discoveries with this new technology than in making pots of money. They

It was a set of images of a horse and jockey in motion, taken by Eadweard Muybridge, that eventually led to the birth of moving pictures.

included that doyen of creative entrepreneurs, Thomas Edison, who in 1887 asked one of his assistants, William Dickson, to devise a suitable camera and a means for projecting the resulting images. The result was the kinetograph, a camera which could capture images at 40 frames a second, and the kinetoscope, a form of "peep show" allowing individuals to watch the resulting film.

The first kinetoscope booths opened for business in 1894, and among the first to visit one was a French photographic materials manufacturer named Antoine Lumiere.

Suitably impressed, Lumiere instructed his two sons, Auguste and Louis, to develop a lightweight camera and projection system capable of making movies that could be viewed by an audience.

In March 1895, they gave the first demonstration of their cinematograph system: a one-minute film of workers emerging from the Lumiere factory in Lyon.

It was not the last word in entertainment, perhaps, but it was the first genuine motion picture, nonetheless, and the beginning of something that has become more than merely a form of entertainment: it has also been an important medium of social and even political change.

Phonograph

Thomas Edison

1877

The phonograph was Thomas Edison's proudest invention. Designed to be little more than a dictation machine, it heralded the start of the home entertainment industry, brought together cultures that would otherwise have remained distinct and paved the way for a new audio history.

In 1855, Frenchman Leon Scott de Martinville used a horn and membrane attached to a stylus to draw sound waves on a spinning cylinder. As a recorder of sound, it worked perfectly, but there was no way of playing back the recording.

Edison, a former telegrapher, was looking for a way to store Morse messages. His first model, built by his mechanic John Kruesi, used tin-foil strips wrapped around a 10 cm drum. When the cylinder was cranked at 60 rpm, a stylus attached to a mouthpiece scratched the foil, converting sound waves into a three-dimensional pattern. To play back the recording, the mouthpiece was replaced with a horn. Edison's first tinny recording was "Hallo, hallo" followed by *Mary Had A Little Lamb*, and he was soon boasting that his phonograph's capabilities went far beyond simple speech. But the foil wore out quickly and it was eventually replaced with a brittle wax cylinder rotated at 90 rpm for four minutes of poor quality speech, or at 160 rpm for two minutes of music.

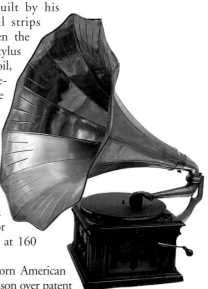

In 1887, Emile Berliner, a German-born American inventor who had already clashed with Edison over patent rights to telephone mouthpieces, invented the gramophone. His records were flat and played at 78 rpm, with the needle moving from side to side in a groove of even depth. As with the cylinder, the sound was amplified with a large cone. The discs, soon replaced by the brittle plastic shellac, were cheaper to press than cylinders and the mass music industry began. Records of the tenor Enrico Caruso made $2million 20 years later.

In 1948, Columbia introduced the vinyl long-playing $33^{1}/_{3}$ rpm record. Squeezing 25 minutes on each side by using 250 grooves per inch where the shellac had accommodated about 80 grooves was seen as a major breakthrough. Victor introduced a seven-inch 45 rpm the following year, carrying as much music as a 78 rpm.

After stereo records were introduced in the late 1950s, the LP and single changed little. But the arrival of the compact disc proved the gramophone to be an almost exclusively 20th-century technology.

Invented Language

Lazarus Ludwig Zamenhof

1878

1880 Oil vapour blowlamp invented.

The first "planned language" was devised by a Polish polyglot ophthalmologist who was convinced there would be fewer wars if all people spoke a common language. In constructing his language, Lazarus Ludwig Zamenhof rejected the most popular European languages (English, French, German and Russian) in order not to give their speakers an advantage. He based it on the Romance languages, using a 28-letter alphabet, a small number of words, six verb endings with no irregularities, and a simple set of grammar rules to devise a language he called Lingvo Internacia, said to be five times easier to learn than Spanish.

By his 19th birthday in 1878, Zamenhof had devised his bridge between cultures and nine years later his first book of instruction was published under the pseudonym of *Esperanto*, which means "person is hoping" in Lingvo Internacia. Today, around two million people worldwide can speak Esperanto, as it has become known.

Hundreds of other languages have been invented, including Basic English, Glosa and Eurolang, but only two besides

Esperanto have been at all successful: Ido (which means "offspring" in Esperanto), devised by a group of French intellectuals that believed Esperanto needed to be reformed, and Volapük ("world language" in Volapük), invented by a Catholic priest in Germany.

Metal detector
Alexander Graham Bell

1881

Alexander Graham Bell, the Scottish-born inventor of the telephone, discovered in 1881 that he could detect the presence of ferrous metals using a coil that produced an electro-magnetic field. It was a primitive form of metal detector.

Described at the time as an induction balance, he used it to search for a bullet lodged in the body of US President Garfield after an assassin had shot him. Sadly for the President, who did not recover, Bell was not able to locate the bullet, only the metal bedsprings beneath him.

Bell did little to develop the device. It was left to radio engineer Gerhard Fischar in 1925 to re-discover the principle by accident. In the course of conducting experiments with portable radio apparatus, he found he was getting interference from a large metal water tank near the receiver. His metal detecting device was sold to the public in 1931.

Fungicides
Pierre-Marie-Alexis Millardet

1882

A new era in agricultural technology was triggered by the chance discovery of the first successful fungicide by Pierre-Marie-Alexis Millardet, a botany professor at the University of Bordeaux.

1882 The electric safety iron and the electric fan invented.

Millardet became famous for saving the entire French vineyards from destruction in 1858-63 by phylloxera, an aphid-like plant lice that was introduced to Europe on American vines intended for grafting. Millardet's solution was swift: he imported phylloxera-resistant American stock and had European vines grafted on to these.

The lice were accompanied by a downy mildew fungus (*Plasmopara viticola*) that attacked grapes. In October 1882, Millardet realized that a mixture that Médoc farmers sprayed on their vines to discourage thieves also kept the mildew at bay.

He spent three years testing the solution of copper sulphate and lime in water, which became known as "Bordeaux mixture". After it was made known, it was widely adopted and became the first fungicide to be used on a commercial scale.

Steam turbine

Charles Algernon Parsons

1884

1884 Lewis Waterman, an American insurance salesman, invents the fountain pen after becoming infuriated by leaks from other pens.

The steam engine that powered the Industrial Revolution was always recognized as a clumsy and inefficient. Engineers knew that by squirting high-pressure steam directly on to turbines, far higher speeds and power outputs would be possible, if only turbines could be made to operate reliably.

This challenge was taken up by a young British engineer, Charles Algernon Parsons, in the 1870s. The son of the President of the Royal Society, Parsons was still a science student at Cambridge when he built a prototype rotary steam engine.

In 1884, after joining the Gateshead engineering company Clarke-Chapman, Parsons patented the first steam turbine system, which used alternating stages of moving and fixed blades to extract the maximum energy with minimal stresses.

Four years later, Newcastle upon Tyne became the first town to benefit from steam turbine electricity generation when the Forth Banks Power Station was equipped with four Parsons engines.

By 1897, Parsons had turned the turbine into a new form of marine propulsion and built the *Turbinia*, a 44-ton launch with an engine developing an unheard-of 1.5 megawatts. The vessel moved at 35 knots (40 mph), almost twice as fast as any other boat. Parsons' claim that steam turbine engines could propel even battleships at more than 20 knots was met by the Admiralty with incredulity and ignored.

On July 4 1897, the Grand Fleet of the British Empire was assembled at Spithead for review by Edward, the Prince of Wales, when suddenly, from among the hundreds of vessels on the sidelines, a small launch darted out and took up position between the giant battleships. Furious officials dispatched two attack boats, but these were completely outrun by the vessel. It was Parsons' *Turbinia*, putting on a display of speed that no one could ignore. His company was immediately commissioned to supply steam turbines to the Royal Navy, and their ability to generate huge power in a compact unit has led to steam turbines remaining at the centre of power generation to this day.

Machine gun

Hiram Maxim

From the moment the first guns were made, armourers were confronted with the demand for a weapon that could fire more than one shot before being reloaded. Early attempts had many barrels mounted together, fired by running a flame across their touchholes, but they took an age to reload.

James Puckle, a London lawyer, devised the first single-barrelled firearm that could fire several shots without reloading. His repeater of 1718 was a large flintlock revolver with the chambers arranged around the rim of a wheel. Turning a handle brought each chamber, in turn, into line with the barrel. Curiously, it was designed to fire square bullets at infidels and round ones at Christians.

To invent the first successful and practical machine gun, Richard Gatling, an American doctor, went back to an earlier idea of using many barrels. Cartridges from a magazine were dropped, one at a time, into the breeches of the 10 barrels, rotated by turning a handle. Patented in 1861, the Gatling gun could fire up to 3,000 rounds a minute and it saw service with many armed forces.

1884

Sir Hiram Maxim, in 1915, shows his grandson the gun he designed. During the First World War the Maxim gun was responsible for terrible casualties and ended the idea of mass attacks by infantrymen.

The first person to make a totally automatic machine gun with a single barrel was Hiram Maxim, an American working in London. He supposedly invented his gun after a fellow American advised him that the route to riches was to "invent something that will enable these Europeans to cut each other's throats with greater facility."

In 1884, Maxim came up with a gun that could handle high chamber and bore pressures created by the explosive in the latest generation of bullet. Cartridges loaded into a belt were fed into the breech and the mechanism of the gun pulled a cartridge free, loaded it into the breech and fired it. It ejected the empty case and then used the recoil action to repeat the loading sequence so that the gun was ready to fire the next round. This sequence was repeated for as long as the trigger was pressed and cartridges were available.

Many armies adopted the Maxim gun and the principle introduced by Maxim is in use today. During the First World War machine guns inflicted terrible casualties and showed that the days of mass attacks by infantry were numbered. For his invention, Maxim was knighted.

Cash register

1884

The bell was heard around the world when the cash register rang.

James Ritty

The first working mechanical cash register, nicknamed the "Incorruptible Cashier", was invented in 1884 by James Jacob Ritty, owner of the Pony House Restaurant at Dayton, Ohio. Ritty's saloon was attracting the celebrities of the day, such as Buffalo Bill Cody and the boxer Jack Dempsey, but much of his profit was disappearing into his cashiers' pockets.

On a trip to Europe, Ritty spotted a device in the ship's engine room that counted the propeller's revolutions. Inspired by this, he built the Incorruptible Cashier. It had two rows of keys with an attached dial that looked like a large clock face and kept a running total of the amount deposited. A bell – later referred to in advertising as "The Bell Heard Round the World" – rang when the register was opened.

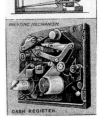

The machine worked, but Ritty did not have the persistence to make a success of marketing it. James Patterson, a businessman with a coal distribution business that had been losing revenue to pilfering had bought two of the machines sight unseen; within six months they reduced his debt from $16,000 to $3,000.

It impressed Patterson so much that in 1886 he bought Ritty's company for $6,500 and went on to invent the first formal salesman training school, becoming a multi-millionaire.

Fingerprint forensics
Sir Francis Galton

I n 1684 Nehemiah Grew, an anatomist, published an investigation into the intricate patterns to be found on fingertips. The usefulness of this knowledge was realized in 1860 by Sir William Herschel, a government official in Bengal. In letters published in *Nature*, the scientific journal, Herschel announced that he had found that the patterns remained unchanged from birth to death.

Sir Francis Galton, an anthropologist and explorer, verified the Herschel discovery and in 1884 he proposed the setting up of libraries of fingerprints for identification purposes, based on his elementary system of grouping fingerprints by the arches, whorls and loops in their patterns.

In 1893, a British government committee reported that no two people had the same fingerprints and this assertion allowed fingerprints to be used as absolute proof of identification in a court of law.

Galton's observations were developed into a method of fingerprint classification and comparison by Sir Edward R. Henry and made public in June 1900 under the title *The Classification and Use of Fingerprints*.

Henry became Commissioner of the Metropolitan Police and the Galton–Henry system was adopted by Scotland Yard in 1901. It forms the basis

1884

Sixteen characteristics must be identified to make a fingertip match.

for most modern fingerprint archives and dactyloscopy, the technique of taking fingerprints.

At first, fingerprints were used to identify individuals to establish their criminal records, but once an extensive database had been formed, prints found at crime scenes could be used to link suspects to the scene.

Matching fingerprints to images on record is almost an art in itself. Generally, fingerprint impressions are classified in three ways: type of pattern, position of pattern on the finger and relative size. The British Standard now states that if two sets of prints share at least 16 characteristics, it is virtually certain that they are both from the same person. Police forces around the world gather increasing amounts of fingerprint data to achieve quick matches, with the donkey work now done automatically by computer.

Motor car

Karl Benz

1885

Karl Benz didn't know how far his patent motor wagon would go until his wife took it on a five-day jaunt.

The growth of the car has been an extraordinary story, as remarkable for its destructive powers as its benefits. Only 100 years ago, horses were the primary transport and speeds across London were faster than they are today in a car. Yet London was choking in a sea of manure, and horses were an expensive luxury, so for the most part, people simply did not travel far. Fifty years later, in 1951, Elma Wischmeir became the millionth American to die on the highway. By the end of the century, the world's car-makers employed nearly four million people to produce 35,801,618 cars in a single year to add to the estimated 500 million in daily use around the globe.

The first self-propelled vehicle could have appeared around 350 years ago. In the 13th-century, Roger Bacon, the Franciscan friar, mentioned the possibility of carriages propelled without the use of animals, and in the 15th-century Leonardo da Vinci sketched a self-moving car, but the first record of a powered vehicle comes from *Astronomia Europæa*, published in 1687. It describes a small steam-powered model carriage constructed under Father Pierre Verbiest, a Jesuit

missionary who spent a large part of his life in China and was appointed Astronomer Royal to Emperor Kang-Hi. The small vehicle, apparently constructed around 1662, was fuelled with a pan of hot coals. Steam turned a simple turbine, which was geared to the front axle of the vehicle.

The first successful steam wagon built by Nicholas-Joseph Cugnot in 1770 to pull cannon also provided a lesson in the dangers of powered transport. His carriage hit the wall of the Paris Arsenal to become the earliest recorded powered passenger vehicle accident.

The motor car's time came in the 19th-century with the emergence of volatile fuels such as coal gas and liquid petroleum spirit. In 1829, Samuel Brown of Brompton, London, produced a vehicle powered by a four-horsepower, 89-litre gas-vacuum engine. This was a true three-stroke, internal combustion engine, which on test climbed Shooters Hill in Blackheath, but it was wildly uneconomical compared with contemporary steam engines and little came of it.

Installing tyres on the Model-T Ford at the Highland Park Plant, Michigan, in 1925. Henry Ford pioneered motor-car production which was so fast the paint barely had time to dry.

1885 Carl Auer,
Freiherr von
Welsbach, patents the
gas mantle.

Karl Benz builds the
first practical
motorcar.

In 1862 the French engineer Alphonse Beau de Rochas proposed a four-stroke explosion cycle, the basis on which almost all modern car engines operate. Rochas's paper coincided with the work of Nikolaus August Otto, who had been developing J-J Étienne Lenoir's first commercially available gas engine of 1860. In 1862 Otto prepared a gas-powered engine that operated on the four-stroke cycle but it promptly blew to pieces. Otto was then joined by Gottlieb Daimler, the son of a German baker, who patented a high-speed petrol engine in 1876.

At the same time, in Mannheim, Karl Benz was working on a petrol tricycle (Frederick Simms, an early motoring pioneer, coined the word "petrol" to dispel public fears about storing inflammable motor spirit in their garages). Benz's tricycle took to the streets in 1885, was patented the following year, and offered for sale as the "patent motor wagon" in 1887. When Benz granted an agency to Emile Roger in Paris in 1887, the modern motor industry was born.

In August 1888, Berta Benz borrowed her husband's 1886 Benz tricycle-car without permission. She packed provisions and her two young sons on board and drove off to visit relatives in Pforzheim, some 75 miles from Mannheim. En route Frau Benz had to refuel with dry cleaning fluid when she ran out of fuel; she repaired a short circuit with garter elastic, unblocked a fuel line with a hatpin and found a cobbler to reline a brake. She arrived safely and five days later she drove home again; it was the first ever long-distance journey by motor car.

Another woman played a major role in the creation of the first recognizably modern motor car, the 1894 Panhard-Levassor. Louise Sarazin secured the first successful export of Gottlieb Daimler's engine and mechanical patents to Panhard-Levassor in France after the death of her husband Edouard, a colleague and old school friend of Daimler who had initiated the deal. The widowed Mrs Sarazin concluded negotiations and then married Emile Levassor, who used Daimler's engine to construct the first modern motor car. With an enclosed, front-mounted, twin-cylinder engine, attached gearbox and foot-operated friction clutch, it was the first successful move away from the horseless carriage layout. Levassor described the primitive gear system as *"brusque et brutal – mais ça marche!"* And so the motor car was born.

In its early years, car development was more rapid than that in today's computer industry. In 1894, the first motor cars were no faster than a horse; 10 years later they were commonly capable of 80 mph and some had been clocked at over 100 mph. Making cars soon became a study in productivity, with Henry Ford leading the way. Taking his example from the Colt revolver factory, Ford invented the moving production line in 1914. His claim that "any customer can have a car painted any colour that he wants so long as it is black" was startlingly honest. His production line made cars so quickly that the only colour that would dry fast enough was black Japan enamel.

Since then, companies such as Volkswagen and Toyota have dominated car making. The VW Beetle became the world's most popular car by combining reliability with a charm that modern car makers have been hard put to rediscover. In January 1978 the last German built VW Beetle rolled off the Wolfsburg lines, but production continued overseas. In May 1980 the 20 millionth Beetle was completed in Mexico, where the car is nicknamed the belly button because everyone has one.

In the 1970s, environmental and safety legislation raised the price of entry into car production and gradually car makers fell by the wayside. Even those old names that remain, such as Jaguar, Aston Martin, Alfa Romeo, Lancia, Saab, Volvo and Ferrari, are part of much bigger automotive groups which themselves are being taken over by even bigger car-making combines.

The 1980s and the 1990s have been a battle for world domination among car makers and a fight for lean production. The industry now spends its time building as many different cars as possible on the same chassis and exporting production to the cheapest wage economies. It has resulted in such oddities as General Motors buying in a Polish-made Suzuki urban car and selling it in Britain as a Vauxhall.

Early motorists recalled the ethereal experience of travelling "like a bird, past workers in the fields, leaving nothing to show our passing." Perhaps we've lost the thrill of it all, but it's been quite a ride.

Motorcycle

Gottlieb Daimler

1885

1886 John Pemberton, an Atlanta pharmacist, invents a health tonic that he sells to soda fountains. It later becomes known as Coca-Cola.

1887 R.W. Brownhill patents first useful coin-operated gas meter.

F.E. Muller, a glass-blower from Wiesbaden in Germany, produces the first workable contact lens.

Alex Crivillé, the 1999 500cc champion: the motorcycle has never been healthier.

Once the internal combustion genie was out of its bottle, inventors started to fit engines to almost everything: cycles, cars and motorboats. Strictly speaking, the first motorcycle, Gottlieb Daimler's 1885, 7.5 mph, 264cc Einspur, was also the first motor car, albeit with tiny training wheels on each side.

Although steam-powered bikes had been tried (such as the 1869 French Michaux-Perreaux Velocipede), motorcycles were peculiarly suited to the compact dimensions and high power-to-weight ratio of the petrol motor. Two years after the Daimler/Maybach Einspur, Edward Butler in Britain built a tricycle, powered by a twin-cylinder, two-stroke engine with rotary valves.

The first production motorcycle came from Germany, the Hildebrand and Wolfmüller, a 1,488cc, twin-cylinder, 28 mph two-wheeler, built between 1894 and 1897. Count de Dion and Georges Bouton, two Frenchmen, pioneered lightweight, relatively powerful engines that allowed the development of the motorcycle worldwide. There was little regard for international patents in those days and the de Dion Bouton designs were widely copied. The Werner brothers were two such plagiarists; their improvements to the de Dion designs set the pattern for 20th-century motorcycle design, which was completed by the invention of the twistgrip throttle control on the 1904 Indian, by George Hendee and Oskar Hedstrom, two Americans. By 1914, there were around 600 motorcycle makers throughout the world. At the beginning of the 1920s there were 200,000 cars in use in Britain and 260,000 motorcycles. The great names of the motorcycle industry – Ariel, AJS, BSA, ACE, Excelsior, Pierce, Norton, Scott – built their reputations in the years before the Second World War, but none of them survives today. In the post-war baby boom, people bought cars as family transport as soon as they were able.

In the past 10 years, however, increasing congestion, environmental legislation, public wealth and leisure time have made people reconsider motorcycles. In the past five years, motorcycles sales have tripled in Britain and new, smaller machines are attracting new buyers. As the motorcycle enters its second century, it has never been in more robust health.

Electric chair

New York State, USA

1888

Once hailed as a new, more humane form of capital punishment, the electric chair is the gruesome product of a bitter commercial battle that raged in America in the 1880s. At issue was whether America should adopt alternating current (AC) or direct current (DC) to power the nation's newly emerging electrical infrastructure. And squaring off against each other in this Battle of the Currents were the two brilliant but ruthless electricity entrepreneurs: George Westinghouse and Thomas Edison.

In 1883 Westinghouse had begun to invest in AC technology, having recognized its many advantages over DC. Chief among these was the ease with which AC could be "transformed" to higher voltages for low-loss transmission over long distances. Edison, in contrast, had committed his company to the simpler DC technology that struggled to power lights just half a mile or so from a generator. Determined to overcome this handicap, Edison spared no effort to ensure a lasting success over his rival.

Edison set up a smear campaign against AC, claiming that it was unproven technology and had already killed several people by electrocution. To back his case, he set up grotesque photo opportunities where newspaper reporters could witness stray cats and dogs being forced on to metal plates connected to 1,000 volt AC supplies, and   duly "Westinghoused".

Edison's most appalling PR coup was dreamt up by a former assistant, Harold Brown, who had joined the state of New York's commission investigating more humane means of capital punishment than hanging. Having personally helped set up Edison's animal executions, Brown directed the commission towards a new means of execution: death by AC.

Seizing the chance to score over his rivals, Edison gave

1888 Marvin Stone patents drinking straws made by winding paraffin-coated manila paper around a pencil. His patented spiral winding process is now used in thousands of electronic devices and components. Theophilus van Kannel invents the revolving door to overcome air pressure differences in tall buildings.

1889 First electrical oven installed, at the Hotel Bernina, near St Moritz, Switzerland.

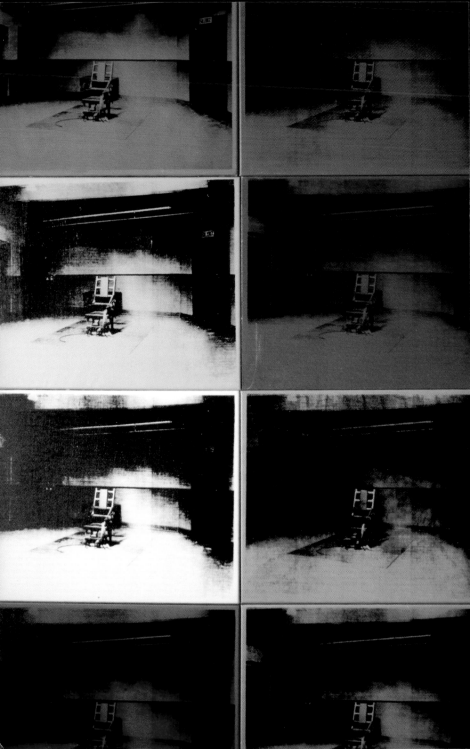

Brown every assistance, including setting up more animal experiments. By the end of 1888, the state of New York formally replaced hanging with death by AC electricity. It took the form of the so-called Electric Chair, a wood-framed chair in which the condemned person sat, while wired up with electrodes before being electrocuted with tens of kilowatts of alternating current.

On August 6 1890, a murderer named William Kemmler became the first to die while in the Electric Chair. But far from the speedy death promised, he took 15 minutes to die in a torment that horrified those who witnessed it.

Since then, more than 4,000 have died in the Chair. Yet despite having been adopted by more than half the US states at some time, it now seems to be fast becoming obsolete. Only three states – Georgia, Nebraska and Alabama – still have "Old Sparky" as the sole means of   execution, and of those only Alabama actively uses it.

Opposite: 'Electric Chair' by Andy Warhol, 1964. The invention was the result of a commercial battle over electricity, and in spite of assurances that it would prove humane, the first victim took 15 agonizing minutes to die.

Jukebox

San Francisco

1890

The first jukebox burst into song in San Francisco's Palais Royal Saloon in 1890. Carousers packed the device named after the Creole slang for dancing – *jook* – with more than $1,000 in six months, a remarkable sum given that it played only one tune.

Much to the undoubted relief of barflies, in 1908 John Danton of Michigan invented the first jukebox with more than one tune. By 1927, the Automatic Music Instrument Company had created the world's first electrically amplified multi-selection phonograph – the jukebox as we know it today.

The Wurlitzer 1100 of 1948: at a peak of jukebox perfection, it played 78 rpm records.

Escalator

Jesse Reno

1891

The residents of Coney Island in America may not have realized they were pioneers in transportation, but in 1891 Jesse Reno built the first escalator there as a novelty ride and soon the idea spread way beyond the local fairground.

Reno didn't get it right first time. The prototype had a simple rubber conveyor belt inclined at 25 degrees. Passengers perched on small platforms called "cleats" and there was no moving handrail.

Charles Seeberger, another American inventor, improved the moving staircase by making one from individual step units linked to a looped chain. The Otis Elevator Company adopted the concept and engineered a version that appeared at the 1900 Paris Exposition.

The improved Seeberger machine was an instant success. The Otis Company registered its version as the "Escalator", a name that stuck when the word was soon adopted for all moving staircases. The invention transformed the practicality of high-rise buildings, and contributed to the opening of the first department stores in the 1920s.

Tractor

John Froehlich

1892

1892 Henry Perky, a lawyer from Denver, Colorado, inventes shredded wheat, the first breakfast cereal.

Most people see travel as the biggest beneficiary of the invention of the internal combustion engine. Yet the oil had hardly been wiped from the first petrol engines when they were being put to use on the fields. In 1892 John Froehlich, an Iowa blacksmith, built the first petrol-powered farm vehicle, and by the turn of the century tractors were taking on a major role on farms.

One big problem with these early tractors was that their brute strength came with brute-like weight. In 1904, Benjamin Holt came up with a famous solution: he swapped the wheels for bands of linked metal tracks. The resulting "caterpillar tracks" spread the load over a broader area, preventing the vehicle from sinking in mud.

Ironically, the brute strength of the tractor was also a liability if the plough it was towing struck a big rock. In trying to move forward, the tractor would just rotate around its back axle and flip on to its back, often killing whoever happened to be driving.

The solution was found by an Ulsterman whose name is known even to townies who have never driven a tractor: Harry Ferguson. In 1916 he was invited by the Irish Board of Agriculture to devise better ploughing methods, and he set about introducing key innovations, including towing bars with linkages that prevented lethal flip-overs. But Ferguson's greatest idea was to discard the massive chassis and use the engine as the main structure, an ideal shape to resist enormous twisting loads. Today's tractors are just the same, with their huge rear tyres (introduced in 1932) to give more grip and comfort, and the power take-offs, which allow other machinery to be driven by the tractor's engine.

Vacuum flask

Sir James Dewar

1892

I t is a common theme running through the history of invention: serious scientists spend a lifetime on dedicated, arcane work, then end up in the history books for coming up with a peripheral but more practical device; something essential for modern living.

Sir James Dewar, a renowned Scottish physicist and chemist, was one such case. His record of important innovations in the field of low-temperature gases is lengthy. He conceived a practical process to liquefy oxygen and hydrogen, and then needed a means to store the liquid gases at very low temperatures.

In response to this demand, in 1892 Dewar came up with the device that made his name: the vacuum flask. Dewar's flask was intended to keep his experimental gaseous liquids and supercooled solids at a consistently low temperature. It made use of a curious property that had been observed in near-vacuums for some time; Torricelli may have known about it in the 16th-century. Vacuums do not have a medium to conduct

Designed by a Scot,
the vacuum flask was
developed by Germans.

or convect heat, so a vessel holding a volume of cold or hot liquid surrounded by a vacuum will remain the same temperature for hours or even days. Coating with silver any surface in contact with the liquid eliminates heat loss by radiation as the mirrored surface reflects radiated heat back into the liquid.

Dewar's invention was taken up in 1904 by two German glassmakers who found a way to mass-produce the flask. They added a tough metal exterior and a shock-absorbing element between the fragile inner and outer flasks, and named their company Thermos after *therme*, the Greek word for hot. And there the design has stuck, because it is simple and it works.

Radio

Popov and Marconi

1895

The story of radio begins with James Clerk Maxwell, professor of experimental physics at Cambridge University. In 1864, Maxwell proved mathematically that an electrical disturbance was capable of producing a remote effect by electromagnetic propagation. His calculations, which concluded that these radio waves move at the same speed as visible light, were put to the test in 1888 by Heinrich Hertz, a German physicist. In his Berlin classroom, Hertz used a condenser linked to a metal loop to create electromagnetic waves. A spark leapt across the loop when the condenser discharged and, a fraction of a second later, a smaller spark leapt across a loop of a smaller circuit two metres away. It proved Maxwell's theories. However, Hertz told one of his pupils: "I don't see any useful purpose for this mysterious, invisible electromagnetic energy."

Two years later, Edouard Branly, a French scientist, noticed that the electrical resistance of a tube of fine metal particles decreased dramatically when a spark discharged nearby, but the particles had to be shaken loose after each discharge in order to detect the next spark. Then, in 1892, Oliver Lodge, an English physicist, noticed that when a spark discharged near two barely touching metal spheres, the spheres fused together and current would flow easily through the junction. Lodge called this phenomenon the "coherer" effect and realized soon afterwards that it could be used to detect the presence of electromagnetic radio waves produced by a distant spark discharging.

Lodge improved on Branly's "coherer tube" of metal filings by using relatively coarse iron chips and attached the tube on to a stand with an electric bell. When the coherer tube detected a radio wave, the bell rang. The mechanical vibrations of the bell loosened the metal particles in the coherer tube and made them ready to detect the arrival of the next radio wave. It was a simple radio receiver.

Guglielmo Marconi was happy to take the credit for inventing the radio, but a Russian got there first.

The birth of radio communications came in 1895, when Alexander Stepanovitch Popov, a Russian, and Guglielmo Marconi, whose father was an Italian nobleman and his mother Irish, separately sent and received radio signals over distance. In May, Popov obtained a distance of 600 metres, while Marconi transmitted and received within the grounds of the family estate at Bologna. Marconi's brother would wave a white handkerchief to signal that he was receiving Marconi's broadcasts. Further success came in September when Marconi attempted to transmit beyond the line of sight and heard the shot of his brother's hunting rifle echo up the valley as confirmation.

Both men used similar equipment, including an antenna, and both had studied the work of Hertz. Popov had read of Lodge's work in scientific journals, and he further improved the sensitivity of the coherer.

Popov thoroughly enjoyed his research but he did not approach it with a sense of urgency. Marconi, in contrast, was determined to develop wireless telegraphy into a profitable technology, lest someone else achieve it first. When the Italian government showed no interest in his apparatus, Marconi set sail for England, where he was granted a patent for "a system of telegraphy using Hertzian waves", dated June 2 1896. From that point on, the future of early radio belonged to Marconi.

Popov felt no personal resentment toward Marconi. In 1902 when Marconi visited Kronstadt, Popov met him and the two had a cordial discussion. Marconi later received a silver samovar and a sealskin coat from Popov as wedding presents. Popov's work won him a Grand Gold Medal at the Paris International Exposition of 1900. (Marconi's gained him a share of the 1909 Nobel Prize for physics.)

Marconi's experiments became increasingly successful. In 1897, the Wireless Telegraph and Signal Company was formed with Marconi as its major shareholder. A year later, the Italian Navy adopted Marconi wireless, the press used wireless, Queen Victoria communicated wirelessly from Osborne House with the Prince of Wales on board the Royal Yacht, and Lord Kelvin sent the first telegram by wireless.

Broadcasting over ever increasing distances culminated in the first transatlantic wireless transmission on December 12 1901 between Poldhu in Cornwall and St Johns in Newfoundland. With the aid of a kite aerial, Marconi and his assistants were able to discern a Morse signal for the letter S – three dots. Many had considered this impossible, saying radio

Guglielmo Marconi with the receiver which picked up the first transatlantic radio message (three dots for the letter 'S') which had been sent from Newfoundland to Poldhu in Cornwall.

waves would not "bend" to follow the curvature of the earth. Instead, the signals were bounced between Cornwall and Newfoundland on the ionosphere, a layer of ionized gas particles 100 miles above the Earth's surface.

The next year, 1902, Valdemar Poulsen, a Dane, invented the first high-frequency generator without moving parts. By generating continuous waves, it enabled transmitters to be fine tuned and minimized signal disturbance between stations. Radio had come of age.

X-rays

Wilhelm Roentgen

1895

I t is hard to imagine a more unexpected, or more spectacular, discovery than X-rays. Certainly Wilhelm Roentgen, the German physicist who first witnessed their amazing powers in 1895, was taken utterly by surprise. At the time, he was just a 40-something academic at the University of Würzburg, doing ho-hum research into the effect of passing electricity through gas-filled bottles.

He was following up reports that some kind of ray streamed out of the bottles, which he called X-rays ("X" being the standard symbol for anything scientists are clueless about). Roentgen discovered that these rays had the miraculous ability to take pictures of things hidden under or inside other objects. In one of the most famous images in science, Roentgen exposed his wife's hand to the X-rays streaming from one of his bottles, and a photographic plate exposed to the rays revealed all the bones in her hand, plus the solid band of her wedding ring. It was a stunning discovery, and one whose potential was rapidly exploited. The British Army used simple X-ray machines to examine soldiers wounded in the Battle of Omdurman in the Sudan in 1898. Dental X-rays and cancer therapy followed a few years later. Roentgen was awarded the first Nobel Prize for Physics, in 1901.

Scientists call something 'X' if they don't know what it is. Wilhelm Roentgen had no idea what he had discovered.

The dangers of X-rays were not recognized for years. In the Spanish-American War, Austrian-born Elizabeth

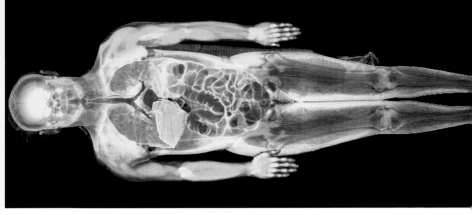

Fleischmann pioneered the use of X-rays to pinpoint bullets in injured soldiers. But she checked the strength of the rays by exposing her hand to them and ended up with radiation-induced cancer. Her arm had to be amputated and seven months later she was dead.

Today's X-ray machines use sensitive photographic films that can give clear images with far lower X-ray doses. A particularly impressive application of X-rays is the computerized tomography (CT) scanner, where images of many X-rays are combined to give cross-sectional "slices" through the body. Its invention won the 1979 Nobel Prize for a British electrical engineer, Godfrey Hounsfield, of EMI.

Lie detector

Cesare Lombroso

1895

The idea that liars can be unmasked by looking for physiological changes is not new. One Far Eastern method involved getting suspects to fill their mouths with rice: those who had the most trouble spitting it out after questioning were deemed guilty on the grounds that the stress of lying had made their mouths go dry.

The first attempt to detect lies scientifically was made by the Italian criminologist Cesare Lombroso in 1895. According to Lombroso, the stress of telling lies inevitably caused changes in blood circulation. These changes in pulse and blood pressure could be detected by the plethysmograph (from the Greek for enlargement), a device invented in 1870 for measuring alterations in blood volume in parts of the body.

In the 1920s, John Larson of the University of California developed a more sophisticated device, monitoring physiological signs such as breathing rate. The resulting Polygraph was adopted by some American police forces in 1924.

Supporters of the Polygraph today claim to be able to detect liars with 80 to 95 per cent success. Critics warn that this is dependent on the way questions are put to suspects. They also point out that effective countermeasures are easily mastered, as demonstrated in 1994, when top CIA official Aldrich Ames, who had passed many Polygraph tests, was shown to be a long-serving Soviet spy.

Safety razor
King Camp Gillette

1895

Whatever else they were, safety razors weren't safe. Trying to make something both lethally sharp and safe has exercised inventors for centuries. In 1771, the French cutler Jean-Jacques Perret developed a razor designed to minimize cuts. But facial trauma was low down the list of priorities of the most famous name in shaving (and what a name), King Camp Gillette of America. He took the advice of a former employer who made a fortune selling disposable bottle-tops: "Invent something that people use once, and then throw away. That way they'll have to keep coming back for more." The story goes that in 1895 Gillette twigged which product could make his fortune while shaving in the mirror. It was a razor fitted with disposable blades. Gillette patented his idea in 1901, and within a year of launching it in 1903 he had sold more than 12 million blades.

1896 The first toothpaste in a tube – Colgate Ribbon Dental Cream – goes on sale.

1897 J.J. Thomson, a British physicist, discovers the electron, leading to the invention of the thermionic diode and the field of electronics.

It's almost impossible to cut yourself with today's razors. But one thing they do is scrape off the dead outer layer of skin. Cosmetic experts call it "exfoliation", and it's supposed to make you look younger.

Aspirin

Felix Hoffman

1899

1899 Lester Allen Pelton patents the impulse turbine, more efficient than conventional turbines.

Aspirin, or acetylsalicylic acid to use its proper name, is related to a treatment known to the ancient Greeks. Hippocrates described a tea made from willow bark that was effective against fevers and which he recommended for treating gout.

In 1763, the Rev Edmund Stone noticed that farm workers in Chipping Norton used a preparation of the bark of the white willow tree to cool fevers and treat aches and pains. Six decades later, the Italian chemist Rafaele Piria revealed the active ingredient of this willow bark: salicylic acid.

Because salicylic acid on its own caused stomach pains and nausea, the German pharmaceutical company Bayer launched a hunt for similar compounds to find one that had fewer side effects. In 1899, Felix Hoffman, a Bayer employee, tried several Bayer compounds on his arthritic father and found the acetyl compound was the most effective. It was soon marketed under the trade name Aspirin. Early samples of salicylic acid were extracted from the meadow-sweet plant Spiraea, so the name was coined from "a" for acetyl and "spiraea".

Aspirin, based on the Spiraea plant.

The drug acts by inhibiting the production of prostaglandins, locally acting hormones that are necessary for blood clotting and sensitizing nerve endings to pain. Because of its ability to maintain blood supply to regions of the heart or brain, aspirin has also been used as an anticoagulant following a minor stroke or a heart attack.

There is evidence that aspirin can also protect against nerve damage, inhibit bowel and colon cancer, and even inhibit a protein that helps the Aids virus to multiply.

Aspirin can block an enzyme that promotes inflammation, pain and fevers but it also affects a related enzyme that is essential for the health of the stomach and kidney. Now scientists have developed compounds that counteract this imbalance.

Paper clip

Johan Vaaler

1899

The paper clip would appear to be an unlikely symbol of protest against an occupying army. But that is what it became during the Second World War, when Norwegians wore paperclips as a protest against occupation by the Germans, who forbade the wearing of buttons showing the Norwegian king's insignia.

Until 1899, when Johan Vaaler, a Norwegian inventor, filed the first patent for a paper clip, papers were fastened together with ribbons or straight pins. Ironically, Vaaler patented his design in Germany, because Norway had no patent laws at the time. Vaaler patented several variations on a length of wire shaped into a rectangle, crossing on one of its two long sides, but did little with his invention.

The next year, an American inventor, Cornelius J. Brosnan, of Springfield, Massachusetts, filed an American patent for a paper clip he called the "Konaclip." In 1907, Gem Manufacturing Ltd., an English company, patented what has become the ubiquitous standard "slide on" double oval design.

With its ability to slip easily on to paper the Gem paper clip is an excellent example of form following function, a design that was perfected further by rounding the sharp points of the wire so they would, as Gem promised, "hold securely your letters, documents, or memoranda without perforation or mutilation until you wish to release them."

Chapter 5
Journey into the atomic age
1900 to 1944

I t is impossible to assess the rapid explosion in the rate of invention between 1900 and 1945 outside the extraordinary stimulation that the two world wars gave to technological development. In times of war, budgetary restraint, technical and methodological conservatism and rivalries between competitors are put aside in the rush to ensure national survival. Wartime is often boom time for invention – and the early 20th-century was a perfect illustration, witnessing more advances over a wider range of activities than in the whole of previous history.

The 20th-century began with two inventions that would shape the century to come: the Wright brothers' mastery of controlled, powered flight, one of the great "firsts". And, secondly, the invention of the vacuum diode and triode, the first electronic components, which made possible broadcasting, long-distance telephony, early computers and a myriad of other devices.

It is hard to believe that, without the world wars, the aeroplane would have advanced as rapidly as it did, from a wood-and-glue construction to the robust, cantilevered, all-metal monoplane of the 1940s. Certainly the jet turbine and the liquid-fuel rocket would have taken longer to reach maturity.

Two other developments would have been far more ponderous without the extraordinary resources, finances and political impetus of military exigency. One was the evolution from valve-based electronics to the transistor (which reached fruition in 1948 but was driven by the demand for miniaturization during the Second World War). The second was the development of digital computers (when the theories of two great British thinkers, Charles Babbage and Alan Turing, united to develop a machine that was used, initially, to break enemy codes).

Many key innovations of this period were significant in a wider sense, in that they were technologically novel. Unlike most previous inventions, they did not build on earlier methods, but were entirely new and unpredictable, mainly because they were the application of knowledge gained from scientific discovery rather than technological investigation. Increasingly, technology took a back seat to science, which was undergoing a revolution of its own as the wars transformed research from small-scale efforts undertaken by a few isolated scientists to "big science", where governments and corporations sponsored and co-ordinated massive research teams working towards a common goal.

The 20th-century was also the century of the motor car, and its rise was accompanied by greater efficiencies in obtaining and distilling oil. The invention of thermal and catalytic cracking techniques enabled the hydrocarbon raw material to be tailored to suit the demands for spirit fractions needed to produce two other key inventions: plastics and pharmaceuticals.

The first plastics – vast chain molecules, called "high polymers" – were made from naturally occurring compounds in the mid-19th century, but true synthetic high polymers did not appear until the invention of Bakelite in 1905. Although plastics have swung in and out of fashion as domestic materials, in industry and commerce their use has steadily grown. Today, materials, objects and clothing made of polymers or artificial fibres are readily accepted.

In the late 19th-century, pharmacy had begun to break free of the empiricism of the herbalist, but it was in the 20th-century that chemists made the first rational, synthetic drugs (where the characteristics of the drug's molecular structure were exploited to induce a particular response in the patient). The first such "magic bullet" drug was Salvarsan, an organic compound containing arsenic, that Paul Ehrlich, a German bacteriologist, developed as an effective cure for syphilis.

The period covered by this chapter ended with an invention possibly more momentous than any before or since, and certainly a dividing point in the century that was as socially significant as it was technologically momentous: the detonation of the first nuclear weapon at Alamogordo in New Mexico in July 1945. With this simultaneous harnessing and releasing of the enormous forces within the atom, science, technology, economics and politics converged, and the last vestiges of technological leadership passed from Britain and Europe to America.

Vacuum Cleaner

Hubert Booth

1901

Many inventions have had a Darwinian-like evolution, with rival companies vying to come up with ever-better products, to the benefit of the consumer. But not the vacuum cleaner. For the best part of a century the basic design remained a dinosaur in the technological Lost World.

The first vacuum cleaner was a fairly silly idea cooked up in response to a really silly one. In 1901 a London-based engineer named Hubert Booth saw a new American device for extracting dust from the upholstery of railway carriages. It used compressed air to blast the dust out, which was all very well, except the dust settled straight back on the furniture again.

Booth suggested that perhaps suction might be better, an idea the inventor dismissed as impractical. Booth later decided to put his idea to the test. He put a handkerchief on the floor of his office and sucked hard. The handkerchief was covered with dust. By February 1902, teams of men from Booth's British Vacuum Cleaner Company were turning up at homes with a five-horsepower engine and a huge hose fitted with a filter for cleaning carpets. Two years later, he introduced an electric-powered machine

From left: the bellows-operated Baby Daisy, 1908; the 1911 hand-pumped Star; Hoover Junior, 1907, the Swedish cylinder Electrolux from 1920 and 1945 (opposite top).

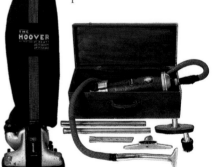

weighing around 40 kg. Booth's design, with the filter in front of the vacuum section, was not the true ancestor of the conventional vacuum cleaner. That accolade (if such it can be called) goes to James Spangler, an American asthmatic janitor who wanted a way of cleaning that didn't stir up dust and bring on an attack. He came up with a machine with an electric fan that dumped the dust into a pillowcase. In 1908 Spangler sold the rights to a leather and saddle maker looking to diversify. The company's name was Hoover.

Put on the market at $70 (around £800 in today's money), the Hoover upright machine turned vacuum cleaners into a mass-market product. A few tweaks were added to the design: Electrolux introduced the cylinder model with a hose in 1913, followed 23 years later by a Hoover with beater-brushes. But that was pretty much it until 1978, when James Dyson found he had run out of vacuum bags at home.

The bagless Dyson DC01.

Fishing the old one out of the bin, he cut it open with the intention of emptying it, only to find it wasn't very full. But the walls of the bag were clogged with dust. The airflow was being choked by the bag walls, causing a dramatic loss of suction.

Determined to come up with a vacuum cleaner with a performance that did not stand or fall by the properties of a paper bag, Dyson focused on using centrifugal force to spin dust out of the air. It sounds simple but it took 14 years and more than 5,000 prototypes to arrive at the final design: a dual cyclone system that accelerates the dust-laden air to high speed, flinging the dirt out of the air-stream. In February 1993, the first Dyson DC01 machines rolled off the production line. Within two years it was the most popular machine in the UK and today Dyson vacuum cleaners are the most popular in western Europe.

1901 Blood types A, B and O first determined. Vlademar Poulsen invents the arc-generator, used to produce continuous radio waves.

Air conditioning

Willis Carrier

I n 1902, a Brooklyn, NY, printer frustrated by the effects that changes of temperature and humidity had on his paper, contracted Willis Haviland Carrier, a local engineer. Carrier's solution was to invent the world's first air-conditioning machinery. It pumped liquid ammonia through a set of evaporation coils. Warm air in the room heated the ammonia which evaporated, absorbing heat and cooling the air until it reached its dew point. The droplets that formed on the coils drained away and a fan returned the cooler, drier air to the printing plant.

Air-conditioned man: a fanciful costume to keep a man cool in summer, warm in winter. But a Brooklyn engineer came up with a more practical idea.

Electrocardiogram

Willem Einthoven

I t had been known since the 17th-century that muscular tissue would conduct electricity, and by the late 19th-century, physiologists had investigated the possibility that the human body and heart, too, had electrical potential. Then, in 1895, a brilliant young Dutch physiologist, Willem Einthoven, used a crude electrical sensing apparatus to establish that the beating heart produced four distinct signals, each one corresponding to a different ventricle. He called these distinct signals the "PQRS" factors. However, Einthoven needed an exact way of measuring the minute amounts of current.

In 1897 a French electrical engineer, Clement Ader, invented the "string galvanometer", containing a tensioned string of quartz. In 1903, Einthoven modified Ader's machine, adding electrodes attached to the patient's limbs and thorax. In use, the string was seen to vibrate in time with the patient's heart. The vibrations were recorded photographically.

With his "electrocardiogram" Einthoven was able to look at each of the PQRS signals and diagnose the health of the heart. For his pioneering work, Einthoven was awarded the Nobel prize in 1924.

Aeroplane
Orville and Wilbur Wright

1904

Credit for the first manned flight was hotly disputed by early aircraft designers and patent holders of the day and it is still a subject for debate among aeronautical historians. However, it is generally accepted that the Wright brothers were the first to achieve sustained and controlled powered flight, on December 17 1903.

Like many inventions, the aeroplane was the result of incremental developments, in this case going back to Sir George Cayley, a British inventor who is regarded as the grandfather of the aeroplane. Cayley devoted much of his life to the study of heavier-than-air flight, establishing the principles of aerodynamics and the basic configuration of a modern fixed-wing aircraft. In 1804, he built and successfully flew a model glider; later models featured propellers powered by twisted rubber bands. Cayley continued to develop and publish his theories, culminating in a full-size glider, which he built in 1853. It flew, carrying his unwilling coachman a distance of nearly 200 metres.

Orville Wright was at the controls of the first powered flight, while his older brother Wilbur looked on.

One of the first patents for a powered aeroplane with fixed wings, an engine and propellers was granted in 1843 to a British inventor, William S. Henson. He called it a "Locomotive Apparatus for Air, Land, and Water", and even had plans to set up an airline. He teamed up with John Stringfellow, a close friend and maker of miniature steam engines in his home town of Chard in Somerset. They built an aeroplane, based on Henson's designs, with a six-metre wing span and powered by one of Stringfellow's compact steam engines. Henson lost interest, and the machine proved too heavy to fly. But Stringfellow went on to complete a lightweight design of his own, with a three-metre wing span, in the summer of 1848. The first test flights took place inside a local lace mill, where the plane was launched from a guide wire to ensure it didn't crash into the walls. The first attempt failed, but eventually it flew for more than 10 metres.

During the 1890s, the French engineer Clement Ader and the American inventor of the machine gun, Hiram Maxim, who was resident in Britain, independently conducted

The first powered flight, on December 17 1903, on the sand flats near Kitty Hawk, North Carolina. Wilbur Wright had the first, unsuccessful flight of that bitterly cold day. His younger brother, Orville, is seen here on the second attempt, at 10.35, with Wilbur running alongside. The plane, Flyer 1, travelled 37 metres at a maximum of three metres from the ground. The flight lasted just 12 seconds.

promising experiments with powered man-carrying craft. Both managed to leave the ground briefly, but uncontrollably, in steam-powered machines.

Between 1891 and 1896, Otto Lilienthal, a German engineer, and his brother Gustav constructed and flew a succession of man-carrying gliders from a hill they built for the purpose. Otto did most of the flying and he discovered how to manipulate the shape of the wings, and shift his body weight to control the flight of his machine, in much the same manner as a modern hang-glider. The Lilienthals made more than 2,000 flights and kept meticulous records of their experiments, providing valuable data for many other pioneers.

In 1896, Professor Samuel P. Langley, an American physicist and astronomer, built a steam-powered model aeroplane that managed to stay aloft for 90 seconds, flying a distance of more than half a mile. Buoyed by his success and a $50,000 grant from the US War Department to develop his design, Langley began work on a full-size machine, powered by a small petrol engine. The machine was called the Langley Aerodrome but on the first test flight, from a houseboat moored on the Potomac River on October 7 1903, the plane snagged the launching mechanism and plunged into the water. Another unsuccessful test followed in December.

Wilbur and Orville Wright had been interested in flight

since childhood, possibly sparked by a toy helicopter bought for them by their father, Milton. In 1896, three years after the brothers opened a bicycle repair and manufacturing business at Dayton, Ohio, Orville contracted typhoid. While nursing him, Wilbur read about the death of Otto Lilienthal. This appears to have rekindled his interest in flying machines and on May 30 1899 he wrote to the Smithsonian Institute requesting information on aeronautical research. The brothers developed a deep interest in flight. They studied books written by Cayley and Lilienthal and corresponded frequently with Octave Chanute, a respected engineer and author of *Progress In Flying Machines*, who became their mentor.

The Wrights' first large-scale glider underwent test flights at the Kill Devil Hills coastal sand flats near the town of Kitty Hawk in North Carolina in 1900. It was selected from a list of sites provided by the US Weather Bureau as an area with steady winds averaging no more than 13 mph.

During the experiments the brothers worked out ways of controlling an aircraft in flight, and they went on to conduct a series of tests on different wing shapes in a home-made wind tunnel. Using information from these tests, they built a glider in 1902 that flew a record 188 metres at Kitty Hawk.

Wilbur and Orville built their first powered aeroplane in 1903 and named it Flyer No 1. It was based on their own experiments and an earlier biplane glider design that Chanute had sponsored. The wings were made from a wooden frame covered in cotton cloth – the first to use a ribbed construction – measuring 12.2 metres from tip to tip. The pilot lay in the middle of the lower wing; to his right was mounted a 12-horsepower petrol engine connected by chains and sprockets to two large wooden propellers. The aircraft was controlled from a cradle that fitted around the pilot's hips and connected to the wing tips by cables; by twisting his hip the pilot could alter the shape of the wings and with luck turn the machine. A small wing at the front of the aircraft, the equivalent of an elevator on a modern aircraft, controlled ascent and descent. Flyer had wooden skids instead of wheels and was designed to take off from an 18-metre rail.

The first flight attempt was on December 14 1903 at the Kitty Hawk sand flats. Wilbur took the controls but the aircraft stalled on take-off. Three days later, the brothers flipped a coin to see who would be next; Orville won and at 10.35 am Flyer 1 lifted off from the ground under its own power. The flight lasted just 12 seconds, during which time Flyer 1 flew a distance of 37 metres and rose to an altitude of three metres, with Wilbur running alongside it.

Three more flights were made that day, the longest of which had Wilbur at the controls; it lasted 59 seconds and covered a distance of 260 metres into a strong head wind. Five locals witnessed the four flights and the news was reported the next day in a newspaper aptly called the *Norfolk Virginian Pilot*.

Knowing that Flyer 1 was, in effect, a glider that had harnessed the wind to get off the ground, with the engine used primarily to steer the aircraft, the Wrights built and flew two more aeroplanes at Huffman Prairie over the next two years. In this pasture, near their home town of Dayton, there was no wind and they could justifiably claim they had used power to take off. From late June to mid-October 1905, they made more than 40 flights on Flyer III (their third aeroplane), staying aloft for up to 39 minutes at a time, banking, turning, flying figures of eight and circling at about 35 mph, proving the structure they had designed and built was strong enough to withstand repeated take-offs and landings.

Then, concerned that rivals would steal details of their invention and unable to hide it any further from the press, they decided to remain on the ground until they had secured patents and contracts for the machine's sale. In 1908, they signed contracts with the US Army and licensed their design to French manufacturers, before leaving for a tour of America and France to make their momentous invention public.

Radar

Christian Hülsmeyer

1904

After discovering radio waves in 1888, the German scientist Heinrich Hertz proved they could be bounced off things. In 1904, the German engineer Christian Hülsmeyer turned this into a patent for a collision warning system for shipping, the first radar system.

Then in 1935, the Scottish physicist Robert Watson-Watt of the National Physical Laboratory near Slough was asked by the Air Ministry to investigate the possibility of using a beam of radio waves as a death-ray weapon.

Watson-Watt found that radio transmitters lacked the power to kill, but that they could trigger echoes from aircraft almost 200 miles away. He was helped by the invention in 1940 of the cavity magnetron, by John Randall and Henry Boot at Birmingham University. This gave a compact, powerful source of short-wave radio waves and thus a means of detecting targets from aircraft at great distances.

Wartime radar experts found their magnetrons could be used as water heaters for their tea. Today magnetrons are the source of heat in microwave ovens.

A wartime radar scanner's electronic eye is marked on a vertical chart.

Colour Photography

Auguste and Louis Lumière

1904

This first colour photograph, of a tartan ribbon, was made in 1861 by James Clerk Maxwell. It was a composite image of three photographs using different colour filters, a process taken on by the Lumière brothers

T he first photographs, seen in 1839, were greeted with wonder and disappointment. How could a process that recorded nature with such faithful detail fail to record its colours? The search for a method of producing colour pictures became photography's Holy Grail.

In 1904, two Frenchmen, Auguste and Louis Lumière, gave the first public presentation of a colour process they called "autochrome". The brothers had been experimenting with colour for many years and had published an article on the subject in 1895, the year that they invented the

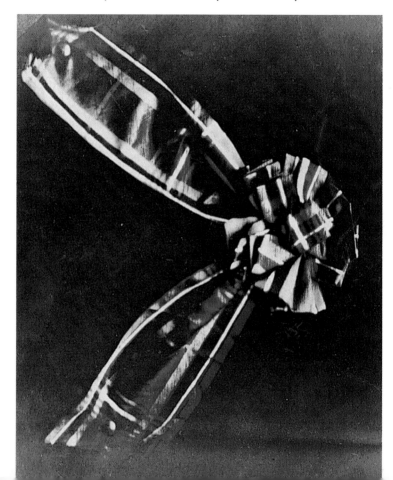

cinematograph. Autochrome plates incorporated a filter screen made from transparent grains of potato starch dyed red, green and blue, the three colours that in 1861, James Clerk Maxwell, a young Scots physicist, had shown could be used to make all colours, a process known as "additive" colour synthesis.

There were about four million microscopic potato starch grains on every square inch of autochrome plate, each of which acted as a tiny coloured filter. The grains combined after processing to produce a full-coloured image. Autochrome plates went into production in 1907 and their success soon prompted the appearance of many similar processes.

At the same time, a number of "subtractive" methods for making colour photographs were under development. These all required negatives to be made through red, green and blue filters. These "colour separation" negatives were then used to make cyan (blue), magenta (red) and yellow prints, which were superimposed to produce a coloured image. Modern colour photography is based on the same concept, but instead of separate negatives, colour film has several layers of emulsion, each of which is sensitive to one of the primary colours.

Kodachrome, the first modern colour film, was invented in America by Leopold Mannes and Leopold Godowsky, nicknamed "Man and God". These classically-trained professional musicians carried out photographic experiments in their spare time, often in hotel rooms when they were on tour. Their musical background was useful when they whistled passages of music to gauge development times in the darkroom.

When Kodak heard about the two Leopolds' work, they were persuaded to give up their musical careers to work full time in the Kodak research laboratories. With the support of Kodak's technical and financial resources, they made rapid progress. Kodachrome cine film first went on sale in 1935, followed by 35 mm film for still photography the following year. In 1936, Agfa introduced a similar multi-layer film, called Agfacolor.

With the perfection of dye-based multi-layer films such as Kodachrome and Agfacolor a new era of colour photography had arrived. All colour film in use today evolved from these pioneering products.

1904 Ice cream cone invented.

Julius Elster invents Photoelectric cell.

Thomas Sullivan uses hand-sewn silk envelopes in New York to invent the tea bag.

Vacuum diode

1904

John Ambrose Fleming

John Ambrose Fleming invented the diode, a vital component in radio and television.

The first electronic device was the vacuum diode, a sealed air-evacuated tube with two metal electrodes, invented by Sir John Ambrose Fleming, professor of electrical engineering at University College, London.

In 1883, Thomas Edison sealed a metal wire inside a light bulb and discovered that a bluish glow could be seen between it and the nearby filament. He had neither explanation nor use for it, but in 1897 Joseph John Thomson, an English physicist, worked out that the glow was due to the passage of tiny particles carrying a negative charge; he had discovered the electron.

In 1904, Fleming substituted a metal plate for Edison's wire and discovered that the electrons travelled to the plate only when it was carrying a positive charge. It meant that the device converted (or rectified) alternating current to direct current. Fleming called his two-electrode device ("diode") a thermionic valve, as it allowed current to flow only in one direction, a characteristic that enabled it to detect radio waves.

In 1906, Lee De Forest, an American inventor, added a grid of fine wire between the filament and plate to invent the triode. He found a small charge on the grid could be used to amplify or reduce a large flow of electrons between the other two electrodes. This was the first amplifier.

Vacuum tube diodes and triodes were at the heart of thousands of electronic devices that followed, such as radios, amplifiers, televisions, long-distance telephony and the first digital electronic computers, until they were replaced in the 1950s by the transistor.

Windscreen wipers

1905

Mary Anderson

A standard fixture on all cars today, the wiper blade wasn't invented until 20 years after the motor car, perhaps because many early vehicles were open-topped, so drivers didn't go out much in bad weather.

On a trip to New York City, Mary Anderson of Alabama noticed how her driver struggled to remove the snow and ice from his windscreen. She came up with the idea of a wiper operated via a lever inside the car, and patented the device in 1905. She never went into production with it but by 1913 it was standard on American cars.

Polymers

Hendrik Baekeland, Wallace Carothers

1905

N o material has changed our lives over the past 100 years as dramatically as plastic. A world devoid of Formica, pvc, nylon, polythene and the other man-made fibres and high-polymer films is hard to imagine. Made by uniting vast numbers of simple molecules to form one massive molecule, plastics exist for the same reason as all organic compounds: the extraordinary ability of carbon atoms to bind with each other in chains and rings.

Human beings are not the only designers of high polymers. Cotton's main constituent is a high polymer, cellulose. This natural product, consisting of hundreds of glucose molecules joined together, became the starting point for the plastics industry.

By the mid-19th century, it was well known that if cellulose was treated with nitric and sulphuric acids, it became a dangerous explosive known as nitrocellulose or "gun-cotton". If it could be persuaded not to explode, nitrocellulose could be dissolved in solvents, such as alcohol or ether, to make a primitive kind of plaster. When the solvent evaporated, the layer of nitrocellulose would keep air from a wound. Until the early 20th-century nitrocellulose was also used in photography and for movie films, but the highly inflammable polymer would not always survive its passage through a hot projector, and it would burst into flames.

In 1862, Alexander Parkes, a Birmingham chemist, invented the first genuine plastic when he discovered that a mixture of nitrocellulose solution and camphor dried to a hard, smooth material that looked like ivory or tortoise-shell. He called it "Parkesine". Promoted as a "beautiful substance for the

1905 Albert Einstein publishes his special theory of relativity.

Hiram Maxian invents th silencer for the gun after working on exhaust sliencers for cars.

Leo Baekeland was playing with chemicals when he stumbled on the secret of the most useful modern material.

The Panton Classic Chair, from the Vitra Design Museum in Germany, proves plastic's versatility and durability. The 1940s camera and clock are made in Bakelite.

Arts", it was used for combs, knife handles and other domestic items, and later marketed under the name xylonite.

Seven years later, John Wesley Hyatt, an American, was intrigued by a competition set up by the New York firm of Phelan and Collender, offering a prize of $10,000 for an ivory substitute for billiard balls. Hyatt had heard of Parkesine and improved the formula to make a material that he called "celluloid". He used it to coat composition billiard balls so that they looked like ivory, although the collision between balls could produce a report like a gun shot. On such occasions, every man in one Colorado saloon is said to have instantly drawn his gun. Celluloid was widely adopted for fancy goods and a multitude of articles, from dental plates and piano keys to cuffs and collars.

The first truly synthetic plastic was invented by Leo Hendrik Baekeland, a Belgian chemist living in New York who had made his fortune in 1898 by selling his invention of the first commercially successful photographic paper to George Eastman for $1 million. Baekeland was therefore free to follow any research he wished and in 1905, while literally playing with formaldehyde and phenol, two common organic chemicals, he found they produced a resinous material that could be used as a binder for all kinds of powders, including asbestos. He perpetuated his own name by calling it Bakelite. It was the first "thermosetting plastic" – a material that once set hard would not soften under heat. It had many useful properties, which led to its being dubbed "the material of a thousand uses". It was water- and solvent-resistant, an

electrical insulator, hard yet easily machined and capable of being cut with a knife. It made the communications revolution of the 1930s possible, being used in electrical insulators, 78 rpm gramophone records, telephones and radio cabinets. Several firms in England made similar resins, often for lacquers – one rejoiced in the name of the Damard Lacquer Company because its products were supposedly "dam' 'ard".

The first synthetic rubber, exploited by du Pont and marketed in 1932, was neoprene, invented by Julius Nieuwland, a Roman Catholic priest. Another polymer was an "organic glass" discovered in Montreal and taken over by ICI as Perspex in 1934. It was first used for aircraft cockpits, then safety glass and is now used for contact lenses. Several new plastics appeared through serendipity, the most famous being polyethylene, or polythene, formed in an ICI experiment that went wrong in 1933; a quite different product had been expected.

The first wholly synthetic fibre was, however, an example of industrial sponsorship. Wallace H. Carothers was a scientist working for du Pont, and in the 1930s discovered how to make nylon just in time to replace the East Asian supplies of silk cut off by Imperial Japan. It took the company 10 years from Carothers's first appointment and about £10 million to develop an effective manufacturing process. Carothers died in 1937 and never saw nylon go into production, but among the initial products made that year were toothbrush bristles and nylon stockings, of which 64 million pairs were produced in 1938. Nylon was reserved for military use for the duration of the war, being used for gear-

Formica, the plastic laminate, enters Britain's most fashionable 1950s suburban homes.

FOR**MICA**

The finest of all the decorative laminates

AUSTRALIAN WALNUT

SUNFLOWER BLUE

SURFACES

Here's beauty–rich, warm colour and patterns always clean and new. Here are surfaces that stand up to wear for ever and a day! Marvellous. 'FORMICA for me!' you say. But–make certain you get it. Look for the FORMICA label before you buy. (The Schneider dining furniture shown here has FORMICA laminates on curved chair-backs as well as flat surfaces.) Ask for FORMICA-surfaced furniture in the shops . . . or ask your local contractor surface your own pieces . . . or Do it Yourself–quite easy!

wheels, bearings, parachute ropes and tyre-cord, but became one of the most versatile of materials after 1945.

Plastic came into its own during the Second World War. Besides Perspex for aircraft cockpits, polythene insulators for radar and synthetic rubber for tyres, four other polymers known from pre-war days now became accepted. Polythene became a major product for insulators and food containers; polyvinyl chloride (pvc) was used for record sleeves, lightweight mackintoshes and billions of bottles; polystyrene was employed as packaging material and as foam fillings for furniture; and Formica, a material derived from formaldehyde and a compound called melamine, became familiar in laminates for furniture and kitchen bench-tops.

Now the hunt was on for many more polymers, and completely new plastics emerged, each with its own special advantages. Polytetrafluorethylene, commonly known as Teflon, has one of the largest molecules known, giving it a stability and water-resistance that makes it ideal for non-stick cooking pans. Lycra, a stretchable polyurethane fabric initially

A tester on an early stocking production line: 64 million pairs were made in 1938, the year they appeared.

made for tight-fitting sports gear by du Pont, has now permeated socks, underwear and fashionable outer garments. Terylene, or Dacron, an artificial fibre that is crease-resistant and rot-resistant, is invaluable for ropes, sails and tents in damp surroundings. Like plastic furniture before it, synthetic clothing was regarded as cheap and nasty only 15 years ago. Nowadays, fleece jackets and trainers made of man-made textiles are the height of street fashion, and many of plastic's early disadvantages, such as its tendency to generate static electricity and therefore attract dust, have been overcome. High polymers have become valued as substances in their own rights, rather than as tacky substitutes for natural materials.

Broadcasting

Professor Reginald Fessenden

1906

The concept of radio broadcasting – simultaneously sending a message to many people at different locations – had been hindered by the technical inadequacies, bulk and expense of early transmitting and receiving apparatus. Early radio equipment was initially used for short-range point-to-point communication.

In 1898, Professor Reginald Fessenden, a Canadian scientist and former chief chemist to Thomas Edison, began work on a radio system that could transmit and receive a full range of sound, rather than just the dots and dashes of Morse code. On December 23 1900, he succeeded in sending a voice signal over a distance of more than 15 metres.

In 1901 Fessenden was granted a patent for a continuous-wave voice transmitter that used a high-speed electrical alternator, the forerunner of today's amplitude modulation, (AM) system used for long-wave radio broadcasts.

On December 21 1906 at Brant Rock in Massachusetts, Fessenden gave a series of demonstrations to interested organizations, including the American Telegraph and Telephone company (AT&T). Three days later, Fessenden used his alternator transmitter for what is now considered the first entertainment broadcast by radio. It comprised an unnamed female voice singing a Christmas carol, Fessenden playing Charles Gounod's *O, Holy Night* on his violin, a speech and an invitation to report on the reception. Wireless operators on board United Fruit Company banana boats in the North Atlantic received it.

Early transmitters and receivers relied largely on crude electromechanical devices. In 1906 Lee De Forest, an American scientist, invented the three-electrode vacuum tube or triode "valve", based on John Ambrose Fleming's diode. De Forest's Audion tube was the first device capable of electronic "gain", or amplification, using a feeble electrical current to control a much larger one.

The Audion tube provided a major catalyst for radio pioneers and in 1913 Edwin Armstrong, an American electrical engineer, developed the "regenerative circuit", which increased the sensitivity of radio receivers by a factor of several thousand.

Armstrong was a major innovator in radio technology and went on to invent the Superhetrodyne circuit in 1917, which enabled precise tuning to a selected frequency, allowing broadcasters to make efficient use of the frequency spectrum. Armstrong also invented the portable radio receiver in 1923, as

Music on the move. An electric filter on early car radios eliminated interference from the ignition system.

a present for his wife, and developed the Frequency Modulation (FM) system in 1933.

In 1952, Armstrong conducted a number of successful experiments into FM stereo radio, but after two years of severe depression brought about by a long-running legal battle over his patents for FM radio, he committed suicide.

Many claims have been made for the first radio station to regularly broadcast an entertainment-based schedule. One of the first was undoubtedly De Forest's experimental station with the call sign 2 XG, based in New York, which began transmitting phonograph recordings and sponsor announcements three times a week from spring 1916. Broadcasting from privately-owned and commercial radio stations began in earnest in 1920 in both America and Britain. In June of that year Marconi opened the first public station in Chelmsford, Essex, with a concert transmitted on the Short Wave band by Dame Nellie Melba. Pittsburgh's KDKA began scheduled broadcasts in November 20 with updates from the returns of the Presidential election.

The success of KDKA started a rush of radio stations in America; by 1924 there were more than 1,400 licensed transmitters and three million receivers in people's homes. In Britain, by contrast, a committee of inquiry advised the Government that the American broadcasting "free-for-all" was an unsuitable model. The experimental British

A wartime valve radio made of Bakelite and chromium.

1906 Lee De
Forest
invents the Audion
and electrical value
later used to amplify
sound.

Broadcasting Company, set up with Post Office approval in 1922, became a public corporation accountable to Parliament and the British Broadcasting Corporation was born in 1925.

The first practical car radio is attributed to an American, Paul Gavin, who in 1929 coined the name Motorola for his company's products. Experimental multiplex stereo transmissions began from KE2XCC, an American station, on April 29 1952. BBC engineers began work on CD-quality Digital Audio Broadcasting (DAB) in 1975, and scheduled services began in 1997.

Electric washing machine

Alva Fisher

1907

Doing the week's washing used to be a major event taking up the best part of a day. It was hard physical work, lifting heavy, sodden clothes with wooden tongs, wringing them out, steering them through a mangle and finally hanging them out to dry, weather permitting, of course.

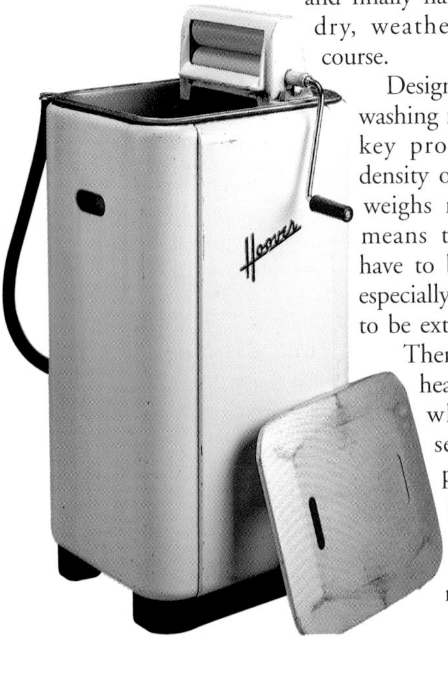

Hoover's first UK product was a washtub with mangle.

Designers of the modern washing machines faced two key problems. First, the density of water (a sink-full weighs more than 20 kg) means that the machines have to be pretty powerful, especially if the water is going to be extracted by spinning. Then there is the huge heat capacity of water, which means that several kilowatts of power are needed to heat enough to wash a single load of washing in a reasonable time.

The American engineer Alva Fisher is generally credited with making the world's first electric washing machine, the Thor, in 1907, while the American inventor Vincent Bendix claims credit for the first automatic washing machine, which washed, rinsed and dried without much outside intervention.

These machines didn't really take off until the 1960s, first in the form of "twin tubs", which were then eclipsed by automatics. Since then, washing machines have been growing in sophistication, able to give decent results at the lower washing temperatures that environmental regulations demand of manufacturers. Spin driers develop hundreds of g-force to dry clothes, and the whole operation is controlled by microprocessors.

Caterpillar tracks

England

1907

Several patents were registered for caterpillar tracks long before anybody constructed a vehicle that could run on a continuous belted tread. The earliest, for "a portable railway, or artificial road, to move along with any carriage to which it is applied", was filed in 1770 in England by a Richard Edgeworth.

Benjamin Holt, the owner of a California heavy-equipment company that later became the Caterpillar Tractor Company, is often credited with making the first caterpillar tractor in 1904, but all he did was to attach tracks, in distinct blocks, to the rear wheels of his steam tractor.

The first caterpillar tractor to use a continuous series of separate links joined by lubricated pins was invented by David Roberts and built by the firm of Ruston Hornsby and Sons. This petrol-driven tractor was demonstrated to the War Office in 1907 but they showed no interest. Hornsby sold the patent rights to Holt and the caterpillar was forgotten until it re-emerged on the tank towards the end of the First World War.

Armour had been fixed to trains by the time of the Boer War when the War Office considered armoured tractors for use as "mechanical horses". In 1900 the first armoured car was built. Twelve years later De la Mole, an Australian inventor,

A First World War tank rumbles over cobbled streets as French troops enter the Rhineland. The first armoured vehicles with caterpillar tracks were disguised as 'tanks' during shipping — and the name stuck.

submitted to the War Office a design for an armoured vehicle with caterpillar tracks. It was rejected. But by 1915 the Great War had become bogged down. What was wanted was something that could crush obstacles, cross trenches and protect the troops. Winston Churchill, the First Lord of the Admiralty, realized that some form of armoured vehicle was the answer and set up the Landships Committee to examine ideas. In July 1915 an armoured car body was fitted on to the chassis of a tractor. It was not a success in trials, but Churchill pressed ahead.

The first tank, called "Little Willie", designed by Major W.G. Wilson of the Royal Naval Air Service and William Tritton, an engineer from Lincoln, was top heavy and its length was insufficient to cross a German trench. Wilson and Tritton then came up with the classic lozenge-shaped caterpillar track on Big Willie, or Mother as it was later called, and the British Army ordered 1,000.

To disguise their identity, the vehicles were shipped to France described as "tanks", and the name stuck. There were two versions, one with two six-pounder guns, the other with machine guns. The first time tanks showed their worth was in the Battle of Cambrai in November 1917, when nearly 400 of them led two platoons of infantry on a six-mile front that penetrated further than any previous attack, taking 7,500 prisoners and 120 guns.

Coffee filter

Melitta Bentz

1908

Gaggia House in Dean Street, Soho. Espresso coffee made by counter-top machines made Italian coffee houses popular.

E uropeans knew nothing about the intoxicating pleasures of drinking coffee until the 16th-century. For the next 300 years, coffee fanatics – J. S. Bach and Immanuel Kant among them – would find their enjoyment tainted by a thick, gritty residue in their final sips.

The gritty coffee-drinking experience was revolutionized on July 8 1908, when the Berlin Patent Office received a registration from a Dresden housewife. Melitta Bentz had invented a prototype for the first coffee filter, with the aid of a piece of her son's blotting paper and a perforated tin can. With her husband, Bentz set up a company, called Melitta, devoted to the filter's development, and in 1936 patented the conical filter that is still used today.

The short, sharp hit of the *espresso* had its origins some 200 years earlier in metal jugs that, when heated on a stove, forced boiling water through ground coffee, but it took a succession of false starts for a mechanical *espresso* maker to be successful.

At the Paris Exposition in 1855, Edward de Santais showed a complicated prototype *espresso* machine with four chambers that forced steam

1908   Henry Ford begins making the model T.

Hugh Moore invents the water cooler and paper cups.

and hot water through ground coffee. An Italian, Luigi Bezzera, patented the first commercial *espresso* machine, based on de Santais's design, in 1901. Bezzera was an inept businessman and his rights were bought in 1903 by Desiderio Pavoni, who produced the *espresso* machine in quantity from 1905 in his factory in Italy.

The best *espresso* was found to be made at a temperature of 90°C and a pressure of about 150 psi (between nine and 10 times atmospheric pressure). In 1938, M. Cremonesi, another Italian, designed a piston-type pump which forced hot, but not boiling water through the coffee. This was installed in Achille Gaggia's coffee bar in Italy. Gaggia saw the value of the design and set up in business in 1946 selling his own version of the Cremonesi machine, called the Gaggia, with great and continuing success.

Neon lamp

Georges Claude

1910

The vivid red glow of neon energized with a strong electrical current was a sensation at its first public showing, the Paris Motor Show in 1910. It had been discovered eight years earlier by the French engineer Georges Claude, who was the first to try the inert gas in a lamp.

Claude set up a company, Claude Neon, to market the invention around the world. Advertizers soon realized that the

Neon lights in Los Angeles, visible in daytime: their potential was quickly realized by advertizers.

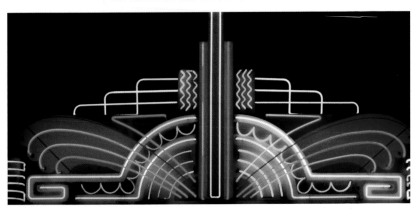

light was highly visible during the day, and the glass tubes could be shaped to form words and pictures.

The vibrant night landscapes of American cities were first lit by neon in 1923, when Claude started exporting his tubes. By using different gases more than 150 different colours can be produced.

Kitchen mixer

L.H.Hamilton, Chester Beach and Fred Osius

1910

T he first patent known to have been made for a mixer belongs to L.H. Hamilton, Chester Beach and Fred Osius. In 1910 the trio formed the Hamilton Beach Manufacturing Company to make a mixer that looked similar to machines used to make milk shakes today.

The first domestic version of a multi-purpose mixer with several attachments is said to be the KitchenAid, invented by Herbert Johnson, an engineer, and sold to the US Navy in 1916 for kneading dough. It was the first mixer in which the beater and the bowl rotated in opposite directions, and it sold for $189.50 for home use from 1919. A more affordable and portable version was introduced in the 1920s and has since become a "design classic".

The KitchenAid mixer, designed in 1916 and making a comeback.

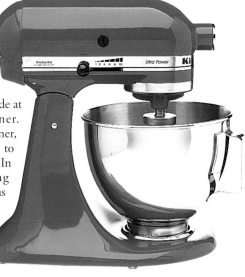

In 1922, Stephen J. Poplawski invented the blender, when he became the first to put a spinning blade at the bottom of a container. Poplawski, a drugstore owner, used his invented appliance to make soda-fountain drinks. In 1935, Fred Waring collaborated with Fred Osius to improve on Poplawski's idea to make the Waring Blender, which is still sold today.

Stenotype machine

1911

1911 Ernest Rutherford proposes modern theory of atomic structure.

Ward Stone Ireland

The problem of how to record speech in courtrooms and conferences quickly, quietly and accurately taxed engineers and inventors throughout the latter part of the 19th-century.

Ward Stone Ireland, an American court reporter and shorthand enthusiast, came up with a machine in 1911. He called it the Ireland Stenotype Shorthand Machine and it was an instant success. Weighing in at five kilograms, it had a fully depressible keyboard, making it possible for typists to record whole words and numbers at a stroke and in tests it was proved faster than the most accomplished shorthand writers of the day.

Ireland founded the Universal Stenotype Company, which went on to improve the machine, making thousands of models for the American government during the First World War. But the government didn't pay for the Stenotypes that they ordered and the company was bankrupted.

The stenotype machines now used have 22 keys (some letters are represented by combinations of keys) that print on to a thin strip of paper. By using abbreviations, stenographers can type up to three words in a single stroke.

Brassière

1913

Mary Phelps Jacob

For centuries the female breast has endured discomfort in the pursuit of the changing fashions for an ideal silhouette.

A "breast supporter" was patented in 1893 by Marie Tucek, a device involving separate pockets for the breasts and straps with hook and eye closures. It did not catch on.

The first practical, popular brassière to receive a patent was invented by the New York socialite Mary Phelps Jacob in 1913. Jacob had bought a sheer silk dress and didn't want to wear the whalebone-stuffed corset of the period beneath it.

She came up with an alternative: two silk handkerchiefs joined with pink ribbon. It freed a generation of women from the confines of the corset, which had reigned supreme since

A seamstress in her cups in Naples in the 1950s. The bra was invented by a New York socialite to wear under a sheer silk dress.

1913 Niels Bohr is
the first to
apply quantum theory
to molecular
structure.

Brillo pad invented by
Miton B. Loeb.

Catherine de Medici's ban on thick waists at the court of her husband, Henri II of France, in the mid-16th century.

Jacob patented the invention in 1914 and went on to sell it under the name Caresse Crosby. It didn't make her fortune: her business collapsed and she sold the rights on to Warner Brothers Corset Company for $1,500. Over the following 30 years they made $15 million from the New York woman's inspired improvization.

In 1928, when flappers' fashion dictated that women should be flat-chested, Ida Rosenthal, a Russian immigrant, came up with the idea for grouping women into alphabetical categories according to cup size.

Ecstasy

Merck

1913

When the German pharmaceutical giant Merck registered the patent for 3, 4 methylenedioxy-methamphetamine (MDMA) in 1913, its destiny as the signature drug for the "Ecstasy Generation" 70 years later was impossible to predict. Merck decided against marketing the drug, developed primarily as an appetite suppressant.

In 1967 the fringe biologist Alexander Shulgin, fascinated by the power of mind-altering chemicals after receiving morphine for a hand injury, rediscovered MDMA. Shulgin was the first known human to have tried it, recording its effects – along with 179 other psychoactive drugs – in his landmark publication *pihkal* (Phenethylamines I Have Known And Loved). Entry 109, MDMA, was described as the closest to his expectations of the perfect therapeutic drug.

MDMA works by inhibiting the uptake of the serotonin, a brain chemical that helps control depression, anxiety, memory disturbance and other psychiatric disorders. Derived from an organic compound, Ecstasy can be made only in the laboratory in a complex process. It acts as an empathogen-entactogen, increasing communication and empathy and promoting a sense of well-being. Between 1977 and 1985, when it was outlawed by the US Drug Enforcement Agency, MDMA was used experimentally on as many as half a million people to treat clinical depression, schizophrenia and even in marriage guidance.

Zip

Gideon Sundback

With an ingenious arrangement of teeth, each machined to tolerances of a few 10-thousandths of a centimetre, with its versatility and ubiquity, the Zip should be hailed as a brilliant invention. But it jams, it corrodes, it falls to pieces, and those teeth can convert any man into a boy soprano in a split-second.

Certainly, it's clever. But it has been asked to do more than it should. When Whitcomb Judson, a mechanical engineer from Chicago, showed his patented "clasp locker" at the Chicago Worlds Fair of 1893, it was intended as a replacement for couplings on high boots. (Elias Howe, famed for inventing the sewing machine, was granted a patent for his zip-like "automatic, continuous clothing closure" in 1851, but never commercialized it.)

Judson's system failed to catch on, and he sold only 20, all to the US mail service as bag fasteners. It was the Swedish engineer Gideon Sundback who, in 1913, got the zipper to work, by trebling the number of teeth and inventing a machine to stamp them out and fix them on a flexible material.

The credit for making the zip a commercial success usually goes to the American company BF Goodrich, which in 1923 incorporated the zip into rubber boots. The company also gave the fastener its name, after the noise that it makes as it opens and closes.

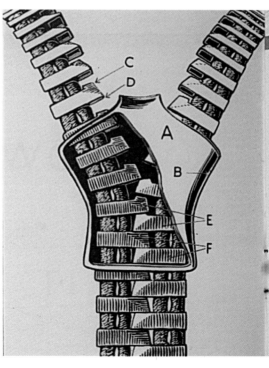

The zip, from A to F: the slider (A) has a closed side (B) and an open side through which the teeth are pulled. Each tooth has a tiny knob on top (C) and a little hollow underneath (D). When the slider draws the teeth (E, F) together, each knob engages with a hollow on an opposite tooth.

1913 Crossword
invented by
the *New York World*
newspaper.

The zip found its way into ever more garments, with one magazine in 1925 trilling: "no fastening is so quick, secure or popular as the zipper". Someone eventually thought it would be a great idea to replace the buttons on the flies of men's trousers with a zip, and claimed that it eliminated the risk of embarrassment through failing to do one's flies up. It is a claim that will no doubt be greeted by any man with total derision.

Sonar

Military scientists

1916

Paul Langevin, the French physicist, was one of several eminent scientists involved in the invention of sonar.

At the outbreak of the First World War, the Admiralty in London was slow to realize the threat posed by the U-boats of the Imperial German Navy. Soon thousands of tons of shipping were being sunk and lives lost, culminating in the torpedoing of the liner *Lusitania* in 1915, on which 1,198 died.

Something had to be done. By July 1915 the Admiralty had established the Board of Invention and Research, with

luminaries such as the nuclear physicist Ernest Rutherford giving advice. The section of the board responsible for submarines and telegraphy was given a great deal more money for research than other departments. The use of dowsing-rods was tried out, as was the possibility of training submarine-spotting seagulls and sea lions.

The Anti-Submarine Division of the Admiralty, together with American and French scientists (notably the physicist Paul Langevin) came up with the first successful submarine detection system in 1916: a line of specialized microphones, called hydrophones, that were towed behind a ship. This was a passive system; it emitted no pulses of sound.

Operators listened for submarine noises and calculated a bearing to guide attackers on to the enemy. The first U-Boat to be detected and sunk in this way was the UC3 on April 23 1916.

British sound detection systems were named ASDIC (Anti Submarine Detection and Investigation Committee). Work was carried out on "active" ASDIC systems which emitted a characteristic "ping" sound when an enemy sub reflected an echo. Anti-submarine warships used the ping to calculate the range, bearing and depth of the sub before an attack with depth charges. Active ASDIC helped to win the underwater war.

In America such systems were dubbed Sonar (SOund Navigation And Radar) in 1942 by the acoustic scientist, F. V. Hunt. The term did not catch on in the Royal Navy until the late 1950s.

Sticky plaster

Earle Dickson

1916

In 1920, a newlywed couple, Josephine and Earle Dickson, were living in New Jersey. Every evening, the accident-prone Josephine would have dinner on the table for Earle's return but she would also have several cuts or burns on her fingers.

Without an adhesive bandage, Josephine had no easy way of bandaging her own cuts, so Earle would cut pieces of adhesive tape and cotton gauze to make a bandage for each wound.

Finally, after several weeks of kitchen accidents, Dickson prepared ready-made bandages by placing squares of cotton gauze at intervals along a strip of surgical tape. He covered them with textured crinoline. Now all Josephine had to do was cut off a length of the strip, peel away the crinoline and wrap the ready-made bandage over her cut.

Dickson worked at Johnson & Johnson, who by the 1920s had become successful selling baby powder and sterile bandages. He mentioned his invention to his superior, who was unimpressed until Dickson demonstrated that he could easily apply a bandage with one hand to himself. Soon the first adhesive bandages were being produced and sold under the Band Aid trademark.

The first handmade Band Aids came in sections 7 cm wide and 45 cm long and did not sell well. But by 1924, Johnson & Johnson had developed a way of manufacturing several sizes by machine, and the sticky plasters caught on. Since then more than 100 billion have been made.

Sticky plasters are still called Band Aids in America, but other brand names, such as Elastoplast, have taken up the idea.

Hairdryer

Germany

1920

The electric hairdryer was invented in Germany in the early 1920s, when compact electric motors developed for use in vacuum cleaners became small enough to fit into a reasonably compact handheld unit. The technology was simple: the motor turned a fan which blew air over an electric heater – usually current-resisting nickel, iron and chrome alloy wire on a mica or asbestos board. Early models were large, heavy and unwieldy. They incorporated large magnets with chrome-plated steel or nickel for the housing and solid wood handles.

The invention of Bakelite, the first synthetic plastic, liberated the hairdryer from its weighty origins and made it a much safer appliance: thermoset plastics would

not distort under heat, and, more importantly, unlike the metal housings, they did not conduct electricity in the event of a short circuit.

The use of plastics also turned the hair dryer into a fashion accessory as models in walnut-effect brown, ivory, bottle-green, jade-green and red also became available. After the Second World War, induction motors replaced the heavy brush motors and the hair dryer became smaller, lighter and almost silent.

The German Sol electric hairdryer from around 1925.

Submachine gun

John Thompson

1920

1920 Hollywood
film director
Preston Sturges
invents kiss-proof
lipstick.

The usefulness of machine guns on the battlefield was limited by their weight and size. They would be much more useful if they were light enough to be handled by one person. The first step to the submachine gun was to use smaller, lighter, pistol cartridges, first seen on the Villar Perosa, an Italian gun patented in 1915, which fired 9 mm ammunition so fast it could empty its magazine in two seconds.

The Villar Perosa had two barrels and a double handgrip and was sometimes mounted on a tray hung around the neck. In 1918, the Germans introduced the Bergman MP18, which used 9 mm Luger bullets loaded and cocked by the force of the explosion of the previous round. It had a shoulder stock and could therefore lay claim to being the first submachine gun.

However, many regard the Tommy gun as the first submachine gun. The American Auto-Ordnance Company, under General John Thompson, who apparently first used the title "submachine gun", developed it in the early 1920s.

The Tommy gun used .45 calibre ammunition designed for Colt pistols, which it fired at a rate of 800 rounds a minute. The first war in which submachine guns were used extensively was the Spanish Civil War of 1936–39. The Allies and the Germans were armed with submachine guns in the Second World War, the British having developed a simple model, called the Sten. Today, the Israeli-made Uzi, the Koch MP 5 and the Heckler (both German) are probably the most widely used.

Robot

Karel Capek

1921

Opposite: Designers
find it difficult to make
a robot with two legs.
This one, the 1.6-
metre-tall Honda P-3,
can walk at one mile
an hour and navigate
around furniture.

The term robot dates to 1921, when   the Czech playwright Karel Capek called downtrodden serfs and slaves "robots" in his play R.U.R. (Rossum's Universal Robots). The play describes a mythical world of leisure, with lucky humans able to enjoy lives of utter pleasure, waited on by machine servants. Paradise turns to purgatory, however, as the robots bring unemployment and cause society to fall apart.

Robotic technology has developed in a benign way. The first commercially available robot was devised in America in 1956 by George C. Devol and Joseph F. Engelberger. Called "Unimates", the robots worked first in a General Motors engine plant. The two engineers formed a company, Unimation, which still makes industrial robots.

Nowadays, millions of robots are used for mundane tasks in factories and even to investigate the surface of Mars, as Nasa's Sojourner did so spectacularly in 1997. However, few conform to the archetypal image of a mechanical humanoid. Making a machine walk upright on two legs remains a tough task for robot designers; nevertheless the designers still strive to make anthropomorphic machines, as much for the engineering challenge as anything else.

Of those robots that do walk on legs, such as the autonomous vehicles that examine pipelines on the oceans' beds, most have six or eight legs, like a spider.

Insulin

Frederick Banting and Charles Best

1921

1922 Gustav Dalen, a blind Swedish scientist, invents the Aga cooker to lessen his wife's workload.

Opposite: Charles Best (left) and Frederick Banting on the roof of the medical building at Toronto University with one of the first diabetic dogs to receive the insulin hormone. In 1923 Banting received the Nobel Prize for his work on insulin, and he shared the money with his colleague.

Inside our bodies, the digestion of carbohydrates results in the production of glucose, known as blood sugar. The amount of glucose present has to be closely controlled. The hormone that regulates this process is called insulin and it is produced in the pancreas.

Too much glucose causes the condition known as hyperglycaemia. In itself, this is not lethal, but it can be a symptom of a disease, diabetes mellitus. People with diabetes have a restricted production of insulin, and the increased levels of blood sugar cause the body to produce poisons, which, without treatment, will result in death.

It was Joseph von Mering and Oskar Minkowski in 1889 who discovered that the pancreas controlled the glucose metabolism. By removing the pancreas of live dogs they could immediately cause the onset of severe diabetes. Although this inferred the existence of insulin, it proved difficult to isolate because digestive enzymes quickly destroyed the hormone.

In Toronto in 1921, following repeated attempts to extract insulin, Sir Frederick Banting and Charles Best discovered that by tying off the pancreatic ducts of a live dog they could cause the pancreas to produce certain cells. When injected into diabetic dogs these cells cured all symptoms of diabetes. This led to the isolation of insulin.

The work of Banting, Best, von Mering and Minkowski meant that diabetics could live a reasonably normal life, either by observing a carefully formulated diet or by having periodic insulin injections.

Quantities of animal-derived pancreatic tissue were required to produce insulin. But by 1981 a process was discovered to   produce it artificially. This became the first protein produced by genetic engineering to be used to treat a human disease.

Hearing aid

Marconi Company

1923

Cupping a hand around the ear can improve hearing by five per cent. In Victorian times, a wide range of trumpets and horns were popular, while hirsute gentlemen could take advantage of an under-beard listener and top hats were available with   discreet earpieces.

Until the advent of electrical microphones and amplification, however, there was no help for the very deaf. In the 1890s Alexander Graham Bell may have been trying to build a hearing aid when he stumbled/p.144 p.142/on the invention of the telephone. His carbon microphone was capable of turning sound waves into electrical   impulses, which were then amplified and turned back into sound at the receiver's end.

The major problem was the lack of suitable batteries. The Marconi Company in Britain came up with what they claimed was the first practical portable hearing device in 1923. Called the Otophone, it used a carbon microphone and a valve amplifier to channel sound to the wearer. Marconi were nearly there – but, with batteries, it weighed 7 kg.

A machine made by A. Edwin Stevens in 1935 was a breakthrough; it used lightweight valves and, weighing 1 kg, could be worn around the neck. The invention of the transistor shrank aids further in the 1950s. Now, in the digital age, selected frequencies can be amplified, giving a much better quality of sound.

Frozen Food

Clarence Birdseye

1924

T he first person to examine whether freezing food might delay its deterioration died in the process. In March 1626, Francis Bacon, lawyer, Member of Parliament, wit and philosopher, was passing Highgate in north London. Observing the snow outside his carriage, he wondered if cold might delay the putrefaction of living tissue. He stopped his carriage immediately, bought a hen, and with his own hands, stuffed it with snow. But he was seized with a sudden chill, which turned to bronchitis, and he died at the Earl of Arundel's house nearby on April 9.

Some three centuries later, Clarence Birdseye, an American businessman, developed Bacon's speculation into a process for freezing foods in small packages suitable for retailing.

Clarence Birdseye got his idea after seeing fish being preserved by ice and wind by the native people of Canada.

In 1912 and 1916, Birdseye travelled to Labrador in northeast Canada to trade fur. He noticed the natives froze food in winter because of the shortage of fresh food. The combination of ice, wind and temperature almost instantly froze fresh fish straight through. When the fish were cooked and eaten, they were scarcely different in taste and texture than they would have been if fresh.

Birdseye, a former government naturalist, realized the fish froze too quickly for ice crystals to form and ruin their cellular structure. Inspired by what he had seen, he experimented with frozen foods on his return to New York. In 1924 he founded the General Seafoods Company, having patented a system that packed fish, meat or vegetables into waxed-cardboard cartons that were flash-frozen between two refrigerated metal plates under high pressure.

Five years later he began selling his quick-frozen foods, which made him wealthy. He introduced refrigerated glass

1924 First transatlantic crossing by an airship the Zeppelin ZR-3.

display cases in 1934 and leased refrigerated railway boxcars to distribute his frozen products throughout America.

Birdseye's invention significantly changed eating habits, but he was not a one-trick pony. He held nearly 300 patents, covering infrared heat lamps, a recoilless harpoon gun for whaling, and a method of removing water from foods.

Preserving food by freezing was an age-old practice in northern Canada

Television

1925

John Logie Baird

I n the early 1920s the name "television" might have been new, but the idea was not. Radiovision, phototelegraphy, telectroscopy, telephonoscope, audiovision, radio movies, the radio kinema, radioscope, lustreer, farscope, optiphone and mirascope were some of the many aliases for a dream chased by experimenters and inventors around the world. In the end, the word television caught on, prompting C. P. Scott, editor of the *Manchester Guardian* to remark: "Television? The word is half Greek and half Latin. No good will come of it." But by the time the century ended, television had developed far beyond the predictions of Scott and the wildest dreams of its innovators.

Most of the component parts of television had been discovered or invented by the late 1890s, including the discovery by Henri Becquerel that light can be changed into electricity, the invention by Paul Nipkow of a device to scan the image and Ferdinand Braun's invention of the cathode ray tube. But the invention of the amplifying triode by Lee De Forest in 1906 was the final piece in the jigsaw. Without it, the signals produced by early experimenters were far too weak to create a picture.

By the mid-1920s there had been more than 50 serious proposals for television. The competition was truly international, with inventors working in Russia, Hungary, France, Britain, America, Germany and Japan. Many had well-equipped laboratories and sufficient funds for staff and equipment, so it is surprising that success was snatched by a most unlikely figure.

John Logie Baird was born in 1888 near Glasgow. He already had a string of business ventures behind him when, at the age of 34, he embarked on his quest to invent television. Baird had narrowly failed to invent a number of items,

John Logie Baird put on his first demonstration of a television in Selfridges department store in London.

John Logie Baird and the large spinning disc that he used to scan the image to be transmitted.

including a process for the manufacture of industrial diamonds and the air-soled shoe. His experiments ranged from a disastrous home-made haemorrhoid cream to a rustless glass razor (with which he badly cut himself before abandoning the project). There were, however, some successes and he had capital left over when he eventually sold his businesses selling socks and soap.

Baird rented modest premises and set about his experiments in television. This was the start of a passion which drove him for the rest of his life. Dogged by ill health, constantly short of funds and usually working alone, he improvized his apparatus from scrap materials. At the heart of his system was a large spinning disc through which the image was scanned and broken down into horizontal lines. Invented by Paul Nipkow in 1883, the disc had a spiral of holes around the circumference through which light passed into a photocell. To increase the sensitivity of this system, Baird replaced the small holes with lenses to let more light through.

Desperate to increase the sensitivity further, he designed larger discs with wider lenses, culminating in one disc nearly three metres in diameter with 20 cm lenses. Spinning at 750 rpm, the heavy

A General Electric console television from 1939 with a nine-inch (22-cm) picture tube.

lenses flew off, bursting into shards of glass as they hit the walls, leaving the unbalanced disc leaping around the room as it destroyed itself. Even so, by early 1925 Baird was able to mount the world's first demonstration of television to the public.

Selfridges department store in Oxford Street, London, was the unlikely setting for these demonstrations. Shoppers were treated to "a recognizable, if rather blurred, image of simple forms such as letters printed in white on a black card".

Meanwhile, other inventors across the world had not been idle. Charles Francis Jenkins was born in Ohio in 1867 and, like Baird, he had grown up with a passion for all things mechanical. In his late twenties, he turned his attention to the newly invented Edison Kinetoscope – a peep-show film viewer that produced a jerky and blurred moving-picture sequence. Jenkins built a much-improved version, which he called the "Projecting Phantoscope". This was the world's first practical movie projector and was so good that it was promptly stolen from Jenkins's workshop, copied and sold around the world as the "Edison Vitascope".

Jenkins was a prolific inventor and by the time he turned his attention to television he had filed more than 400 patents. Rejecting the Nipkow disc favoured by others, Jenkins devised the "prismatic-ring" to scan the image. As this wedge-shaped glass ring turned, it varied the angle at which light was bent as it passed through. Throughout the 1920s Jenkins had some success but his system, like Baird's, was doomed to fail with the coming of electronic television.

The champions of electronic television were Vladimir Zworykin at RCA, Isaac Shoenberg's team at EMI and a 14-year-old farm boy from Idaho called Philo T. Farnsworth who conceived the basic requirements for electronic television while still at high school. It took him six years to take his system from first ideas to prototype and by 1929 he was the first to demonstrate a complete electronic television system with no moving parts. But over the next 10 years, despite a number of stunning successes with his "Image Dissector" camera, he was overtaken by other inventors and faded from the scene.

Vladimir Zworykin, who had fled the Russian revolution, once designed a car radio, but the local police chief thought it

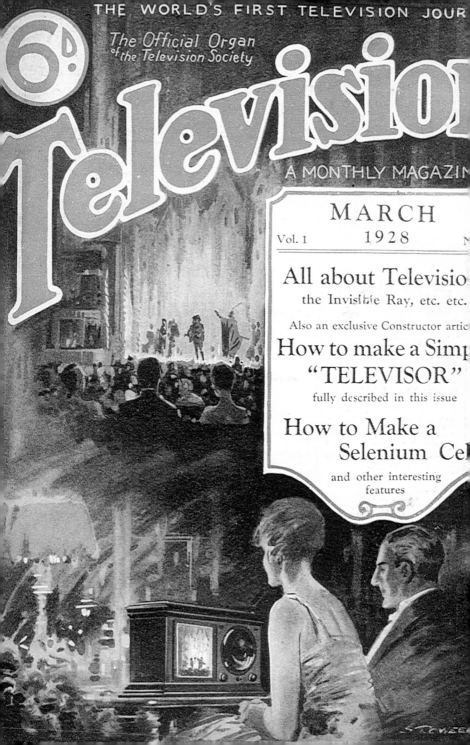

would distract the driver and promptly put him out of business. Later, working for RCA, he developed an electronic receiver tube he called "the Kinescope" and an electronic camera tube called the iconoscope. This heralded the true start of modern television.

While Zworykin was still developing his system, a team of inventors at the EMI factory at Hayes, Middlesex, led by Isaac Shoenberg, turned their thoughts to television. By the mid-1930s they, too, had developed a version of the iconoscope. Called the "Emitron", this was the world's first practicable television camera.

The outbreak of the Second World War suspended most television operations in Europe, although Germany continued regular transmission of programmes. In the post-war period, the tremendous developments in electronics, radio and radar were embraced by television-set makers who turned out cheap, reliable sets.

By the 1950s television had become so popular that it threatened the future of the cinema. Manufacturers vied with each other to corner the market, offering sets with big magnifying glasses on the front to enlarge the picture, tinted screens, portable sets and more. In a taste of things to come, the Philco Predicta had a screen that could be detached on a long umbilical cable leaving the receiver box next to your armchair. The remote control television had arrived.

Since then, colour television and video recording have been perfected and 3D-television has been tried. All these developments were demonstrated by Baird back in 1927. Indeed today's television system is fundamentally the same as the one developed at EMI in the 1930s when Isaac Shoenberg turned to his team and said, "Well, gentlemen, you have now invented the biggest time-waster of all time. Use it well."

Opposite: Television brings entertainment into the home.

Geiger-Müller counter

Geiger and Müller

1925

Many who grew up during the Cold War will be familiar with a nagging feeling of doom whenever the crackling, insistent din of a Geiger counter filters over the radio or television. It was the soundtrack of the nuclear paranoia that enveloped much of the world in the post-war years.

The Geiger counter is, in fact, an extremely useful measuring device – able to scan the environment for background radiation of many different kinds. The brilliant German physicist, Johannes Wilhelm Geiger, gave his name to a machine he devised to prove his own ideas about atomic science.

On finishing his PhD in 1906 at the University of Erlangen in Germany, he went to work for Ernest Rutherford at the University of Manchester. Rutherford was engaged in his pioneering work about the nature of nuclear matter.

Geiger thought that gases would conduct electricity only when their atoms had been ionized – or split up into free electrons or ions. Radioactive materials emit such free electrons and ions and, in Manchester, he set out to build a machine capable of sensing these particles.

Rutherford and Geiger's work revealed more of the nature of the atom.

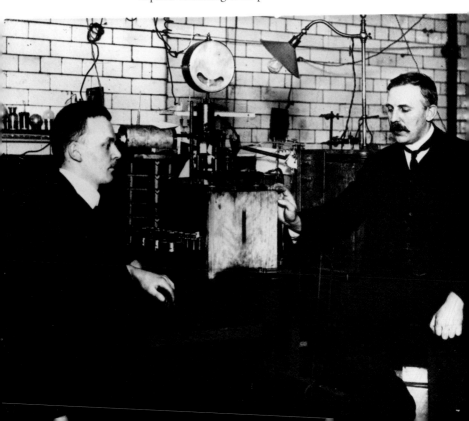

At the heart of the first counter, completed around 1911, Geiger had a small metal tube containing a thin wire between two electrodes. The assembly was sealed into a gas-filled glass bulb. A high-voltage current passed through the wire between the two electrodes. Free ions or electrons passed into the tube and interfered with the ionized gas around the wire, creating pulses of current. These were recorded or used to drive a loudspeaker, hence the familiar sound of the Geiger counter.

This first version of the Geiger counter was used to identify alpha particles as the nuclei of helium atoms, a vital step in Rutherford's determination of the nature of the atom.

The next year, Geiger returned to Germany, developing the counter with another physicist, Walther Müller.

In 1925, the pair produced an instrument capable of detecting a wide range of radiation and particles, such as electrons and ionizing electromagnetic photons, the matter-penetrating radiation that provokes such fears at the tick-tick sound of the counter.

Pop-up toaster
America

1926

Bread has been toasted to prolong its edible life since Roman times – *tostum* is Latin for roast or burn – but, for thousands of years, toasting was accomplished simply by holding the bread close to a flame, often with a fork or tongs.

The first electric toaster, invented in 1893 by Crompton and Co, a British company, became possible when current-resisting elements from electric fires and irons were built into a unit with a wire cage that could sit safely on the table. Its appeal was immediate.

The first refinement came in 1918, when sprung doors were added. Provided they were opened with a snappy movement, these doors turned over the toast automatically, but still required a watchful eye to prevent the toast burning. That changed in 1926, with the pop-up Toastmaster, an American invention.

The 1930s saw an eccentric British version ejecting the toast sideways and pop-up toasters did not become popular

Burnt toast ruined many a breakfast until the arrival of the pop-up toaster, an invention to rival that of sliced bread.

in Britain until Morphy-Richards introduced a model in 1948.

Equally important was the invention of sliced bread in 1928 by Otto Frederick Rohwedder, an American jeweller who made a machine that wrapped bread immediately after slicing it, keeping it moist and fresh.

Liquid-fuel rocket

Robert Goddard

1926

Born in 1882 in the small city of Worcester in Massachusetts, Robert Goddard was inspired by Jules Verne and H.G. Wells about flying to the moon. He developed an instinctive feel for pyrotechnics and was intrigued by the black powders that provided the chemical violence for TNT. But as professor at Worcester Clark University, he discovered that to launch a rocket any distance would need more sheer power than conventional solid fuels could afford.

If the fuel was to explosively expel itself from a rocket tail at a high enough speed to push the rocket forward (the Newtonian basis for all of rocket science) it needed energy to spare, with more fuel to sustain flight.

Goddard knew he needed the thrust of a vast amount of fast-burning fuel to overcome the weight of a moon rocket of any reasonable size. Static solid fuels were hard to control and lacked efficiency, but a liquid fuel, such as hydrogen, might be a different matter. He believed that if hydrogen could be piped into a combustion chamber sufficiently fast and burnt with liquid oxygen, enough power would be produced to get to the moon.

He used solid-fuel rockets to refine his techniques and prove basic principles. By 1915 he had taken solid-fuel rockets to their limits by carrying simple instruments to investigate the atmosphere. At the same time, for the war effort, he invented the bazooka, which was taken up by the US Army shortly before the end of the war.

Goddard first published his work in a 1920 report to the Smithsonian Institute entitled *A Method of Reaching Extreme Altitude*. It concluded that the best way to test rockets was to fly them to the moon and to use explosive flash-powder to show their arrival.

The paper was picked up by the *New York Times*, which gave him a pasting. As anyone knew, the January 13 editorial explained, travel to the moon was impossible, since without an . atmosphere to push against, a rocket could not move so much as an inch, and Goddard lacked "the knowledge ladled out daily in high schools".

Robert Goddard forced the *New York Times* to eat its words about the possibilities of space travel.

Robert Goddard (on the left) checks fuel pumps on a rocket at his base in Roswell, New Mexico, in 1940. His ideas were taken up by the Germans and used in their V-2 rockets against Britain (following page).

Goddard was furious and he immediately embarked on a quarter-century of research. It began with a first test flight on March 16 1926. Having solved the problems of fuel delivery to the combustion chamber and stabilizing the rocket's course, he built a top-heavy three-metre missile called Nell. A team of four horses wheeled it out to his aunt Effie's farm where his wife took notes. An assistant lit the oxygen-gasoline mix using a blowtorch on a long stick, and for a short moment the rocket did nothing. Then it leapt, screaming, into the sky at 60 mph, climbing to a modest 14 metres before crashing in a cabbage patch.

Still smarting from the press attacks, Goddard kept quiet about his trials, but the launch of a fourth rocket brought attention. "Moon rocket misses target by 238,799 miles" read one headline.

More importantly, however, Charles Lindbergh, the transatlantic aviator, was impressed, and he introduced Goddard to Harry Guggenheim, the millionaire financier who provided sufficient funds for Goddard to move to Roswell, New Mexico, where for nine years in arid scrubland Goddard continued his research in great privacy.

Few took any interest in his work, but now and then German engineers would contact him with technical questions. In 1939, with the outbreak of war, Goddard became concerned and he contacted the US Army to show

Opposite: The Saturn V rocket takes the first men to the moon, though the inventor of its means of propulsion, Robert Goddard, did not live to see it. The rocket blasted off with enough power to light up New York City for 75 minutes, and would reach 25,000 mph to escape the Earth's pull.

1926 Aerosol sprays invented by Norwegian scientist Erik Rotheim. In 1943 they were first used as insect repellents by Americans in the Pacific.

Moisture-proof cellophane invented by DuPont chemists and used to seal cigarette packets.

them movies of the Nell rockets and their potential for destruction. But the Army paid no attention to his warnings. Five years later, the first German V-2 rocket blasted off for London.

When Goddard finally had a chance to examine a V-2, he recognized his own handiwork. The Nazis had taken his papers, read his 200 patent applications and twisted his peaceable Nells into weapons of war.

Goddard died in 1945, never seeing his dreams of space travel come true. His work was taken up by American and émigré German scientists to produce the Redstone rockets that sent the first Americans into space and, in 1969, placed man on the moon. It prompted an apology from the *New York Times*: "Further investigation and experimentation have confirmed the findings of Isaac Newton in the 17th-century, and it is now definitely established that a rocket can function in a vacuum as well as the atmosphere. The *Times* regrets the error."

Talkies

1927

America

S ound is as much a part of cinema's beginnings as pictures. Attempts to combine the two were made from the 1890s, but these experiments involved actors or musicians miming to previously recorded soundtracks rather than the simultaneous recording of pictures and sound.

A recently-restored 20-second section of a film made around 1895 for Thomas Edison, who wanted to invent motion pictures so he could "do for the eye what the phonograph does for the ear", provides intriguing evidence that Edison might have been first to record pictures and sound simultaneously. It shows Edison's British assistant, W. K. L. Dickson, playing a violin into a huge phonograph horn as two men dance, but we do not know if Dickson was miming or being recorded.

Simultaneous filming and recording became feasible only when camera and gramophone motors could be run at exactly the same speed or sound could be recorded on the film itself. The breakthrough came in the 1920s when electrical recording

1927 Buckminster Fuller invents his Dymaxion Dwelling, a prefabricated six-sided house.

Charles Lindbergh crosses the Atlantic by aeroplane.

technologies superseded acoustic recording methods. Western Electric Company synchronized sound with picture by using self-synchronous motors to drive camera and recorder. This led to the first commercial sound system, Vitaphone, which was adopted by Warner Brothers in 1925.

Each reel of a Vitaphone film came with a 16-inch (40-cm) disc that was played on a turntable linked to the projector. The projectionist aligned start points on the disc and film, and then ran the two in synchronization. On October 6 1927, the public was first treated to this invention when Al Jolson's catchphrase 'You ain't heard nuthin' yet' became the first words spoken in a feature film. But even as they were heard at *The Jazz Singer's* premiere, another process was waiting to replace Vitaphone.

Warner's rival, the Fox Film Corporation, had a system that recorded sound on the film, alongside the pictures. The Movietone system was based on work by Lee De Forest and Theodore Case, who had developed light valves that glowed in response to microphone signals. Light from the valve, passing through a slit, produced an optical track along the film as it ran through the camera. During projection, light was focused through this soundtrack on to a photoelectric cell to produce variations in electrical current that the loudspeakers converted into sound. Several test films were made in 1924 using the Case and De

Forest Phonofilm system, among which is the first true "talkie" – a speech by President Coolidge.

Fox used the Movietone system for newsreels as well as features and in March 1929 became the first studio to abandon silent films. Audiences responded enthusiastically to the 'talkies' though they brought disaster for some in the industry. The careers of stars with heavy accents or poor voices ended, as did the livelihoods of thousands of musicians who had accompanied the silent movies in cinemas.

Sound also changed how films looked. In the silent days, the camera dominated the studio; now, it was incarcerated in a sound-proof booth to prevent its whirring being heard by the microphone. Talkies resembled filmed plays until the

Al Jolson and Mae Clark in 'The Jazz Singer'. the first feature film with spoken words. It was recorded on Vitaphone and each reel came with a disc that had to be played on a turntable linked to the projector.

blimp, a sound-proof cover for the camera, was developed. Some directors immediately seized the opportunity to use sound creatively, like Hitchcock in a scene in his 1929 thriller *Blackmail*, where the word "knife" was emphasized every time it was spoken, to convey the murderer's sense of guilt.

For 20 years, cinema sound quality scarcely altered, but behind the scenes engineers worked on improvements. In 1940, Walt Disney premiered *Fantasia*. Audiences had never seen such animation, nor heard such sound: it was stereophonic.

A further development occurred in 1952, when Cinerama appeared with its wide, deeply-curved screen. RCA had brought out the magnetic film recorder in 1948 and Cinerama's sound engineer, Hazard Reeves, exploited it to the full with seven separate tracks fed to speakers behind the screen and surrounding the audience. Other widescreen formats soon appeared, such as Fox's CinemaScope with stereo sound for *The Robe* in 1953 and the first of the 70 mm formats, Todd-AO, with six magnetic tracks for *Oklahoma!* in 1955. The era of large-scale epics had begun.

By the 1970s cinema was losing its battle with television and most local cinemas had reverted to monophonic sound. Audiences heard better sound on their home hi-fis. A new talent, Ray Dolby, applied noise-reduction techniques, similar to those he had invented for the compact cassette, to optical sound tracks. He widened the dynamic range to match the best magnetic tracks, and in 1975 introduced Dolby Stereo. This was encoded with sound information for centre screen and surround speakers.

In the 1980s, as cinema's fortunes improved, the race was on to produce a digital optical system. Kodak's Cinema Digital Sound in 1990 proved unreliable. Dolby's system, first used for *Batman Returns* in 1992, placed the digital soundtrack between the film's perforations, leaving the analogue track to be printed alongside as normal so the same print can be used in both digital and analogue cinemas.

In future, pictures will be projected from a computer disc using a high-definition video projector. Attention is focused on how pictures are generated and projected, with sound almost being taken for granted.

Videophone

Bell Laboratories

Despite its obvious appeal, the videophone has failed to catch on significantly since it was invented in 1927 – because videophones have to be bought in pairs for the caller to be able to see the person they are calling and, even then, the pictures are jerky and fuzzy.

The American telecommunications company AT&T demonstrated the first experimental videophone, but it worked in only one direction. A two-way system was unveiled in 1930, linking AT&T's head office and its research department, Bell Laboratories, both in New York.

In 1956 Bell demonstrated its "Picturephone" system, but it needed up to 125 telephone circuits to produce a reasonable picture. By 1968 Bell had perfected a system that required a relatively narrow bandwidth and it went into service in 1971, but was eventually discontinued.

It wasn't until the early 1990s that phones with moving pictures became a practical reality, the key development being digital image processing.

Quartz timekeeping

Warren Marrison

Wearing a watch on the wrist became fashionable for women following the late 19th-century craze for cycling, when leather wrist "converters" were designed so that women could securely hold their small fob watch on the wrist. The fashion soon outgrew the need of an accompanying bicycle and spread to men, who overcame their initial hostility to wristwatches. Soldiers found them useful in the First World War.

In 1925, Hans Wilsdorf, a German-born English national living in Switzerland, purchased a Swiss patent for a waterproof winding button, and a year later the first successful waterproof watch, the Rolex Oyster, was born. Two years earlier, John Harwood, an Englishman, patented a system of automatic winding that led, in 1928, to a successful automatic wristwatch.

However, the most significant development in modern watchmaking happened in 1927, when Warren Marrison, a Canadian-born engineer working at the Bell Telephone Laboratories in America, built the first quartz-crystal controlled clock. The quartz in the Marrison clock vibrated 50,000 times a second; by using a combination of electronic and mechanical gearing at a ratio of three million to one, this oscillation was used to move the second-hand through one rotation every 60 seconds. Capable in observatory conditions of losing only one second in 10 years, quartz technology produced a leap in accuracy, but needed the corner of a small room to house the workings.

Research into electric-powered watches began in the 1940s and the Hamilton Watch Company of America launched the first, the Ventura, in 1957. It was a tremendous success and other manufacturers in France and Switzerland soon followed. These new electric watches were still part mechanical and culminated in the first quartz wristwatch. The first 100, made by Seiko, went on sale on Christmas Day 1969. Advances in miniaturization technology allowed the quartz workings to be squeezed into a watchcase, a feat that cost the consumer $1,250 a piece at the time, the same as a medium-sized car. Within 20 years quartz watches would be given away free at petrol stations with a tank of petrol.

In 1972, the Hamilton Company developed the first all-electronic wristwatch, the Pulsar. It had flashing red light emitting diode (LED) digits in the place of hands on its dial. Its main disadvantage was that a button had to be pressed to read the time.

By 1977, permanently visible liquid crystal display (LCD) had become the most popular way of showing the time. Sufficiently powerful small batteries were a big problem with the early models and much research went into their improvement. Watches with small electric generators powered by wrist movement were invented, as were watches powered by body heat. Solar-powered watches appeared in 1973, and battery design has improved so that many can run for two years or more.

A 1970s Lasser digital watch, forerunner of the electronic digital watch.

In 1990, Junghans produced a wristwatch controlled by super-accurate caesium atomic clocks in Germany, England and America. It is accurate to one second in a million years, as long as the watch can receive a radio signal from the atomic clock.

Antibiotics

1928

Alexander Fleming

One of the most important advances ever made in medicine began with a Monday morning clean-up in a hospital laboratory in central London.

On September 3 1928, Alexander Fleming, a professor at St Mary's Hospital Medical School, was sorting through a heap of glass plates coated with staphylococcus bacteria as part of a research project. Fleming noticed one of the plates had mould on it: spores of *Penicillium notatum* had landed on the plate. "That's funny," said Fleming to a colleague, pointing to a strange ring around the mould that seemed to be free of staphylococcus.

Alexander Fleming made his life-saving discovery when he was tidying up the laboratory one day.

Many scientists would probably have thrown the plate away. But Fleming had a long-term interest in ways of killing bacteria, which are responsible for a vast range of diseases from food poisoning to the Black Death. The bacteria-free region around the *Penicillium* suggested it was exuding some substance that was killing the staphylococcus.

Fleming made a careful drawing of what he saw and then set about trying to find out more about the mystery substance. Filtering some of the mould, he discovered that it killed some other bacteria apart from staphylococcus, and could be given to mice and rabbits without side-effects.

However, Fleming's investigation of the mould's medical potential for treating human diseases was not encouraging, and within a year he had moved on to other things. A decade passed before Howard Florey and Ernst Chain, an Australian pathologist and a German biochemist working together at Oxford University, made the key breakthrough of isolating the bacterium-killing substance produced by the mould: penicillin.

The Petri dish, used for the culture of bacteria, is among those Fleming used to illustrate his discoveries.

In 1941, Charles Fletcher, a doctor at Oxford's Radcliffe Infirmary, happened to hear about their work, and told them about one of his patients, a policeman named Albert Alexander who had scratched his face on a rosebush. Both streptococci and staphylococci had invaded Alexander's body, and he was close to death. Fletcher gave Alexander 0.2 grams of Florey and Chain's penicillin, followed by smaller doses every three hours. Within a few days, the patient underwent an almost miraculous recovery and his wounds began to heal up. Tragically, Fletcher did not have enough penicillin to finally rid his patient of infection, and the bacteria fought back, killing the patient a few weeks later.

Even so, it was an impressive demonstration of the powers of penicillin. The problem now was to mass-produce the substance so that more patients could benefit. Florey succeeded in persuading American drug companies to work on the problem, and by the time of the Normandy landings in

Alexander Fleming as
a student.

1944 there was enough of the "wonder drug" to treat all the severe cases of bacterial infection that broke out among the troops.

For their brilliant work in turning a chance discovery into a life-saving drug, Florey and Chain shared the 1945 Nobel Prize for Medicine with Fleming. Today, almost half a century later, penicillin-based antibiotics are still the safest and one of the most widely used in medicine.

Iron Lung
Philip Drinker

1928

For centuries the polio virus was a feared killer, especially of young children. Ingested via contaminated food or water, the virus targets the nerves in the spine responsible for movement. Paralysis can follow within hours, and if the virus reaches the nerves responsible for respiration, victims are unable to breathe, and slowly drown on their own secretions.

Doctors had long recognized that they might be able to spare many polio victims this terrible fate if a way could be found of keeping the lungs working until the nerves recovered from infection.

Iron lungs have saved the lives of thousands since the 1920s.

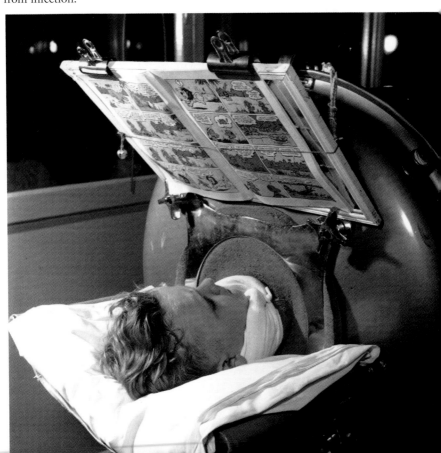

In 1926 the Rockefeller Institute in America set up a commission to investigate this possibility. One of its members, Harvard physician Philip Drinker, had an idea after talking with his brother, who was studying the respiration process in cats. To measure the volume of air breathed in, Drinker's brother had devised a sealed box, the air pressure within being recorded. As the cat inhaled, its extra volume caused a pressure rise, and vice versa.

Drinker realized that by flipping this process around – using pressure to control lung-volume – he could maintain respiration in paralyzed polio victims. He had a craftsman build a sealable box, with the air pressure being controlled by a vacuum cleaner. In 1928, the first patient, an eight-year-old girl on the brink of death from respiratory failure, was put in the box. Within a few minutes she was conscious, and asking for ice cream.

The girl later died of pneumonia, but the principle had been proved. Drinker's machine, dubbed the "iron lung", went into production in 1931. Iron lungs went on to save the lives of countless thousands. Today sophisticated portable ventilators have largely replaced them, but some still remain in use.

Artificial life

John Bernal

1929

The Irish crystallographer John Bernal anticipated in 1929 the possibility of machines with a lifelike ability to reproduce themselves. He wrote of this "postbiological future" in *The World, The Flesh and The Devil:* "To make life itself will be only a preliminary stage. The mere making of life would only be important if we intended to allow it to evolve."

Two decades after Bernal's vision, the great computer pioneer John von Neumann performed the first demonstration of the possibility of artificial life, in his work on self-reproducing automata. Though he conceived his automaton some years before the structure of the genetic blueprint (DNA) had been unravelled, he laid stress on its ability to evolve.

Modern computers have brought von Neumann's realization of the logical nature of life closer to reality. Some examples of weak "ALife" forms were born during Core Wars, a game played by computer addicts. The idea was to create programs that compete against each other for processing time and space in computer-memory – rather like animals competing for food and territory. Now there are unintended versions of Core Wars, better known as computer viruses, existing in computers around the world.

Core Wars programs and computer viruses are capable of self-replicating, but not of evolving in an open-ended f ashion. For there to be any chance of creating genuine artificial life within a computer, ways must be found to introduce novelty by mutations and then to select the "fittest" objects.

Thomas Ray, while at the University of Delaware, was thought to have created the first example of artificial evolution, based on Darwinian principles, in 1990. His computer contained digital organisms which fought for memory space in the central processing unit just as living organisms might fight for food and energy sources.

Ray's simulation of evolution, which he called *Tierra*, provided a metaphor for biological complexity, a tool to understand why it is seething with diversity. Computer scientists now "breed" software along genetic models, with the hope that the most efficient software will evolve. And there have even been computer games, such as Creatures, that exploit "ALife" ideas.

The field took a new direction recently when Dr Craig Venter of the Institute for Genomic Research, Rockville, Maryland, reported that he had whittled down the smallest known genetic code – of a simple bacterium – to a bare minimum set of essential genes.

It might even be possible, said Venter, to build the cell from scratch, making a chromosome and adding cell ingredients to see if it "gets its act together and starts replicating."

Jet engine

Frank Whittle

1930

1930 Mechanical toothbrush invented, using water pressure to turn brush head.

Scotch tape, the first sticky tape, is invented by 3M engineer Richard Drew to help a mechanic two-tone spray-paint a car.

1931 Jacob Schick, a retired U.S. army colonel, designs the electric razor.

The success of the steam turbine in achieving direct rotary action and thereby giving a new lease of life to the steam engine at the end of the 19th-century encouraged inventors working with the reciprocating internal combustion engine to believe that a comparable technology could be found in that field.

The design problems, however, were formidable as they required the construction of a chamber capable of maintaining continuous combustion at a very high temperature, and a satisfactory technique for securing the compression of the fuel-air mix. Little progress was made with gas-turbine designs during the 1920s, although the young Frank Whittle, then an apprentice with the Royal Air Force, produced a design regarded as promising. The Air Ministry was not convinced and rejected the idea, but Whittle went ahead and took out a patent for his "turbojet engine" in 1930.

The increasing international tensions of the 1930s brought a change of heart in official circles and in 1936 Whittle was encouraged to resume work on the jet engine. He had been sent to the University of Cambridge to take a degree and do postgraduate research, but left to set up a company called Power Jets Ltd. to pursue the development of a viable engine. With new alloys available to construct the combustion chambers, Whittle produced the first engine to be bench-tested, in 1937.

With 10 combustion chambers and a turbine to drive the compressor, the engine produced thrust in direct reaction to the stream of hot gases that it generated. The results were impressive, and two years later an experimental aircraft was ordered from Gloster Aircraft Company. This flew on May 15 1941, and the Gloster Meteor, which was developed from the prototype, entered service with the RAF in 1944.

Meanwhile, a parallel development occurred in Germany, where Hans Joachim Pabst von Ohain took out a patent for a jet engine in ignorance of Whittle's work in 1935 and was recruited by Heinkel the next year to build a jet aircraft for the Luftwaffe. The result was the Heinkel He 178, which first flew on August 24 1939, thus beating Whittle's machine into the air.

Frank Whittle took out a patent for his turbo-jet engine in 1930.

Other German companies developed the design further, and by the end of the European war in 1945, several types of jet aircraft were in combat service, although not enough to make a difference to the outcome of the war. Ohain emigrated to the United States, where he worked for the Air Force. Whittle, on the other hand, was frustrated by what he took to be government indifference to his idea once the conflict was over.

Even though the jet engine made little impact on the Second World War, it had developed sufficiently to be available for wider applications after the war was over. Outstanding among such applications was the production of civil aircraft. The new power units radically transformed the experience of flying, giving longer and quicker flights to larger and more robust aircraft, so that there was a tremendous expansion in the number of people using these services. The De Haviland Comet made a brilliant start in jet transport in 1952, but failed disastrously two years later with a series of tragic accidents due to unanticipated airframe weakness.

American manufacturers then took a commanding lead in jet airliners, and the first Boeing 707 entered service in 1958. With its facility for making distant places accessible to a large public, the jet engine is one of the great life-transforming inventions.

Radio Telescope

1932

Karl Jansky

1932 Carlton Magee invents the parking meter, installed in 1935 in Oklahoma City.

Harry Jennings invents the folding wheelchair for his friend Herbert Everest, a paraplegic, following a mining accident.

In 1932 an American radio engineer, Karl Jansky, was investigating odd interference on transoceanic telephone lines when he received a shock: he encountered a signal, the origin of which he was at a loss to explain. Eventually he found out: the signal was coming from outer space.

His explanation for this phenomenon was that it was being created in space by the interaction of stray electrons and ions. As far as Jansky was concerned, the signals, which sounded like radio static, were emanating from the centre of our own galaxy.

Astronomers the world over were galvanized into action by the findings. In 1937 Grote Reber, an American amateur astronomer, began building the first radio receiver aimed up at space. Reber confirmed Jansky's findings about the direction of the radio waves and he went on to map the whole sky, using his dish to pick up naturally occurring radio transmissions from stars, galaxies, quasars, and other astronomical objects. For six years it was the only purpose-built radio telescope.

But as Europe returned to peace in 1945 there was an upsurge of interest in radio astronomy. Resourceful astronomers in Britain pressed captured German radars and radio receivers into peacetime service. One of these experts was Bernard Lovell, Professor of Astronomy at Manchester University.

Lovell set up his first radio telescope in 1945 at Jodrell Bank, just south of Manchester. But this wasn't enough. He started planning a truly giant radio telescope, consisting of a steerable dish receiver 75 metres (250 feet) across. Construction of the Mark 1a began in 1952 and took five years.

Lovell's project soon hit the headlines in Britain. In October 1957 his team picked up a stream of bleeps that told the world the Russians had become the first nation to launch a satellite into orbit around the Earth.

In the 1960s radio astronomers began for the first time to build up detailed pictures of distant features in space, far beyond the reach of optical telescopes. Pulsars, quasars and exotic phenomena such as Black Holes were discovered.

The Manchester dish, now renamed the Lovell Telescope, is one of the biggest steerable receivers in the world. It is a vital part of the SETI project, a worldwide effort to listen out for signs of intelligent life elsewhere in the universe.

Electron microscopy
Ernst Ruska

1933

E ven scientists have trouble accepting some of the tenets of quantum mechanics, the laws of the sub-atomic world – Einstein couldn't bring himself to accept them all. Certainly one of the hardest to credit is that bullet-like particles such as electrons can sometimes behave as if they are waves, like light.

1933 Anglepoise lamp invented by George Carwadine, a car suspension designer from Bath, who sets out to build a lamp with the same versatility as the human arm.

Hard to believe, maybe, but the electron microscope proves it's true. Just as ordinary microscopes focus light waves to magnify images a thousand-fold, the electron focuses the wave-like properties of electrons to produce massive magni-fication. With wavelengths far shorter than those of ordinary light, electrons can reveal far more detail. A magnification of 50,000 times is routine.

It took less than 10 years for the idea of electrons behaving like waves – put forward by the French theorist Prince Louis de Broglie in 1924 – to be turned into a working electron microscope by Ernst Ruska, an engineer at the Technical University of Berlin.

Ruska began by making an electromagnetic "lens" that brought the electron waves into sharp focus. By 1933, he had found a way of stacking several of these lenses above each other and accelerating electrons to such high energies that their wavelength was around 100,000 times shorter than that of light. Shone through a very thin section of material, the electrons fell onto a photographic plate, where they left a highly magnified image of the material.

This first "transmission electron microscope" could magnify only a few hundred times, but its performance was quickly improved. By the 1960s, electron microscopes were achieving magnifications of several million times, enabling individual molecules to be seen.

But while electron microscopes are excellent for studying inanimate objects, they are useless for studying living, breathing things. Electrons have trouble passing through air, let alone biological tissue. Those vivid, apparently three-dimensional electron microscope images of fleas, bed bugs and the rest are taken by coating their lifeless carcasses with metal, and detecting the electrons reflected off them.

Insect spectacular: a common ant carrying a rose aphid, portrait courtesy of the electron microscope.

Catseyes
Percy Shaw

British roads in the 1920s were much more primitive than the smooth well-lit carriageways we expect today. They were mostly unlit: in urban areas moonlight shining off tramlines guided motorists at night – if there were any tramlines.

A young Yorkshire road contractor called Percy Shaw thought he could find a solution to the dangers of driving in the dark. His moment of inspiration came one night when he was driving home across country to Halifax. His headlights caught a cat on a fence and its eyes "lit up" as they reflected the dismal six-volt illumination back to Shaw. He at once realized that reflectors down at the level of the Tarmac could guide motorists safely home.

Shaw set to work, trying to create a reflector out of different combinations of glass lens, mirror and rubber cover to hold the assembly together. By 1934, still only 23 years old, he had perfected the first model, cleverly designed to self-clean the reflectors when the lenses were forced down into the road by a car passing over them.

Transport officials were slow to see the potential of Catseyes: not until the war years, when drivers had to use blacked-out headlights, was the device seen as invaluable. Catseyes were doubly useful in wartime blackout conditions because they reflected light back in a very narrow field of view; there was no danger of their attracting the attention of the Luftwaffe.

In 1947, during the post-war Labour government, junior Transport Minister James Callaghan instituted a massive plan to install Catseyes all over Britain's roads. The basic design is little changed today.

1935 Robert Watson-Watt, a Scottish physicist, develops radar.

Adolph Rickenbacher invents the electric guitar, using the pick-up invented in 1923 by Lloyd Loar.

Percy Shaw was driving home from Halifax one night when a cat crossed his path and gave him a rewarding idea.

Helicopter

Heinrich Focke

Leonardo da Vinci is often credited with the invention of the helicopter on the basis of his famous sketch of an "airscrew" machine from 1480. Hand-launched helicopter toys had, however, been in existence for several hundred years, probably originating in China. The earliest record of a rotary-wing craft dates from the 4th-century *AD*, in a book called *Pao Phu Tau*, which describes a flying car kept aloft by spinning blades.

Da Vinci's craft could never have flown, neither did the many and ingenious man-, steam- and even electrically-powered designs that appeared over the following 427 years.

On November 13 1907, a French bicycle dealer, Paul Cornu, became the first person to ascend vertically in a powered, man-carrying, rotary-wing aircraft. A 24-hp petrol engine powered his twin-bladed machine and it achieved an altitude of 1.8 metres. Unfortunately, it proved to be almost uncontrollable and was stabilized by men with sticks on the ground. It managed to say aloft for 20 seconds and was destroyed on landing.

A Spanish aeronaut and autogyro designer, Juan de la Cierva, solved one of the fundamental problems in helicopter design in 1923 with the invention of the hinged, flapping rotor-blade, giving the pilot control over both forward and vertical movement.

Juan de la Cierva in the front seat of his 'autogiro'. His articulated rotor-blade paved the way for Heinrich Focke (below) to build the first free-flight helicopter in 1936.

Cierva's invention, which he went on to produce in quantity for Britain and the US, paved the way for the first practical helicopter, the German Focke-Achgelis FA61. It was designed by Professor Heinrich Focke and made its first free flight on June 26 1936. The machine was based on the fuselage of a small

biplane trainer. Two outriggers supported the contra-rotating rotors, which were powered by a 160 hp engine. The FA15's impressive handling abilities were famously demonstrated by a German test pilot, Hanna Reitsch, in the Deutschlandhalle Stadium in Berlin; she claimed to have had only three hours' experience on the machine prior to the historic stadium flight.

Three years after the FA15, Igor Sikorsky, a Russian-born engineer who emigrated to the US in 1919, designed, built and flew the VS-300, using the now familiar single main lifting-rotor and boom-mounted tail-rotor layout.

Russian-born Igor Sikorsky's design went into production for the US Navy.

The purpose of the tail rotor was to counteract the torque of the main rotor and provide directional control. This was not a new idea; another Russian, Boris Yuriev, had proposed such a system in 1911. The VS-300 was powered by a four-cylinder 75-hp air-cooled engine, driving a three-blade main rotor, 8.53 metres in diameter. The craft was an open-cockpit design, built around a welded tubular frame chassis.

Igor Sikorsky in his
75-hp helicopter in
Bridgeport,
Connecticut, about to
make the first fully
controlled vertical
flight.

The first tethered flight of the VS-300, with Sikorsky at the
controls, took place in September 1939. The first free flight was
on May 13 1940 and with each successive ascent the prototype
established new flight and endurance records. On May 6 1941
Sikorsky created a new world record by keeping the
VS-300 in the air for 1 hour 32 minutes and 26 seconds. The
VS-300 underwent another year of development,
establishing the design for the Sikorsky R4, the world's first
production helicopter, which flew in January 1942
and went into operational service with the US Navy on
May 6 1943.

1938 Chester
Carlson, an
American scientist
and lawyer, invents
photocopying.

Ballpoint pen

Laszlo José Biró

There's a world of difference between having a great idea and getting it to work successfully. Take the ballpoint pen.

The "biro" was born in response to the frustrations of Laszlo José Biró, a Hungarian journalist who was all too familiar with the failings of conventional pens, with their leaks and smudges. However, the ink actually used to print his words, he noticed, dried quickly and was much less given to smudging.

Experimenting, Biró found that printer's ink clogged up a fountain pen, so he and his brother Georg developed a new type of nib: a ball-bearing arrangement with the ink acting as a lubricant. They patented it in 1938. Two years later, they fled Hungary for Argentina, and the first commercial version of their pen appeared in 1945, launched under licence by Eterpen of Buenos Aires.

Their next target was the US market, but they were beaten by Milton Reynolds, a Chicago businessman, whose pen design dodged patent problems. Marketed as the Reynolds Rocket pen, 8,000 of the $12.50 pens (equivalent to £70 today) were sold on the first day.

But it took innovative mass-manufacturing methods devised by Baron Bich of France to give the ballpoint its present ubiquity: around 14 million Bic ballpoint pens are sold every day.

1939 Viable form of cloud-seeding to trigger rainfall developed by Bergeron, a Swedish meteorologist.

Atomic Power

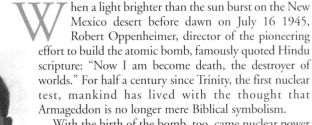

Enrico Fermi

Enrico Fermi construced the first nuclear reactor in a former squash court at the University of Chicago.

When a light brighter than the sun burst on the New Mexico desert before dawn on July 16 1945, Robert Oppenheimer, director of the pioneering effort to build the atomic bomb, famously quoted Hindu scripture: "Now I am become death, the destroyer of worlds." For half a century since Trinity, the first nuclear test, mankind has lived with the thought that Armageddon is no longer mere Biblical symbolism.

With the birth of the bomb, too, came nuclear power and the promise of electricity "too cheap to meter". But public confidence was shaken first by the 1979 accident at Three Mile Island in Pennsylvania, then by the 1986 explosion at Chernobyl in the USSR.

The Trinity test marked the culmination of an unprecedented undertaking called the Manhattan Project. An army of physicists led by the best brains in the field exploited advances made in the first decades of the century, when scientists came to understand that an atom consists of clouds of electrons swarming around a tiny nucleus of protons and neutrons.

The electrons carry negative electric charges and their behaviour dictates the chemistry of an element. Their numbers are balanced by protons in the nucleus of the atom, which carry positive charges, and which weigh roughly the same as the chargeless neutrons. All atoms of a given element contain the same number of protons; but different "isotopes" contain different numbers of neutrons.

The first leap forward had come in late 1938, when Otto Hahn and Fritz Strassmann bombarded uranium with neutrons at the Kaiser Wilhelm Institute for Chemistry in Berlin. Hahn was puzzled by the results and wrote to the Austrian chemist Lise Meitner, who had worked with his group in Berlin until she fled Germany in 1938. Within a few weeks, Meitner and her nephew, Otto Frisch, explained that the uranium nucleus was splitting into two roughly equal parts. Borrowing the term used by biologists to describe the behaviour of dividing cells, they called the nuclear process "fission".

The total mass of the products was less than the mass of the original nucleus. This missing mass was given off as an enormous quantity of energy, according to Einstein's famous 1905 equation that connects energy and mass: $E = mc^2$. In other words, energy equals mass multiplied by the speed of light squared (the latter being a vast number).

The neutrons given off during fission (when atoms come apart) could split more atoms in turn. The first to envisage this was Leo Szilard, a Hungarian theoretical physicist who had also fled Hitler's Germany. He had been struck by the thought of a self-sustaining chain reaction while looking at traffic lights on a road in Bloomsbury, London, one dull September morning in 1933.

Control the chain reaction and you could boil water to make steam and drive the turbines of a power plant. Allow it to run amok and you could create a bomb that would reduce cities to rubble. "There was very little doubt in my mind," Szilard recalled later, "that the world was headed for grief." Because fission had been first discovered in Germany, there were fears, notably among Szilard and other physicists who had fled to America, France, and Britain, that Nazi Germany might develop just such a bomb. President Franklin D. Roosevelt was persuaded by a letter from Albert Einstein to initiate at least a token effort devoted to this purpose.

The attitude of the US government changed rapidly a few months later when news arrived of the work of Rudolf Peierls and Otto Frisch, German physicists working in Britain. Their brief memo outlined how to produce uranium-235 industrially, how it could be used to make a bomb, and what the grim consequences of dropping it would be. Backed

The effects of the bomb on housing is shown in tests at 'Survival City' in Nevada.

by large amounts of money, the Manhattan Project was born.

The Project included reactor development under the supervision of the brilliant Enrico Fermi who, as the 25-year-old professor of theoretical physics at the University of Rome, had come close to discovering fission back in 1934. His Chicago Pile No 1, CP-1, was constructed in a makeshift laboratory in an unlikely venue, the squash court under the grandstands of Stagg Field Stadium at the University of Chicago.

The reactor consisted of uranium in a stack of graphite to make a cube-like frame of fissionable material. There were control rods of cadmium, a metallic element that absorbs neutrons. When the rods were pulled out, more neutrons were available to split atoms. The chain reaction sped up.

At 3.25 pm on December 2, 1942, shortly after Fermi had returned from his customary 11.30 am lunch, the first self-sustaining chain reaction was created by his team, showing it was possible to unlock atomic energy in a controlled way. The world had entered the nuclear age and Fermi allowed himself a grin. Szilard, who worked on CP-1, shook hands with Fermi and remarked: "This will go down as a black day in the history of mankind."

Four years after the discovery of fission, a laboratory was set up to harness it as a weapon of war at Los Alamos, which Oppenheimer had come across while on holiday. Many of the free world's finest physicists were recruited, not so much to develop new theory as to solve the detailed mathematics required to turn science into technology.

Oppenheimer (left) and General Groves of the Manhattan Project.

The familiar mushroom cloud of a nuclear bomb.

It had turned out that normal uranium was not suitable for use in making bombs. Most of it was an isotope called U 238, which had a tendency to absorb neutrons, making an explosive chain reaction impossible. But about 0.7 per cent of natural uranium was a lighter isotope, U 235, which fell apart when a neutron hit it. With uranium enriched in 235 you could make a bomb.

Plutonium, an element not found in nature and produced when uranium nuclei absorbed neutrons, was found to be fissile and also good for making bombs. By mid-1943, programmes were under way to breed plutonium at Hanford in Washington state, and sift U 235 from raw uranium at Oak Ridge in Tennessee. Each was expected to yield enough raw material for a bomb. Los Alamos was to design it.

The most primitive nuclear weapon created a critical mass by firing a slug of uranium 235 into a target of similarly enriched uranium. This approach was used in Little Boy, the

Calder Hall in Cumbria, Britain's first nuclear power station, opened in 1956.

bomb dropped on Hiroshima on August 6 1945. The team was so confident, that Little Boy was never tested. In any case, they lacked the uranium to make more than the one bomb.

The alternative, called Fat Man, was dropped on Nagasaki. The granddaddy of all today's weapons, this design contained a sphere of plutonium that is compressed to form a critical mass. First put forward by Seth Neddermeyer, a physicist from the National Bureau of Standards, this design could assemble the critical mass quicker than a gun.

But it was a headache figuring out how to use explosives to collapse the sphere in a neat way. It took the genius of John von Neumann, a dazzling Hungarian mathematician, to model the complicated physics. But a test was crucial. The site selected for Trinity was near Alamogordo, south of Los Alamos. The bomb was winched to the top of a tower and scientists in emplacements miles away took bets on the explosive yield. It was just shy of 19 kilotons.

Today, if you open up the "re-entry vehicle" of a warhead you will see a sphere of high explosive. When detonated, the shock waves compress an inner shell of plutonium until it is "supercritical" for a period of a millionth of a second or so.

Bomb scientists strive to avoid "pre-initiation", when the chain reaction starts before the plutonium is compressed to the optimum density, resulting in a fizzle rather than a bang. For this and safety reasons, bomb plutonium is so highly enriched that it contains little of the isotope that generates neutrons (plutonium 240). Thus there are relatively few neutrons to start the chain reaction.

Instead, when the critical mass is achieved, the chain reaction is kick-started with an "external neutron initiator", a tiny particle-accelerator – like a Van de Graaff generator – to bombard a target to generate neutrons. The explosive yield of this fission reaction is enough to get up to the 10 million °C temperature range in which nuclear fusion occurs, which is the opposite of fission and is the basis of the hydrogen bomb.

Pumping fusion fuel – deuterium and tritium, heavy isotopes of hydrogen, in gaseous form – into the heart of the weapon results in what is called a boosted device. At the high temperatures inside the A bomb – similar to those in our local fusion reactor, the sun – fusion can occur. Under ordinary conditions atomic nuclei would repel and it is only at these high temperatures that positively charged nuclei can gain enough speed to overcome their mutual repulsion to fuse.

To push the magnitude of the explosion into the megaton range, the primary stage is used to implode a secondary stage, which contains more fusion fuel. Edward Teller and Stanislaw Ulam and others developed the first hydrogen bomb, which was tested at Enewetak atoll on November 1 1952, in Operation Ivy. The USSR first tested a hydrogen bomb in 1953, followed by the United Kingdom in 1957.

Bouncing bomb

Barnes Wallis

The German anti-aircraft gunners on duty on the calm, clear night of May 16–17 1943 at the Mohne Dam in the Ruhr Valley might have been expecting a quiet watch, after all they had a quiet war so far. But the gunners were unlucky to be defending one of three dams chosen to be attacked by RAF Bomber Command.

The 19 Lancasters that were launched on the daring raid were carrying a unique weapon: the "bouncing bomb".

Barnes Wallis was the British engineer responsible for the imaginative plan. He was convinced that a devastating hammer blow could be dealt to the German war effort if the dams of the Ruhr Valley were destroyed and he conceived the idea of a skipping or bouncing weapon, able to leap over torpedo nets that protected the dams.

Barnes Wallis:
Churchill intervened
to give his bomb its
chance.

Using models at home and at the National Physical Laboratory at Teddington, Middlesex, he refined the idea further: the cylindrical bomb would be spun at up to 500 revolutions a minute just before release to make it bounce across the water. When it finally hit the dam, the spin that remained in the bomb would drive the explosive-packed cylinder down to the base of the dam, where the water would amplify the blast.

Weighing more than 4,320 kg, it could be carried only by the four-engined Avro Lancaster, which needed to fly at exactly 220 mph, at 20 metres above the water, releasing the bomb at 150 metres from the target.

It took the personal intervention of Winston Churchill to persuade the War Office to go ahead with the plan. As few Allied aircrews were at that time getting within five miles of their targets in night operations, a special squadron was handpicked from all over the Commonwealth. It was called 617 Squadron.

In a raid later celebrated in the film *The Dam Busters*, the attackers took off from RAF Scampton in Lincolnshire at 9.30 pm and flew at 30 metres or less across the North Sea towards the Dutch coast. In complete radio silence they screamed over rooftops toward their targets. Three dams were hit, but only two were breached.

Of the 19 Lancasters that left Scampton, 10 returned, 53 aircrew were lost, three were taken prisoner. Wing Commander Guy Gibson, leading the raid, made repeated passes over the dams giving covering fire as his fellow fliers made bomb runs. He was awarded the Victoria Cross; 35 other medals were awarded.

The raid, the first true precision air attack of the war, was an immense propaganda success. Vast tracts of the Ruhr Valley, the industrial heartland of Germany, were flooded. More than 1,300 were killed on the ground, including Ukrainian women and children in a prison camp downstream of the dams.

Aqualung

Jacques Cousteau

1943

The idea of using bottles of compressed air to breathe underwater dates back to at least 1825, when William James, an English engineer, sketched out the basic requirements of an aqualung.

The key problem lay in designing a valve capable of delivering air at the correct pressure to the diver on demand. This was the achievement in 1943 of Emile Gagnon, a Parisian engineer, and Jacques Cousteau, a French naval officer who then used the aqualung to show the wonders of the undersea world to TV audiences of millions.

Kidney dialysis

Willem Kolff

1944

Working in the municipal hospital of Kampen, in Holland, Willem Kolff, a medical doctor who had previously invented the blood bank, constructed the first, crude artificial kidney in 1944 from wooden drums, Cellophane tubing, and laundry tubs.

1944 Chromatography invented by A.J.P. Martin and R.L.M. Synge enabling chemical analysis of insulin.

The device drew blood, removed toxins that deficient kidneys were unable to excrete, and pumped the cleaned blood back into the patient. Working with comatose men and women, Kolff managed to prolong the lives of his first 15 patients for a few days.

In 1945, a woman Nazi collaborator was brought to him for treatment. Although many in the town wanted to see her dead, Kolff's "haemodialysis" apparatus saved her life. It enabled Kolff to continue developing his dialysis machine, which he shipped free to researchers in England and America after the war. Kolff went on to invent an implantable artificial heart.

Chapter 6
Re-inventing the world
1945 to 2000

Periods of invention can easily be categorized by the dominant technology – the age of the wheel, of steam, of industrialization, of electricity, of radio and so on. But the post-war 20th-century was remarkable because it was the age of many technologies simultaneously. The atomic age, the space age, the computer age, the information revolution, the genetics era – all these terms have been coined to describe the past 55 years.

The world underwent a remarkable acceleration in technological achievement after the Second World War. A second industrial revolution revolved around two key developments, both of which occurred at sub-atomic or molecular levels: the invention of the transistor in 1947 and the exploitation of the gene after the discovery of its structure in 1953.

By the end of the 20th-century, these two innovations resulted in technology's setting the agenda in most spheres of society. Breakthroughs in genetics and reproductive medicine have had a profound effect on our attitudes to fundamental issues. They meant we can now tinker with our own existence and that of other life forms. The moral and social ramifications of delaying the onset of menopause, changing the nature of foodstuffs, altering inherited characteristics or even cloning ourselves far exceed the technological complexities involved in their execution.

The consequences of the invention of the transistor in 1947 have been similarly profound. Stimulated by the demand during the Second World War for a much smaller and more reliable version of the thermionic valve, scientists at Bell Laboratories in New York unveiled a tiny scrap of germanium that, in housing the transistor, made possible many of the century's ensuing inventions.

As a result of the transistor and its evolution into the silicon microchip, we can now carry more computing power in our pockets than was collected in all of the room-sized computers across the world in 1950. The transistor and microchip made possible the mobile phone, the internet, desktop computers, space travel, colour television, the video recorder, satellites, portable calculators, MRI scanners and many other arithmetic-intensive applications, including the mapping of the human genome, our entire genetic make-up.

Throughout the latter half of the 20th-century, technology often stood accused of inflicting harm on humanity; certainly the harnessing of nuclear power was accompanied by the destructive power of the atomic bomb. But technology was also central to man's single greatest extension of his environment, the exploration of space. Without the liquid-fuel rocket, invented in 1926 by Robert Goddard, and the silicon microchip, man would not yet have walked on the Moon or gazed back through time using the Hubble space telescope.

Though invention is acknowledged as one of the most formative and persistently creative themes in our development, at the start of the 21st-century it is still regarded with suspicion, its contribution to our improved living standards weighed against the threat of technological self-destruction, ecological imbalance and silicon-chip hegemony. Certainly, if computers continue to increase in power at the current exponential rate (a doubling of power every 18 months, according to one silicon microchip pioneer), then by 2025 artificial intelligence systems could rival human intelligence.

It has thus become commonplace to blame inventors, scientists and engineers for many of society's ills. While this position rightly reflects the central role of science and technology in today's society, it often disregards the fact that scientific discovery and technological invention are merely tools in our hands, and their misuse a reflection of human frailty.

Invention is the one characteristic that sets us apart from other species yet it continues to be devalued in cultural terms. And as Christopher Cockerell, inventor of the hovercraft, famously remarked, "But for the silly chaps we would still be living in the Stone Age."

Microwave oven

Percy LeBaron Spencer

1946

1945 Fleming,
Florey and
Chain awarded the
Noble Prize for their
work on penicillin.

I n spring 1946, Dr Percy LeBaron Spencer, an engineer working on a radar-based research project for the Raytheon Corporation of America, felt something sticky in his pocket. It turned out to be a molten peanut bar. Spencer guessed that it had been affected by high-frequency radio emissions from a device called a magnetron, the key component in radar, which the British had developed during the Second World War. While working at the Massachusetts Institute of Technology, Spencer had played an important role in devising improved production methods for the device.

The heating effect of high-frequency radio waves had been documented by other scientists but dismissed as unworthy of investigation. (Stories of partially cooked birds found at the base of early radar installations had a basis in fact but appear

A Radarange, the first kind of microwave, developed from wartime radar technology. The chicken took two minutes and 20 seconds to cook

to be largely apocryphal.) Intrigued, Spencer immediately wanted to learn more, and he sent a boy to buy a packet of popcorn. When he placed it close to the magnetron, the popcorn exploded.

The next morning, Spencer cut a hole in the side of a kettle and mounted the mouth of the magnetron next to the hole. Into the kettle he placed a raw egg, still in its shell. He then switched on the power. When a curious colleague looked into the kettle to see what was happening, the hot egg exploded in his face.

Raytheon engineers immediately began work on Spencer's invention. The company filed a patent in 1946 and in the same year the first prototype micro-wave oven was built and installed for trials in a Boston restaurant.

The first commercial microwave ovens, known as the Radarange, were sold in 1947, for between $3,000 and $5,000 (£28,000 to £46,000 in today's money). They were the size of a refrigerator, almost 1.6 metres high, weighing more than 340 kg and requiring a cold-water supply to cool the magnetron.

Sales were slow at first but smaller, more efficient models with air-cooled magnetrons followed and, before long, microwaves were being used in a wide range of food processing applications.

Following the acquisition by Raytheon of Amana Refrigeration in 1965 the company produced the first domestic model, which went on sale for just under $500.

Automation

Henry Ford

1946

There is something mesmeric about watching lines of industrial robots swinging this way and that as they make things. But to some it's a sinister sight, seeing the work of thousands of people performed by machines that never need a tea break or crave a fortnight's holiday in the sun.

Automation was a natural progression from the application of science to manufacturing. The first major company to embrace automation was one synonymous with mass-production: Ford. By 1946, a significant part of Henry Ford's

1946 Tide, the
first
synthetic laundry
detergent, developed.

Louis Reard, a French
designer, launches a
two-piece swimsuit
named after Bikini
atoll when the
Americans had tested
an atom bomb four
days earlier.

Earl Silas Tupper, a
New Hanphire tree
surgeon, invents
Tupperware after
experimenting with
polyethylene.
Brownie Wise, a
single mother from
Detroit, becomes a
multi-millionaire
after instigating the
Tupperware party.

Detroit production was undertaken by automata, with engine blocks being made by machines that could automatically adjust themselves to the task in hand.

As electronic computers became more powerful in the early 1950s, so-called numerical control (NC) methods started to emerge, allowing machines to be programmed to deal with a range of tasks, rather than just one. The first robot, the Unimate, which had mechanical arms, was devised in 1956 and first used in an American die-casting operation in 1961.

Increased automation led to an interest in the idea of putting the whole process of manufacturing under computer control. The first to succeed was a company which specialized in the manufacture of cigarette-making equipment: the Molins Machine Company of Deptford, south London. Designed by a British production engineer Theo Williamson, the Molins System 24 began operation in 1964, and over the following 20 years various parts of this pioneering "flexible manufacturing system" (FMS) were taken up and improved upon by leading companies in both Britain and America.

Automation has since spread far beyond the factory floor, from "hole in the wall" bank machines (first demonstrated by Barclays in 1967) to the 1996 Sharp Logicook microwave cooker, which was "trained" to cook automatically whatever was put in it.

Much has been made of the impact of automation on jobs, and the failure to plough the resulting increased profits and tax revenues into retraining. But it is also true that the growth of automation has relieved many people of hard, boring and dangerous work.

Transistor

Bardeen, Brattain and Shockley

1947

1947 Chuck
Yeager
breaks the sound
barrier in the X-1
rocket aeroplane.

Transistors make it possible to squeeze into today's camcorders and mobile phones more electronic components than existed in the whole of Britain at the end of the Second World War.

By forming the fundamental switching components of a vast array of electronic equipment, the transistor made thousands of other inventions and products possible, from pocket calculators, computers and radios to industrial robots and satellites.

The transistor was invented in 1947 by three American physicists at the Bell Telephone Laboratories in New Jersey: John Bardeen, Walter Brattain and William Shockley.

The trio had been recruited by Bell to find a way out of the conundrum that the thermionic valve faced: it was an essential component in thousands of electronic devices, but the technology that it helped to advance came to require a more compact and reliable device.

The researchers examined the peculiar electronic characteristics of materials such as germanium and silicon, which, because they are neither electrical conductors nor resistors, are known as semiconductors. They discovered that some semiconductors comprised two distinct regions that conducted electricity in completely different ways. In the n-type region, the charge carriers were electrons; in p-type the charge carriers were positively charged "holes".

They also discovered that they could reproduce the regions at will, by adding certain impurities such as arsenic or boron to the silicon or germanium semiconductor.

The trio experimented with various set-ups including one in which, to Brattain's surprise, the amplification jumped dramatically after he dumped his contraption into a thermos of water. Bardeen, the theorist of the group, worked out they needed two contacts fractionally apart, resting on a sliver of germanium. Brattain, an ingenious engineer, built the contacts by folding gold foil over the point of a plastic triangle, which he sliced at its tip with a razor blade.

On Christmas Eve 1947 they invited their managers to witness the amplification of an electrical signal using this sliver of germanium; a small signal through one contact controlled a larger one through the other gold contact. With the help of John Pierce, a colleague who wrote science fiction in his spare time, the team came up with the name "transistor".

By 1953, hearing aids, the first commercial products using transistors, were on sale. Each transistor cost between £2 and £20 to make; now the transistors on a microchip are, for all practical purposes, free.

Hologram

1947
Denis Gabor

The field of holography has a vast range of applications: to highlight the vibrations in a musical instrument, in embossed holograms on credit cards, magazines and sweets; and to create lenses for supermarket checkouts. It may one day figure in computers that run on light.

Holography, creating what appear to be three-dimensional images in a two-dimensional medium, was invented in 1947 by a Hungarian-born physicist, Denis Gabor, who was struggling to improve the resolution of electron-microscope images.

Holograms may one day enable computers to run on light.

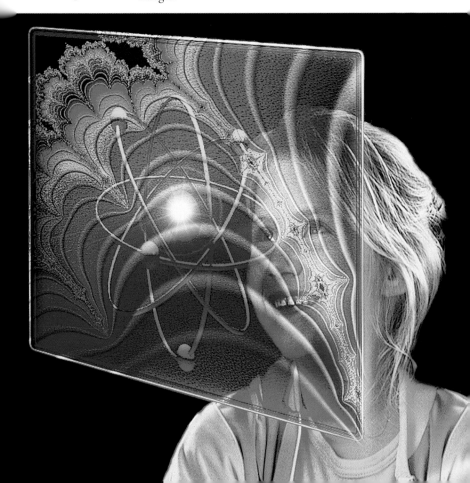

Between 1948 and 1951 Gabor published papers on his work but he had to wait for the construction of the first laser in 1960, and other work at Michigan University, before the first "transmission" holograms of a toy train and bird were produced in 1962. He coined the term from the Greek: *holos* meaning "whole" and *gramma* meaning "message".

A conventional hologram is made from special film that captures the pattern of light waves from an object illuminated by a laser. As light shines through this film, the original pattern is reproduced, giving the eye the impression that it is looking at a three-dimensional object.

The pattern captured is the result of "interference": when two light waves travel in synchrony they add together to become a bright region; when they are not synchronous they cancel each other out to become a dark region.

Early holograms relied on laser light, in which light waves are all in step, or "coherent". However, soon after the pioneering work at Michigan, Uri Denisyuk of the USSR found a way to make holograms visible in normal daylight.

Artificial intelligence
Alan Turing

1947

By using clever techniques in computing – from sophisticated software to computers based on the architecture of the human brain – machines may one day be able to learn, reason, self-correct and adapt to changing environments, just as we do.

The father of artificial intelligence was Alan Turing, the mathematician who devised a plan for a computer called the Automatic Computing Engine (ACE). In 1947, he argued that the brain could itself be regarded as a computer. "In working on the ACE, I am more interested in the possibility of producing models of the action of the brain than in the practical applications to computing. "

In 1950 Turing suggested a practical definition – the "Turing Test" – for intelligence, with arguments why it should be attainable by computers. But his inspiration eluded proponents of early AI, called "good old-fashioned AI" (GOFAI).

Theirs was a "top-down" approach which tried to compartmentalize intelligence into discrete "modules", such as perception, planning, and executing actions, which interact by rules enforced by an "inference engine" to produce intelligent behaviour in a robot, such as grasping or avoiding objects. GOFAI got off the ground when Allen Newell and Herbert Simon of the Rand Corporation showed towards the end of the 1950s that the strings of ones and zeros within a computer could be used not only to represent numbers but also more general symbols.

Since human brains also appeared to manipulate symbols, whether musical notes, pictures or text, and carry out thoughts of infinite variety, it seemed it could only be a matter of time before GOFAI embraced anything a human could do. GOFAI experts placed faith in digital cramming, arguing that a machine can be made clever by stuffing it with facts. In 1956, at Stanford University, California, Edward Feigenbaum, worked with Nobel laureate Joshua Lederberg to develop dendral, a system that could infer the structure of a molecule from a spectograph readout.

It was soon discovered, however, that expert systems have no "common sense". In the 1960s, the ELIZA program was

By 2001 the wildest dreams of 20th-century futurologists, scientists, writers and filmmakers had been exceeded. This female robot face, from the Science University of Tokyo, has 'shape-memory' electric actuators beneath its silicone skin which can alter its facial expressions

designed to reproduce the conversational skills of a psycho-therapist. A dialogue with a human interrogator looked impressive but it was programmed to respond inquiringly to any mention of mother, father or dreams. A statement such as "Nelson Mandela is the father of the new South Africa" would be met with: "Tell me about your father."

There was an alternative bottom-up approach to AI based on networks of "neuron" processors, analogous to nerve cells in the brain. The psychologist Frank Rosenblatt, of Cornell University, produced one "neural net" in 1960. His "Mark 1 Perceptron" could be trained to recognize letters of the alphabet – though it took 20 years before shortcomings were resolved. One of the first neural nets to reach the market was Wisard (Wilkie, Stonham and Aleksander's Recognition Device), which could recognize objects.

Today, there are a host of applications: for instance, identifying a submarine target from a sonar trace and recognizing handwritten postal codes. Efforts are under way to

The world as seen through the 'eyes' of Cog, an android with artificial intelligence developed at the Massachusetts Institute of Technology. Cog has a human-like face with four cameras for eyes.

develop a brain with 40 million neurons, equivalent to that of a kitten, even to "breed" neural nets by mimicking evolution in a computer. Artificial intelligence is seen as a subset of Artificial Life.

Some people dismiss the notion of artificial intelligence, no doubt because they don't like the idea of one day being tested by a smart machine. But it is something we'll all have to get used to. The Dyson robotic vacuum cleaner has vision-like sensors whose input is processed and acted upon every quarter of a second. That's about twice as fast as a human can manage. The robot also knows to within a millimetre where it is in the room. No human could aspire to such accuracy, which explains why they can't vacuum a room as well.

Disposable nappy

Marion Donovan

1947

Marion Donovan, born in 1917 into a family of inventors, thought she could find a better kind of nappy for her two children: something waterproof in a way that the terry towelling diapers had never been. Terry nappies leaked everywhere and created mounds of washing.

Rubber pants, worn outside cotton nappies to prevent leakage, were uncomfortable in summer and trapped too much moisture inside, creating the perfect conditions for nappy rash. So, in 1947 Donovan, a former assistant beauty editor on *Vogue* now living in Connecticut, cut up her plastic shower curtain to make a square-shaped reusable waterproof outer layer to cover the cotton nappy. It worked. Her baby's bed remained dry.

Following experiments on her sewing machine with other kinds of fabric, she eventually settled on parachute nylon as the best material for the re-usable outer liner. Her prototype nappies were fastened with an easy-to-close plastic clip instead of a safety pin.

Flushed with the success of her idea, which she called the "boater" because it helped the baby stay afloat, Donovan visited New York to market her invention. Saks Fifth Avenue was the first to sell the new nappy in 1948. They were an instant hit and other stores across America were soon clamouring for stocks.

The resourceful Donovan was already working on a new design: a patented all-paper disposable diaper with an outer layer that had similar waterproof properties as the nylon prototype, keeping the baby's skin dry. She sold the patents in 1951 for $1million, but it took another 10 years before Victor Mills, a Proctor and Gamble chemical engineer, came up with a way to mass-produce the disposable nappy.

In 1956 Proctor and Gamble had taken over a company with a redundant wood pulp processing plant. Mills had a hunch that the clean pulp could be used to make nappies along the same lines as Donovan's.

As a result, the Pampers brand was introduced in 1961, launching what has now become a £2 billion-a-year industry, to the disgust of many environmentally-conscious parents. Millions of ripening nappies are clogging up landfill sites around Britain, and the terry nappy is having a minor renaissance as "green parents" turn in small but significant numbers to reusable cotton nappies.

Mobile phone

Bell laboratories

1947

Such is the present ubiquity of the "mobile", that it is hard to imagine a time when hand-held portable telephones did not exist. Of even more surprise is that the mobile phone, forever associated with the yuppie of the 1980s, first appeared in 1947. However, it was not until the advances in electronic technology of the 1980s and the creation of an infrastructure of local base stations that mass mobile wireless communication became possible.

The earliest practical use of mobile radio was in 1921, when the Detroit Police Department installed an experimental two-megahertz system in its patrol cars. Channels soon became overcrowded, however. So in the 1940s the Federal Communications Commission in America was given the authority to manage the radio spectrum, enabling the police, emergency and utility services to become the pioneers of mobile radio.

In 1946, commercial mobile radiotelephony began in earnest when AT&T and Southwestern Bell introduced the

A mobile phone brings a worshipper in France into contact with the Wailing Wall in Jerusalem to join in prayers.

first service in Saint Louis, Missouri. It was a complicated system that used a central 250-watt antenna to contact mobile phones and deliver voice calls. The 20-watt mobile sets returned calls on a different frequency, not to the central antenna but to one of five receivers stationed throughout the city. A mobile telephone service combined the signals, amplified them and passed them on to a switchboard that linked the two parties. Operation was by push-button, and only one party could talk at a time.

In 1947, AT&T introduced a radiotelephone "highway service" on the road from Boston to New York, but problems arose when the signals interfered with other distant services. The solution was the cellular concept, devised the same year by Bell Laboratories. Instead of one large transmitter, it used a network of small geographical areas, or "cells", each with a low-powered transmitter. A central switch located the telephone, which was passed from one cellular transmitter to the next as it moved through the network.

It was 30 years before the first cellular network was operational, but there were several interim systems along the way. In 1948, the Richmond Radiotelephone Company introduced the first fully automatic radiotelephone service, in Richmond, Indiana. In 1964, the Bell companies introduced the Improved Mobile Telephone Service (IMTS), which dispensed with the push-button and allowed simultaneous two-way conversation for the first time.

Finally, in 1978, Illinois Bell began operating the first cellular system, called Advanced Mobile Phone Service, in

Chicago. With a dedicated computer and switching system, it proved that cellular technology worked. It used 10 cells to cover 21,000 square miles and operated in the 800-megahertz band, setting the pattern for the development of commercial cellphone services worldwide.

Europe's first cellular service was the Nordic Mobile Telephone System of 1981, introduced in Sweden, Finland, Denmark and Norway. The UK followed in 1985 with Total Access Communications System (TACS), an analogue system that was easy to eavesdrop and often cracked by criminals, who cloned customers' phones to make calls on their accounts. Secure digital networks, introduced in the 1990s, have now replaced all the analogue networks.

More than 24 million Britons now own mobile phones, some of which use WAP (wireless application protocol) to enable their users to access a customized version of the web and send and receive e-mail.

Frisbee

Morrison and Fanscioni

1948

Anger still occasionally erupts among Frisbee fanatics about where the name came from: but most now agree that pie tins of the Frisbie Baking Company of Bridgeport, Connecticut, were the origin of the name.

According to American folklore, college students in the 19th century found that empty Frisbie tins made superb missiles: something like a cross between a discus and a boomerang. Soon the tins were flying across campuses all over the country.

In 1948, two men in Los Angeles, Walter Frederick Morrison and Warren Franscioni, invented a plastic version of the Frisbie tin flyer. Morrison had spent long hours trying to improve the aerodynamic qualities of the pie tin. His rounder and smoother version flew further, faster and straighter.

Morrison's partnership with Franscioni foundered but he eventually got the patented plastic plaything on to the market. He called it the Pluto Platter, hoping to hitch a ride on the mania for anything tinged with the mystique of outer space.

The business never really took off for Morrison, so he sold up in 1957 to Wham-O Toys, who renamed the disc "Frisbee" and began to build the idea of it as a sport. It was an immediate success. Wham-O even introduced a "professional" version in 1964, although no figures exist about how many sportsmen have ever earned a living from the Frisbee.

Velcro

George deMestral

1948

Intrigued by the sticking power of the cocklebur seeds adhering to his jacket after a walk in his native Switzerland, George deMestral, an electrical engineer from Lausanne, examined the seeds under a microscope and found they were covered with tiny hooks that had caught in the loops of his jacket's fabric.

Recognizing that this "hook and loop" principle might make a new type of fastener, deMestral developed two tapes: one, made from loops, felt like velvet, while the other was covered with crochet-style hooks. Pressed together, the hooks fitted into the loops, and the "Velcro" tapes held tight. Velcro was launched in 1948 and has since become the simple fastening of choice on garments and shoes.

General-purpose computer

Freddie Williams

1948

The visionary English mathematician Charles Babbage first floated the idea of a general-purpose computing machine in 1834, but a century would pass before it began to turn into reality. Until the 1930s, scientists and engineers used either mechanical calculators to solve their arithmetic problems, or purpose-built "analogue computers" to tackle more complex problems such as the prediction of tides.

1948 The modern
contact lens,
made of plastic and
resting on a cushion
of tears, invented by
Kevin Tuohy.

Yet within five years of Babbage's death in 1871, Lord Kelvin, the Scottish physicist who invented a mechanical tide-predictor, had raised the idea of building analogue machines able to cope with more general problems.

It took another 50 years for Kelvin's idea to become a working machine. Known as a differential analyser, it was built by a team at the Massachusetts Institute of Technology under Vannevar Bush, who wanted a speedy way of solving differential equations arising in electrical engineering. Completed in 1930, the machine filled a room and could solve a fairly broad range of problems, printing the solutions as curves on a drawing board.

Even so, it still fell a long way short of Babbage's vision. The real breakthrough came from scientists and engineers exploiting electrical technology unavailable to Babbage, plus a simple yet incredibly powerful idea that was known to mathematicians as far back as Leibniz. This was the binary system, which allows every number to be expressed as a series of ones and zeros. Binary is ideally suited to electrical devices, which can express binary digits as "on" and "off". Binary was, in short, the perfect language for any would-be general-purpose digital computer.

The first person to act on this fact was a German engineering student, Konrad Zuse. Fed up with the mathematical drudgery of his course, he wondered if a machine could do the donkeywork. So in 1934, the 24-year-old Zuse drew up a plan to build a mechanical computer capable of tackling a broad range of mathematical problems. Four years later, Zuse and some friends completed the Z1. It featured many of the components of a modern computer – input, processor unit and memory – and had a binary-based operation that could cope with floating-point calculations. Zuse had even given the machine the ability to deal with logical problems, re-inventing work by the Victorian mathematician George Boole, who had shown that logical reasoning could be reduced to manipulations of binary digits.

Some historians regard Z1 as the world's first programmable computer. It lacked one key feature, however: it could not perform "conditional branching", in which operations are carried out only if some condition is met. So the Z1's claim to primacy is not universally accepted.

Z2, based on electromechanical relays, worked a little better. Then, in 1941, Zuse's team completed the Z3, another all-relay device programmed to use old movie film as punch-tape. It is now regarded as the world's first programmable calculator.

The following year, Zuse and his colleague Helmut Schreyer sought funding from the German High Command for a faster, all-electronic version based on valves. Their request was rejected

Although Colossus, built at Bletchley Park, could crack the toughest Nazi codes, it was not considered a fully-fledged general-purpose computer.

on the grounds that the war would be won before they finished the machine. Had they been funded, Zuse and Schreyer might have won the race to build the first genuine computer. Instead, their lead was eroded and then lost to researchers in America and Britain.

In America, Howard Aiken and colleagues at Harvard University had begun work on a electromechanical machine, the Mark I, in 1937. Completed in 1943, with the help of a team of engineers, it weighed five tons and could perform basic arithmetic functions, producing the results on punch cards.

The Mark I was the realization of Babbage's plan to build a machine capable of churning out mathematical tables and printing the results. Yet, once again, because it could not perform "conditional branching", the Mark I cannot be regarded as a genuine computer. It did, however, inspire the machine-making company that had helped Aiken to take a greater interest in electronic computation: its name was International Business Machines, or IBM.

Just as Aiken was starting on the Mark I in 1937, two other Americans began work on two key components of a genuine computer. George Stibitz, a physicist at Bell Laboratories in New York, showed that the binary concept of "on" and "off" was all that was required to turn telephone relays, batteries and lights into a device capable of performing arithmetic functions and even simple logical reasoning.

That same year, a physicist at Iowa State University named John Atanasoff drew up plans to build an all-electronic device that used binary arithmetic and Boolean logic (allowing arguments that contain qualifiers such as "and", "or" and "not") to solve mathematical problems. By the summer of 1939, Atanasoff and some colleagues had a device that could manipulate 16-digit binary numbers. It was the first all-electronic digital calculator ever made.

A few months later the Second World War began and, with it, the most crucial period in computing history. Its course was determined by the contrasting attitudes of the opposing regimes to the likelihood of victory. While Nazi Germany's ludicrous confidence in victory led to the death of Zuse's plans, the Allies had no such confidence and had a very specific need for computing power.

If Charles Babbage was the grandfather of the computer, then Alan Mathison Turing can justly lay claim to being its father. At King's College Cambridge, he worked on probability theory and mathematical logic, which led to the concept of the Univeral Turing Machine the cornerstone of the modern theory of computation.

A year before the war, Britain had begun a secret project at Bletchley Park, near present-day Milton Keynes. Its goal was to break the Nazi military codes, and some of the country's most brilliant minds were recruited to help, including a 26-year-old Cambridge mathematician named Alan Turing.

Some years earlier, Turing had conceived a plan for a "universal machine", capable of performing all the operations of mathematics. The idea was perfectly suited to code-breaking, and by December 1943 some of its power was realized in an all-electronic device designed to attack the toughest Nazi ciphers.

Known as the Colossus, it was not a fully-fledged computer. It did, however, show that machines could do more than handle purely numerical problems, and it gave military insights that some historians claim helped shorten the war by a year or more.

While Colossus took shape at Bletchley, American engineers Presper Eckert and John Mauchly began work on the Electronic Numerical Integrator Analyser and Computer (ENIAC). Originally designed to calculate artillery tables, the 30-tonne machine was finished in November 1945, its 18,000 valves giving it 1,000 times the speed of its predecessors. Although still just a calculator, ENIAC proved that vast arrays of electronics could work reliably and it led engineers to focus instead on the final step needed to make a real computer: finding ways of storing instructions in a machine so it could cope with conditional branching.

The first to succeed was a team at Manchester University under Dr Freddie Williams, who had found that cathode ray tubes could store binary digits in a form that a machine could use. In June 1948, a prototype computer was programmed to find the factors of a number. It projected the answer on a cathode ray tube. The British government seized on the success and commissioned Ferranti to turn it into the first commercial computer. The Ferranti Mark 1 was installed at Manchester in February 1951.

Despite its success, the limited abilities of the Manchester prototype have led to the accolade for the first serious stored-program computer generally being awarded to another British machine, the Electronic Delay Storage Automatic Computer (EDSAC). Completed by a team at Cambridge University in

May 1949, it was routinely used for scientific research until 1958.

After being snarled up in a patent dispute over the ENIAC, Eckert and Mauchly regrouped and in 1949 delivered BINAC, America's first stored-program computer. It was followed in 1951 by UNIVAC, America's first commercial computer. The following year, UNIVAC was used to predict the outcome of the US presidential elections, but its human operators refused to believe its prediction of a landslide for Eisenhower and reprogrammed it to come up with a more "sensible" result. The actual result was indeed a landslide, prompting one wit to proclaim, "The trouble with machines is people".

Credit card

Ralph Schneider

Until credit cards became ubiquitous, queuing up at foreign exchange counters to get currency or traveller's cheques was a regular chore whenever you went abroad. Today, you can exit a French motorway tollbooth quicker if you hand over a credit card than if you proffer cash, which has to be counted.

American hotel companies were the first to introduce credit cards as a means of keeping customers loyal to their chain during the 1920s. The first credit card available to everyone – well, the credit-worthy, anyway – was invented by Ralph Schneider, an American businessman, who launched the Diners' Club card in 1950; American Express, perhaps the most famous credit card, followed in 1958.

1950

Singer Patti Page opens a Diners' Club exhibit in 1964.

The first card to be accepted across America emerged in 1966 with the launch of the BankAmericard, which became Visa in the late 1970s. This prompted different types of credit card to merge under a single banner, completing the worldwide coverage needed to make then truly convenient.

The emergence of internet shopping has created a new market for credit card companies, but it is not quite the success expected. People are worried about giving card details down an encrypted telephone link – though they don't mind handing over their card to some total stranger in a foreign restaurant.

Junction transistor

William Shockley

1950

1950 Bill Bloggs invents the Bloggoscope, a tilt finder to help make accurate maps from aerial photographs.

Upset by the lack of credit for his work on the invention of the point-contact transistor (see page 416), William Shockley decided to concentrate on his own ideas to refine the technology. On New Year's Eve 1948, while staying in a Chicago hotel (he was attending a Physics Society meeting), Shockley wrote 30 pages of notes for a new type of transistor that would improve on Bardeen and Brattain's device.

Shockley returned to his notes on January 23 when, sitting at his kitchen table, he conceived the idea of a three-layer "sandwich" of semiconductor materials. The middle layer would act like a tap to control the flow of electrons through the other layers.

On February 18 1949, work by colleagues at Bell Laboratories proved Shockley's ideas should work, prompting him to share them for the first time.

Much work had to be carried out to produce the sandwich for his transistor. Initially, slices of germanium crystal were joined together, but this idea failed.

Bell Labs chemists Gordon Teal and Morgan Sparks devised a way to grow germanium crystals in layers, changing their electrical characteristics by "doping", or adding impurities as they were drawn from the melt.

The first device (nicknamed the "persistor" because persistence was what made it eventually work) was tested on

April 12 1950. It worked, but not very well due to the thickness of the middle layer. Improvements were made but Bell concluded that the transistor would develop much faster if its research was made available to other companies.

In April 1952, at the Transistor Technology Symposium, companies that had paid a $25,000 licensing fee were given all the information necessary to make transistors, bound in a two-volume book set that later became known as *Ma Bell's Cookbook*.

Self-cleaning house

Frances Gabe

1950

D riven by her hatred of housework, Frances Gabe began developing a self-cleaning house in the 1950s at Newberg, Oregon. "Housework is a nerve-twangling bore. I was determined that there had to be a better way," she said.

Placed on each ceiling was a 10-inch spinning sprinkler that emitted a powerful spray of water. Walls were coated with waterproof paint and the floors and furniture were varnished. Her 36 patents included waterproof covers for books and a material to cover upholstery.

The washing cycle took 45 minutes: "You open a valve, punch a button and it washes your ceilings, walls, floors, windows, curtains, your furniture, your dishes and your clothes. Then it dries them."

Water-emitting ducts washed the floors before the water ran through her Great Dane's kennel, giving it a bath at the same time. Attempts to market the invention were not successful.

Spring clean: Francis Gabe, aged 83, protects herself from a downpour she created in her self-cleaning kitchen in Oregon.

The Pill

Carl Djerassi

1951

The contraceptive pill marked a turning point for science as well as for women. By providing control over fertility, it demonstrated that science had reached the point where a biological process was understood well enough to tinker with it. Before this, the action of drugs was only elucidated after their discovery. But the Pill was designed to mimic a hormone, progesterone, that was already well known.

Progesterone is an important part of the dialogue between the brain and the reproductive system during the monthly menstrual cycle. Imitate it and you can butt in. The main message carried by progesterone is "we're pregnant". It also, crucially, tells the brain not to ask the ovaries to release more eggs.

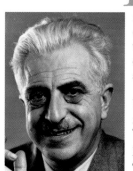

Gynaecologist Gregory Pincus carried out the first tests of the contraceptive pill in Puerto Rico.

1951 Tetrapak milk cartons invented by Ruben Ruasing

But progesterone was expensive to extract. A chemist named Russell Marker discovered an alternative source, however, in yams. He quit his job at Penn State University, went to Mexico, set out by mule to collect tons of yams and returned to synthesize progesterone.

Then, in 1949, Carl Djerassi, a novelist and chemist who also invented antihistamines, discovered a compound more potent than progesterone. It had to be modified to survive administration by mouth rather than injection, and the first successful synthetic progesterone, unveiled in 1951, was used to treat menstrual problems. Its contraceptive application was only later realized in tests in Puerto Rico in 1956. Approval was granted to the company G.D. Searle, in 1960 for the first oral contraceptive pill, and marketing in Britain began two years later.

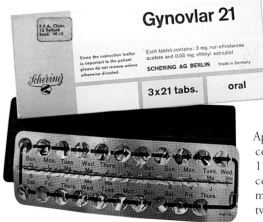

Widescreen cinema
Fred Waller

1952

The proportions of 35mm movie film dictated the shape of early cinema screens until the 1920s, when experiments with 56, 65 and 70mm film were conducted. They proved too expensive. But in the late 1940s, a New York photographer, Fred Waller, developed a widescreen cinema system by adapting a device used to train aerial gunners during the war. "This is Cinerama" used three projectors to produce a wide curved image, and was first shown at the Broadway Theatre on September 30 1952.

Transistor radio
Texas Instruments

1953

In 1953, when Texas Instruments was manufacturing transistors under licence from Bell Laboratories, the company's president, P.E. Haggerty, commissioned the design of an all-transistor radio.

Established manufacturing companies took little interest, so IDEA, a small Indianapolis firm that made aerial boosters, was selected. In collaboration with industrial designers Painter Teague and Peteril, IDEA developed the Regency TR1, the first commercially produced pocket-size four-transistor radio. It went on sale in November 1954, selling for $49.95 (£285 in today's money).

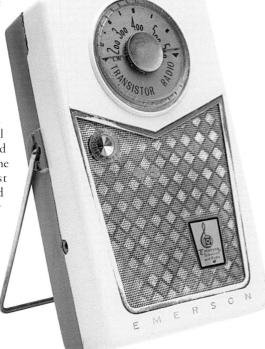

Flying saucer

Avro

1954

A s unlikely as it sounds, the flying saucer has been invented and built as the result of fertile human imagination rather than alien intelligence. To date there have been more than 50 designs for aircraft with circular-shaped bodies and wings. There is even a US Patent category for such vehicles: Class 244, Aeronautics; Subclass 21.2 Airplane, Circular.

The first recorded design dates back to 1716 when a Swedish scientist, Emanuel Swedenborg, published a description of an aircraft with oval flapping wings in the scientific journal *Daedulus Hyperboreus*. It was never built and the concept remained unproven until the early 20th century when a number of aviation pioneers produced aircraft with circular and ring-shaped wings. These included the Annular

The sci-fi view, from 'Earth Vs The Flying Saucers', 1956. Meanwhile, back in real life, 50 patents for circular aircraft have been lodged.

Quadruplane (1904), the Ring Rite (1906) and the Umbrella Plane (1912).

During the Second World War, the US Navy commissioned a circular-wing aircraft designed by aeronautics engineer Charles Zimmerman. A prototype made more than 100 successful flights but by the end of the war the Navy had lost interest in the "Flying Flapjack" and the project was abandoned.

One of the first circular flying machines using ducted-air or jet-propulsion systems was the British-designed Avro VZ-9V Avrocar, taken over by the US Air Force in 1954. Saucer-shaped, six metres in diameter and powered by three turbo-jet engines, the prototype became unstable more than two or three metres above the ground.

The Moller XM-2 was another saucer-shaped design, built by Paul Moller, a professor of engineering at the University of California in 1966. It reached one metre on the first test flight. A second design, powered by eight engines, reached three metres in 1968 and subsequent versions have risen to more than 12 metres. Moller is currently working on a wingless aircraft called the Volantor.

The most bizarre design is the subject of a patent application filed by the British Railways Board in 1970. It describes a circular spacecraft powered by laser controlled pulsed thermonuclear fusion. It has yet to be built.

1954 Electronic music synthesizer invented by Harry Olsen and Herbert Belar, scientists working at RCA.

Solar cell

Bell Laboratories

1954

The possibility of using the sun to generate electricity had been discovered in 1839 by the French physicist Edmund Bequerel, and 47 years later the American inventor Charles Fritts developed the first selenium solar cells. They were only one per cent efficient, however, and no one understood why they worked. However, by the 1930s, quantum mechanics explained to physicists why electrons would flow between the junctions of a sandwich of thin slices of selenium left under light.

The crucial discovery came in 1954, when Darryl Chapin, Carl Fuller and Gerald Pearson at Bell Laboratories found that coating selenium with a layer of impurities increased the flow of electrons, improving the efficiency of the cell to 15 per cent.

In the 1960s, photovoltaic cells went into space to power exploration, but then, as now, more energy was needed to produce a voltaic cell than it would produce in its lifetime.

Hovercraft

Christopher Cockerell

Once described by his father as "no better than a garage hand", Christopher Cockerell was a prolific inventor. He is best known for the hovercraft, which began life as an empty tin of cat food in 1955.

During the Second World War, Cockerell worked at Marconi on the development of radar and airborne navigation systems, which he always maintained were among his most important achievements. While there, he filed 36 patents, for which he was paid £10 each, a foretaste of his future as a largely unrewarded genius.

With a legacy from the death of his father-in-law, Cockerell left Marconi in 1950 and bought a small run-down boatyard in Norfolk. His fertile mind turned to boat design and he began to investigate ways of reducing the friction between the hull of a boat and the water surface. His idea was to support the craft on a cushion of air, not a new concept, but his refinement was to contain the low-pressure cushion within a "curtain" of high-pressure air.

Sir Christopher Cockerell used an empty cat food tin and a vacuum cleaner to develop the Hovercraft.

In 1955, Cockerell built a test rig to confirm his theories. It consisted of a tin of cat food, inside a coffee tin, connected to a vacuum cleaner with the airflow reversed, mounted above a set of kitchen scales. It worked.

Cockerell set up the Ripplecraft company and, in 1956, with the help of a fellow boat-builder, he set about designing and building a radio-controlled working model. This was demonstrated to the Government at the Ministry of Supply and was immediately classified as secret, but nothing further happened. No one knew what to do with it: the Admiralty said it was a plane, the RAF maintained it was a boat and the Army was just not interested.

When Cockerell discovered that the Swiss were working on a similar project in 1957, however, he informed the Ministry of Supply and was immediately given a £1,000 grant by the Government-backed National Research and Development Corporation on the understanding that they would have first refusal on development rights.

Two years later, the Ministry of Supply agreed that the idea could be declassified and developed commercially. The NRDC formed a company called Hovercraft Development Limited to oversee the project and find investment. Cockerell became a director and five companies were licensed to build hovercraft. The contract for the first machine went to Saunders Roe (later to become part of Westland Helicopters) and work began on a man-carrying vehicle.

The SRN-1 was completed on May 31 1959 and the seven-ton craft made its first public appearance on June 11. On July 25, the 50th-anniversary of Bleriot's first cross-Channel flight, the SRN-1, with Cockerell aboard as "ballast", crossed from Calais to Dover in just over two hours.

Despite the success of his designs, Cockerell had to fight for another 10 years to receive £150,000 for his patents. He was knighted in 1969, but he never recouped his development costs and at one stage had to pawn his mother-in-law's engagement ring to survive.

He died in 1999, in relative obscurity, eking out his retirement on a state pension and believing that everything was stacked against inventors. Yet, but for the "silly chaps", he once said, we would still be living in the Stone Age.

A new kind of ferry service was offered by Christopher Cockerell's hovercraft, seen here arriving at Gosport from the Isle of Wight in 1985. Hovercraft still serve the island.

Atomic clock

UK National Physical Laboratory

1955

The realization that an oscillation can be harnessed to create a clock came to Galileo Galilei in 1583, supposedly while daydreaming in the cathedral of Pisa. Using his own pulse, he found the swings of an altar lamp to be regular enough for timekeeping.

Almost 350 years later, these oscillations had been replaced by those in a piece of quartz, which can be controlled by applying an electric field. Today, time is measured by using the vibrations within atoms and molecules: every atom and molecule absorbs and emits electromagnetic radiation (such as light or radio waves) at characteristic frequencies. The first molecular clock was built by Harold Lyons at the US National Bureau of Standards in 1949 and was based on the vibrations of an ammonia molecule.

The first successful atomic clock was unveiled in 1955 by the National Physical Laboratory in Teddington, Middlesex, and relied on the outermost electron of the caesium atom, which loiters in an orbit much further from the nucleus than the rest. There, the electron's energy can have two slightly different values, depending on an electronic property called "spin". The energy difference between the two spins corresponds to radio waves of a frequency of 9,192,631,770 cycles a second.

A quartz crystal controlled a radio tuned to the correct frequency to flip the spin of the electron. As atoms in the lower energy state drifted by, the perpendicular beam of radio waves nudged them into the higher energy state. By maximizing the numbers of atoms elevated this way, the crystal was fine-tuned to produce an incredibly accurate timekeeper.

Harold Lyons with a molecular clock he later developed for Nasa. Its tubular heart housed a stream of ammonia molecules generating a highly stable current.

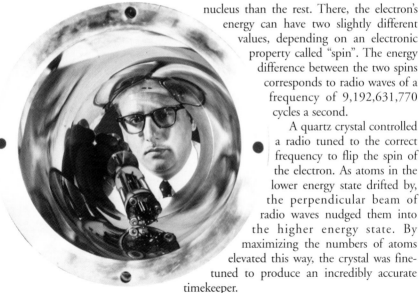

To allow a more accurate measurement of the all-important transition between electronic spin states, scientists created the caesium "fountain". About 10 million caesium atoms were collected in a magnetic field, then a laser was used to push the atoms so they rose and fell slowly under gravity. Measurement by radio waves enabled scientists to fine-tune a crystal clock, as they had the conventional atomic clock. These atomic fountains have achieved accuracies of one second in around 15 million years. By launching the apparatus into space, the atoms, no longer under the influence of gravity, may linger longer in one spot and accuracy could reach one second in 150 million years.

An even more advanced clock is now under study: the trapped ion machine. Ions are atoms that have gained or lost an electron. They are, as a result, electrically charged so they can be held in one place by magnetic fields. Using this technology, scientists expect to push accuracy to one second in 10 billion years.

Polio vaccine

Jonas Salk

1955

Poliomyelitis used to be called infant paralysis because the virus attacked young children, destroying the nervous system. Sometimes this resulted in permanent paralysis and occasionally in death. During the 1940s and 1950s, the acutely infectious virus struck thousands of infants. In 1950, in America alone, there were 33,344 cases.

Jonas Edward Salk, an immunologist who had worked on influenza vaccines for the US Army during the war, was convinced he could combat polio. Funded by the government, he worked from a purpose-built laboratory at the University of Pittsburgh. He published his first papers on polio just after the end of the war, creating enough of a stir to get a generous research grant from the US National Foundation for Infantile Paralysis.

Jonas Salk was a great fund-raiser and publicist, but he received nothing for his life-saving vaccine.

Not all immunologists were pleased at Salk's rapid success at raising funds. Some, working quietly away in hospitals and universities, felt that Salk had not "paid his dues" as a researcher and that his success in fundraising masked his true ability as an immunologist.

One such was Dr Albert Sabin, who had his own views about how the virus could be conquered. Sabin and many others in the field were using live cultures of the virus to develop a vaccine. Salk disagreed with this approach and made a breakthrough at the end of the 1940s with a "killed virus" vaccine. It was based on three strains of the disease that others had isolated, but which he had corroborated. Monkeys injected with the vaccine produced antibodies, but did not develop the disease.

In 1952, Salk successfully conducted field tests with the vaccine on children who already had suffered polio, and on a sample of children who had not. None of the children contracted the disease from the vaccine. In 1954, after mass trials, Salk declared that a safe vaccine had been found and a vaccination programme was instigated. By 1961 American cases of polio had diminished by 95 per cent.

Controversy followed. Salk had neglected to submit his triumphant discovery to his peers, instead splashing his findings at press conferences and on radio. He never received the Nobel Prize because he was adjudged to have developed a technology and not made new breakthroughs. Salk was unable to patent the vaccine and made no money from his polio work.

In 1955, following Jonas Salk's breakthrough, a massive polio inoculation programme was undertaken in America. In Jay County, which suffered one of the worst polio epidemics, schoolchildren were given shots of Salk's vaccine at the rate of 300 an hour.

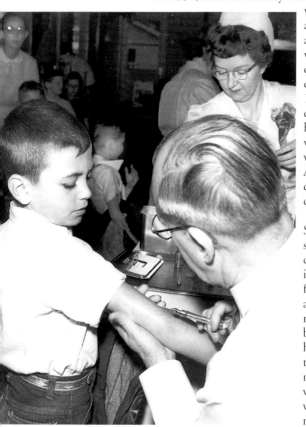

Structurally modified antibiotics
Lloyd Conover

1955

I n 1955, Lloyd Conover, a researcher at Pfizer, revolutionized pharmaceuticals by synthesizing tetracycline, a "broad spectrum" antibiotic used to fight dozens of types of bacteria and other micro-organisms. It was the first antibiotic to be improved by changing the chemical structure of a compound produced by microbial metabolism and it sparked a hunt that resulted in most of the important antibiotic discoveries made since.

Structurally modified antibiotics have recently fallen out of favour, however, amid concerns about the proliferation of tetracycline-resistant "superbugs".

Video tape recorder
Ampex Corporation, USA

1956

A s well as perfecting the mechanical television system, John Logie Baird also devised the first video recording system, using 10-inch (25-cm) wax discs, called Phonovisors. The system was patented in 1926 and demonstrated in September 1927, reproducing a crude 30-line picture.

However, the first steps towards the use of tape to record video images began in 1931, when engineers working for AEG in Berlin developed a magnetic device called the Magnetophone for sound recording. It used a tape coated in an iron-oxide formulation produced by BASF.

Video recording comes full circle. The latest cameras use a disc in place of tape.

The Magnetophone was used for regular broadcasts by German radio from 1941 until after the war, when captured machines were brought to America by Jack T. Mullins, an American Army Signal Corps major whose mission was to ferret out German technology that might prove useful to the Allies.

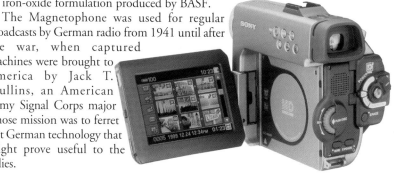

Returning to civilian life in 1946, Mullins refined the machines and demonstrated the device's capabilities to Bing Crosby in August 1947 at the NBC/ABC studios in Hollywood. Crosby immediately realized its potential.

In 1950, Mullins joined Bing Crosby Enterprises, founded to conduct research into recording technologies to alleviate the anxiety of live broadcasts. With help from colleague Wayne Johnson, Mullins began work on a magnetic video recorder. They developed several prototypes that used fixed recording heads, but the high tape speeds needed to produce decent images limited the maximum recording time.

In 1952, the BBC began work on a fixed-head system called VERA (Vision Electronics Recording Apparatus), which was later used for on-air broadcasting in 1958. But the project was abandoned when it became clear that "transverse" recording systems, using spinning tape-heads instead of fast-moving tape, were the way forward.

Toshiba was known to be working on such a system as early as 1953, and this later became the "helical scan" system now used in domestic VHS VCRs. But Ampex Corporation won the race to develop the first successful transverse system. Work started in 1951 and progressed slowly, but, in early 1956, funding was approved for a crash development programme to get a promising prototype machine ready for the National Association of Radio and Television Broadcasters Convention in Chicago in April.

Giant spools of magnetic tape were used to record sound and pictures by the BBC in the 1950s.

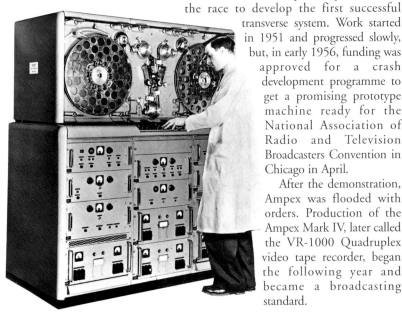

After the demonstration, Ampex was flooded with orders. Production of the Ampex Mark IV, later called the VR-1000 Quadruplex video tape recorder, began the following year and became a broadcasting standard.

There were several valiant but unsuccessful attempts to develop a home video tape recorder during the 1960s. One of the first was an open-reel, fixed-head machine called the Telcan, made in 1962 by the Nottingham Electronic Valve Company. Sony introduced a black and white home model, called CV-200, in 1965, and in America the CBS Electronic Video Recording system made a brief appearance in 1967.

In 1972, Philips launched the first true video cassette recorder, with a television tuner, timer and TV aerial connection. The tape, housed in an easy-to-use cassette, employed a format called VCR. The first video-recorder from Sony, launched in Japan in 1975, used a format known as Betamax and a year later JVC introduced its VHS system. Thanks to the more aggressive marketing policy employed by JVC, it was VHS that became the industry standard.

Satellite

Soviet Union

1957

When the Soviet Union launched the first artificial satellite, Sputnik I, on October 4 1957, science fiction became a matter of urgent political reality. It stimulated a space race that led to the first step by human kind into space, when Yuri Gagarin made the first orbit of Earth on April 12 1961, and to Neil Armstrong's first step on to the Moon on July 20 1969.

In order to develop these ambitious and costly programmes, America and the Soviet Union conducted extensive experiments with satellites in Earth orbit. From these they began to acquire a wealth of new resources, information and communications capacity. Some of this derived from the facility provided by satellites for observing Earth. It included vastly improved weather forecasting from meteorological satellites and a range of useful information about geology and land use from photographic and remote-sensing equipment in space.

Also, enormously valuable scientific information was obtained from instruments placed in satellite orbits to observe the sun, the planets and the stars without the interference of the Earth's atmosphere. This culminated in the

1957 Soriet Union shoots Sputnik, the first manmade satellite, into space.

First flexible fibre endoscope invented by Basil Hirschowitz at the University of Alabama.

stunning pictures taken by the Hubble space telescope, launched in 1990 but not fully functional until 1993.

The most immediate impact of satellite technology has been through the great expansion of communications which it facilitated. In 1945, Arthur C. Clarke, the science-fiction writer, was first to envisage the possibility of using three geo-stationary satellites placed in synchronous orbits, of 22,000-mile radius over the equator so that they appear fixed in space from the surface of the Earth, in order to provide complete world coverage for telecommunications.

NASA took up the idea. It launched the first synchronous satellite, Early Bird, in 1965. It was adopted for a range of telephone, television, and information technology services that made possible as never before a worldwide, instantaneous exchange of communication. There are now dozens of these satellites in synchronous orbit, serving military, commercial and public operators, bringing practical benefits that have helped to justify the great expense of space research.

Nikita Kruschev (centre) with Sputnik I model at the 1959 Leipzig Trade Fair.

Microchip
Jack Kilby

1958

The alchemist's dream of turning base metals into gold seems positively unambitious compared to the silicon microchip, in which ordinary sand has been turned into an electronic brain. It is hard to credit how these tiny slivers of metal have transformed our lives, finding their way into everything from toys to teraflop computers. No wonder few took seriously the idea of such a device when it was first suggested by a British electronic engineer almost 50 years ago.

A reliability expert at the Royal Signals and Radar Establishment in Malvern, Worcestershire, Geoffrey Dummer, had noticed that as valves, relays and wires gave way to transistors, diodes and interconnects after the Second World War, reliability had improved dramatically. In 1952, Dummer delivered a paper at a conference in Washington in which he envisaged the use of a solid block of material such as silicon, whose various layers would act as all the key components needed by electronic systems. It was a stunningly prescient description of a monolithic integrated circuit, or microchip.

Dummer was unable to find people in the UK willing to fund further research, and others began having similar ideas. During the summer of 1958, Jack Kilby, an electronics engineer at Texas Instruments, Dallas, built the world's first integrated circuit (IC), using slivers of semiconductor material connected by tiny wires soldered under a microscope.

Patented in 1959, Kilby's IC was a major advance on all previous attempts at miniaturization, but it was delicate and easily damaged. The real way forward was found by a team of physicists led by Dr Robert Noyce, who in 1957 had set up Fairchild Semiconductor

Jack Kilby invented the first integrated circuit, the microchip that started the miniaturization of technology.

Microwires bonded to a silicon chip on an integrated circuit.

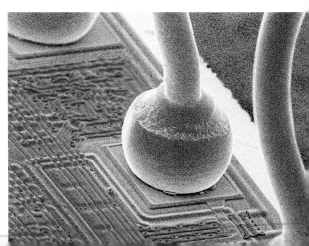

in California with the aim of mass-producing transistors on tiny patches of silicon.

In late 1958, Dr Jean Hoerni of Fairchild had found a way of chemically embedding all the key components of a transistor into silicon. By the summer of 1959, Noyce and his colleagues had used this so-called planar method to make an integrated circuit that could be mass-produced far more easily than Kilby's flimsy design.

As transistors can act as on-off switches, and thus perform binary arithmetic and logic, the IC allowed enormous computing power to be packed into tiny spaces. So ubiquitous is the microchip that its prices fluctuate dramatically. Microchip factories costing £2 billion are frequently built yet there is never a happy convergence of supply and demand.

Video game

Willy Higinbotham

1958

What might be loosely interpreted as interactive video "games" were devised separately by American and German scientists during the mid-1940s to predict the trajectories of missiles on an oscilloscope display. However, the first purpose-built video game, created purely for entertainment, appeared in 1958. Physicist Willy Higinbotham's idea was to provide visitors to the Brookhaven National Laboratory in New York with a "hands-on" demonstration of his department's work.

The image was generated by an analogue computer and a bank of relays and initially displayed on an oscilloscope, later on a 15-inch (37 cm) monitor. The game was based on table tennis and presented a side view of the table, with the net represented by an inverted "T" and a moving spot of light for the ball. Each player held a box with a knob, used to alter the angle of the "bat", and a switch to register a "hit". Higinbotham never patented his idea and it was largely forgotten until 1985 when games manufacturer Nintendo called him as a witness in an unsuccessful lawsuit to establish rights to key patents.

In 1961, the Digital Equipment Corporation lent a massive mainframe computer, called the PDP-1 to the Massachusetts

Institute of Technology. It was the first computer to use a screen and a typewriter keyboard instead of punchcards for viewing and inputting data. The computer company intended it to be used for serious research, but Steve Russell, a young computer programmer at the Institute, spotted it and, inspired by the work of sci-fi writer E. E. "Doc" Smith, wrote the first computer video game called Spacewar.

Spacewar allowed two players to fire torpedoes at each other from their spaceships spinning weightless in space. At first it was against a black background, until Pete Sampson, a fellow research scientist, devised a programme called Expensive Planetarium. In those days, there was a family of "Expensive" programs, such as Expensive Typewriter and Expensive Desk Calculator, a word processor and adding machine so called because the computer time needed to run them was much more costly than buying an actual typewriter or desk calculator. Expensive Planetarium added a moving background of stars, towards which the Spacewar spaceships were attracted by gravity.

Russell and his colleagues soon tired of using the computer keyboard to move the spaceships around the screen and fire buttons, so they devised a separate handheld unit with four buttons. It was an embryonic joystick.

Virtual-reality video games, such as the one above in Ohio, are moving from amusement 'palaces' into the home.

The game, played on a computer the size of three refrigerators, was an immediate hit. Despite the enormous cost of computer time in the 1960s, it was copied by hundreds of other engineers and programmers. Within a short time, it could be found on almost every research computer in America. But no one could imagine that computers would become sufficiently small, powerful and affordable for video games to be played outside universities and military facilities.

In 1962, engineering student Nolan Bushnell encountered Spacewar at the University of Utah. Three years later, while in charge of an amusement parlour for a summer job, he came up with the ideal of a self-contained arcade computer video game, called Computer Space and based on Spacewar. Nutting Associates, a mechanical arcade-game manufacturer, purchased the game and Bushnell was hired to oversee development. He built the first video arcade games in 1971 but the public was unimpressed, finding them too difficult to play.

Disillusioned with poor sales of Computer Space, Bushnell left Nutting to set up his own company, Syzgy. Finding that the name was already used, he renamed the company Atari, after the Japanese equivalent for the word "check" in the game Go. Bushnell hired Al Alcorn, a computer programmer, to develop a video tennis game called Pong but arcade-game manufacturers showed little interest it, so Bushnell manufactured and marketed it himself. A prototype was test marketed in a local bar and within a few days the game broke down because the coin box was full. By the end of 1973 more than 25 companies were manufacturing video arcade games.

Mario, an Italian plumber from Brooklyn, was the champion of the first globally successful video game. He went on to become the leading icon of an industry that makes more money than Hollywood.

In 1974, Atari started work on a home console version of Pong which became the fastest-selling game in America. But in 1976, there was public concern about the effects of violence in games on children following the release of Death Race 2000, based on a film of the same name, and the game was withdrawn.

Atari introduced its first programmable game, the Video Computer System, in time for Christmas 1977; it dominated the market and children's bedrooms for the next six years. Atari's market share was then whittled away by a succession of more advanced game systems and the emergence of the home personal computer. The Japanese entered the market, reviving interest in consoles, in 1985 with the launch of the Nintendo Entertainment System and an arcade game favourite called Super Mario Brothers. The game triggered the video-game craze.

Nintendo's run of success continued with the Game Boy, a hand-held game launched in 1989, but by end of that year it was faced with a new generation of advanced "16-bit" consoles providing better and faster graphics. It fought back in 1991 with the Super Nintendo Entertainment System (SNES) and the third update of the Super Mario Brothers.

Sega responded by introducing its star character, Sonic the Hedgehog, and though others tried to enter the market, it was essentially a two-horse race. Sega and Nintendo battled for market supremacy until 1994, when Sony introduced the PlayStation. Rival manufacturers have tried to knock the PlayStation from its leading position, but success has become self-fulfilling: the best game designers release their products first on the most popular system, ensuring the dominance of the market leader.

1958 The computer modem enabling computers to communicate with each other over telephone lines, introduced.

1959 Volvo invents the car seat belt.

Cardiac pacemaker
Wilson Greatbatch

1960

Anyone unlucky enough to suffer from cardiac problems 50 years ago might have been hooked up to an early version of the cardiac pacemaker, a cumbersome device about the same size as a large television set. Such machines supplemented the electrical impulses that triggered the heart, but were too large to be of practical use anywhere other than a hospital.

By 1958, American medical engineer Wilson Greatbatch had been working for several years on a pacemaker design in his garden shed, and was inspired one afternoon to invent one of the most significant medical devices of all time: the implantable cardiac pacemaker. While trying to use the recently-invented transistor technology to build an oscillator in order to record the beating of the human heart, Greatbatch

Wilson Greatbatch took the wrong resistor from his tool box — an error that would benefit thousands.

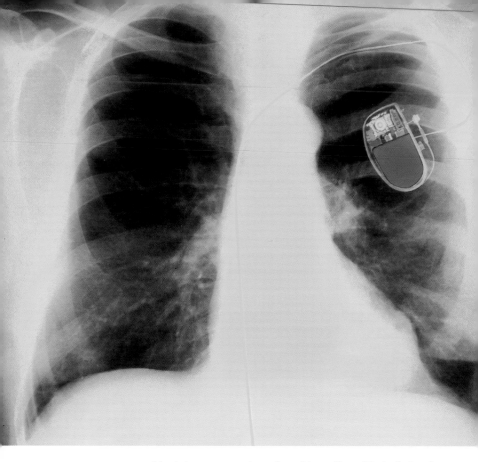

An X-ray of a patient fitted with a battery-powered pacemaker. The short yellow lead connects it to the heart, which receives an impulse if it misses a beat.

grabbed the wrong resistor from his toolbox. He had already put it into the circuitry when he realized the mistake. Out of curiosity he switched on the circuit and found that it produced a regular blip of current, co-incidentally at the same frequency as the average human heart. Greatbatch realized that it could be used to control a cardiac pacemaker.

In May 1958 the first working model was successfully implanted in a dog. Greatbatch then worked at perfecting the tiny electrical generator, powered by a small battery, fitted with electrodes to carry pulses of current to different parts of the heart.

In 1960 surgeons at Millard Fillmore Hospital at Buffalo, in New York state, implanted the first production Greatbatch pacemaker into 77-year-old Henry Hannafeld. It worked perfectly for 18 months.

Greatbatch went on to develop a battery capable of long life and safe operation. His breakthrough in the 1970s was to adapt the corrosion-free long-life lithium oxide battery for pacemaker use. He patented the technology, founded a company to make the cell, and today his company sells or licenses production of 90 per cent of the world's pacemaker batteries.

Laser

Theodore Maiman

1960

Although there was bitter controversy for more than 30 years over who invented the laser, there is no doubt that Albert Einstein laid the groundwork. In 1917, he theorized that, under the right conditions, atoms could be stimulated to emit photons (units of light, or radiant energy) of a single frequency, in a single direction. Thirty-seven years after the phenomenon had been described by Einstein, it was observed when Charles H. Townes and Arthur L. Schawlow at Columbia University, New York, developed the Maser (Microwave Amplification by the Stimulated Emission of Radiation).

Soon after the publication of Townes and Schawlow's paper, researchers began investigating the possibility of using light instead of microwave radiation for transmitting energy. Early efforts concentrated on getting gases to "lase" but the first device to emit laser light was a ruby crystal. It was built in 1960 by physicist Theodore Harold Maiman at the Hughes Research Laboratories in Miami.

Maiman's laser comprized a cylindrical crystal with precision-polished ends, parallel to within a third of the wavelength of light. Both ends were silvered, to act as mirrors, with one end less reflective to allow some of the radiation to escape. Surrounding the crystal was a helically wound flash tube, to "pump" photons into the crystal. The device was encased inside a polished aluminium cylinder through which cooling air flowed.

Although Maiman produced the first working laser, based on Townes and Schawlow's research, a lengthy legal battle resulted in many of the key patents being finally granted to American scientist Gordon Gould.

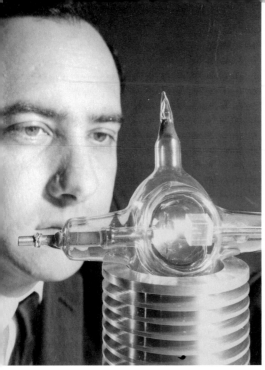

In 1957, Gould was a graduate student carrying out research into optics and microwave spectroscopy while working towards a PhD in physics at Columbia University. Gould became interested in laser research. He conceived a device that he predicted could heat a substance to the temperature of the sun in a millionth of a second.

Gould kept detailed notes, which he had checked by a notary with a view to applying for a patent, in the mistaken belief that he had to have a working prototype before one would be granted. Gould left Columbia to join Technical Research Group Inc in Melville, New York state, where he completed his notes and submitted his patent application on April 6 1959. However, Schawlow and Townes had already published their research and applied for patents. Several lengthy court cases followed, which Gould initially lost, but in 1977 he was finally granted the first of several basic patents for the laser.

Theodore Maiman and the first working laser, based around a cube-shaped ruby crystal that emitted a light 'brighter than the centre of the sun'

Space travel

Yuri Gagarin

1961

1961 Valium invented

The launch of the first man into orbit, the Soviet cosmonaut Yuri Gagarin on April 12 1961, marked the start of the Moon Race, the most ambitious technical project in history. Just six weeks after Gagarin's flight, President Kennedy committed America to putting a man on the Moon and returning him safely to Earth by the end of the decade – just 447 weeks away.

Gagarin's flight thus marked the start of a technological battle to overcome two key hurdles facing manned space travel. First, escaping from the Earth's gravity demands colossal amounts of energy and wielding such energy levels around humans has to be done with extreme care. Second, rockets are a woefully inefficient form of propulsion, in which propellants take up around 95 per cent of the total vehicle weight, leaving precious little left for a useful payload, such as a passenger capsule.

The best-known outcome of these challenges was the development of huge rockets, most famously Saturn V, used in the Apollo lunar landings. Around 120-metres tall, weighing 2,900 tonnes and primed with as much energy as an atomic bomb, a total of 15 Saturn Vs were built, including the one that launched Neil Armstrong, Edwin "Buzz" Aldrin and Michael Collins towards the Moon for the first lunar landing on July 20 1969.

Yuri Gagarin was the first man in the space race, which brought inventions from the cordless drill to silver foil.

The Apollo missions ended in December 1972, having safely delivered 12 astronauts to the Moon and back. Since then, many of the innovations developed to meet the twin challenges of manned space travel have entered our everyday lives.

More than 1,000 cases of space "spin-off" have been documented, and examples have turned up in unexpected places. Some of the most important examples centre on electronics. The development of miniaturized computer technology was accelerated by the need to provide astronauts with on-board computers. Today's powerful cordless drills were developed for use by Apollo astronauts on the Moon.

Nasa doctors wanting to monitor the well-being of astronauts developed the lightweight sensors now routinely attached to patients in hospitals. Cool suits, which pumped water through tiny capillaries to keep Apollo Moon walkers comfortable, are today worn by a range of people from racing car drivers to patients with multiple sclerosis and rare skin disorders.

Powdered drinks and freeze-dried and dehydrated food – plus the silver foil they are wrapped in – were all developed for the space programme. Modern smoke alarms, which use a radioactive source to detect smoke particles, were also a by-product, as was the fire-retardant metallic clothing now regularly used by fire-fighters.

Some of the most memorable spin-offs, however, are neat gadgets such as ballpoint pens which write upside down. Developed in the mid-1960s by the Fisher Pen Company of

Buzz Aldrin steps down on to the Moon in July 1969. Racing drivers and patients with skin diseases have benefited from the cool suits developed for the Moon walkers.

Forest Park, Illinois, these pens used nitrogen gas to drive the ink towards the tip, and were first tried out on the 1967 Apollo 7 mission. Ironically, the often-cited example of space spin-off, the non-stick frying-pan, is one product that did not emerge from the Moon Race. The first products coated with polytetrafluoroethylene (PTFE), or Teflon, were in the shops a year before Kennedy made his famous speech.

While the Moon Race is long since over, manned space exploration continues, as do efforts to overcome the two original key challenges. One major step forward took place in April 1981, exactly 20 years after Gagarin's first flight, with the launch of the US Space Shuttle.

The first reusable spacecraft, the Shuttle, was conceived in 1969 to service an orbiting space-station, and to make spaceflight "routine". A fleet of four Shuttles was built to reach a target of around one flight a week, but in January 1986 one of the four, *Challenger*, exploded on take-off, killing all seven crew. The disaster dealt a blow to the concept of "routine" spaceflight, a setback from which it will probably never recover.

The Shuttles have since proved their worth in launching, retrieving and repairing various satellites and space probes,

1962 The first communications satellite, Telstar, is launched.

including, in 1993, rectifying the defective optics of the orbiting Hubble Space Telescope.

In 1999, the Shuttles finally began work on the construction of an orbiting international manned space station, but the entire project has been dogged by continual delays and cost over-runs. Whether Alpha, the new space station, produces even a fraction of the useful spin-off that came from the manned missions of the 1960s remains to be seen.

Lava lamp
Craven Walker

The lava lamp, a fixture and symbol of 1960s and 1970s living rooms, was invented by Singapore-born Craven Walker, who was inspired by an egg-timer-cum-lamp he saw while drinking a pint in a pub in 1948. "It was a contraption made out of a cocktail shaker, old tins and things," he recalled.

Fifteen years later, Walker unveiled his creation. It had a light bulb in the base to warm a patented wax that softened, became less dense and rose to the top of coloured water, before cooling and dropping to the bottom.

1963

1963 The compact tape cassette for dictation machines introduced by Philips.

A 1960s icon, the lava lamp began life as a tin and a cocktail shaker.

Computer Mouse
Douglas Engelbart

When it comes to ease of use, there's not much to admire about the typical home PC. Oddly enough, its most user-friendly feature, the mouse, predates even the microchip itself.

This brilliant device is an inspired invention in every sense. In 1945, Doug Engelbart, a US Navy radar technician, came across a magazine article by Vannevar Bush, America's chief scientist, on the future challenges for science. It argued that the sheer wealth of scientific information was threatening to overwhelm society. A better means of managing it must be found, Bush declared.

1964

1964 André Courreges invents the mini skirt.

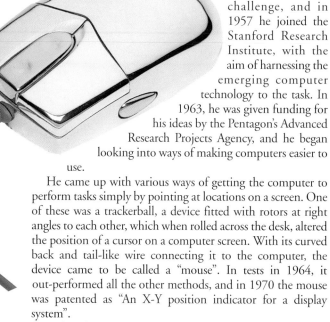

Engelbart decided one day to take up the challenge, and in 1957 he joined the Stanford Research Institute, with the aim of harnessing the emerging computer technology to the task. In 1963, he was given funding for his ideas by the Pentagon's Advanced Research Projects Agency, and he began looking into ways of making computers easier to use.

He came up with various ways of getting the computer to perform tasks simply by pointing at locations on a screen. One of these was a trackerball, a device fitted with rotors at right angles to each other, which when rolled across the desk, altered the position of a cursor on a computer screen. With its curved back and tail-like wire connecting it to the computer, the device came to be called a "mouse". In tests in 1964, it out-performed all the other methods, and in 1970 the mouse was patented as "An X-Y position indicator for a display system".

Today the mouse is found on more than 350 million computers worldwide. Engelbart never made any money out of his invention, but you only have to use a trackerball to realize how brilliant the mouse really is.

Optical disc

James Russell

1965

The shiny, five-inch (12 cm) disc that has become ubiquitous in CD players and computers has its origins in several patents filed by American inventor James T. Russell dating from 1965.

Frustrated by stylus wear and reduction of sound quality on vinyl records, Russell set about developing a method of recording digitally-encoded sounds on a disc. To ensure the disc would not deteriorate through mechanical wear, Russell devised a system of using a finely focused laser beam to read the information off the disc optically.

In 1969 Klass Compaan and Piet Kramer, physicists working for the Dutch manufacturer Philips, began development of an optical disc for storing video. The VLP (Video Long Player, later called Laservision) system encoded analogue video and sound signals as a spiral of microscopic reflective pits on a spinning disc, read by a laser beam. The first prototype player, using a 12-inch (30 cm) disc made of glass, was built in 1972.

Development of an optical disc technology for high-quality audio was begun in 1975 by a Philips engineer, Lou Ottens, who had worked on the compact cassette 15 years earlier. The Compact Audio Disc or CAD as it was first known, used Russell's concept of recording sound encoded in the form of digital data, an error-free series of ones and zeros.

Akio Morita, founder of Sony, also wanted CDs to be capable of playing his favourite symphony

The prototype 115mm-diameter discs were enlarged to 120 mm to simplify manufacture and to increase their capacity to 74 minutes on the insistence of Akio Morita, the chief executive of Sony, which had joined Philips in developing the new system. Morita, a personal friend of Herbert von Karajan, then conductor of the Berlin Philharmonic, stipulated that the disc must be capable of carrying the whole of Beethoven's ninth Symphony.

The CAD, later renamed the compact disc, was demonstrated publicly in March 1979 and the first players and discs reached the shops in autumn 1982. The computer industry soon seized on the CD's ability to hold large volumes of digital data; computer hard drives at the time typically held 10 or 15 megabytes of information whereas the CD could hold 650 megabytes. It led to the development of CD-roms, and recordable and interactive CDs.

The latest variant of the optical disc is Digital Versatile Disc, or DVD, a high-capacity system that can hold as much information as 26 CD-roms. Ideal for movies, it is expected to supersede videotape as a home-recording system.

1965

Microbiologist Benjamin Rubin invents forkshaped vaccinating needle. It helps eradicate smallpox.

Ted Nelson invents hypertext which becomes the basic building block of the worldwide web.

Baby buggy

Owen Maclaren

1965

Anyone born in the 1950s would have first seen the outdoor world from the luxurious surroundings of a coachbuilt pram, perhaps with four large spoked and sprung wheels.

But Owen Finlay Maclaren, a retired aerospace engineer, test pilot and grandparent, felt these prams were a handful and he thought he could make something better using his engineering knowledge. Having worked on the design of the undercarriage of the Supermarine Spitfire, he was used to designing ultra-light structures in aluminium, making sure just the right amount of strength was built in.

His first buggy weighed in at just six pounds, less than many new babies. It had a cunningly designed double joint in the middle, and could fold in half or be clipped together, like an umbrella, with just one hand.

Maclaren patented it on July 20 1965 and the first models were rolling off the production line in Maclaren's converted stables by 1967. The Maclaren buggy was an instant success across the world.

Internet

America

1969

Possibly the most liberating invention of the late 20th century has its origins in the prosaic-sounding technology of packet switching, invented in the early 1960s. Until its invention, computer users could be connected to only one other computer at a time, using a long-distance telephone connection. It was expensive and unreliable.

Donald Davies, at the National Physical Laboratory in London, and Paul Baran, at the Rand Corporation of Santa Monica, invented packet switching independently. It works by converting data into packets which, rather like telegrams, contain the address of the sender and recipient. The packets are sent via a package-switched network and directed to a destination by small, message-processing computers, nowadays called routers.

Using packet-switched networks, it was possible to attach any heterogeneous collection of computers and users cheaply and reliably, and allow access to any computer.

From the invention of packet switching, it took 30 years for the first primitive computer networks to evolve into today's ubiquitous information structure, a progression that began with Arpanet, a packet-switched network that in 1969 connected just four "host" computers funded by the US Department of Defence's Advanced Research Projects Agency. Development of the Arpanet was contracted out to a group of American universities and this led to a democratic, occasionally anarchic, culture.

By 1971, Arpanet had 23 computers attached to it. Originally, the network was designed so that users could make use of specialized computers remote from their place of work. However, it turned out that the main use of the network was for electronic mail. In the next five years, thousands of Arpanet users began to communicate by e-mail.

Another unplanned activity was the Usenet news system. Like a giant electronic bulletin board, Usenet enabled tens of thousands of users to access and contribute to newsgroups, free of charge. From 1975 to

1969 Neil Armstrong is the first man to set foot on the Moon.

Luther Simjian, a Turkish inventor, devises the cash dispenser.

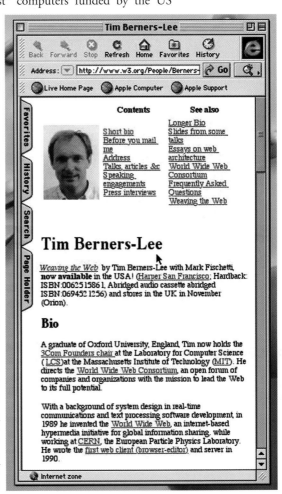

1985, other computer networks sprang up, including online services such as CompuServe and America Online, or AOL, for home computers. But these were all military, scientific or private networks that could not communicate with each other.

In the late 1970s, the sponsors of Arpanet began to address this problem, which they called inter-networking or, simply, the internet. They devised a set of rules (protocols) for communication between networks. This was the Transmission Control Protocol/Internet Protocol, known better as TCP/IP.

Gradually, many of the world's non-military networks connected with one another. Thus, the internet became simply a network of networks. It was a miracle of co-operation, each network adding to the telecommunications infrastructure piece-by-piece without payment from any authority.

By 1988, there were 50,000 host computers attached to the internet. Three years later there were a million. The early 1990s saw the first commercial Internet Service Providers (ISPs) – such as Demon, founded in the UK in 1992 – providing inexpensive commercial and domestic access to the internet.

A model tries out a DigiLens Display unit at an internet fashion show in Beijing. The show was part of the China Internet World 2000 exhibition.

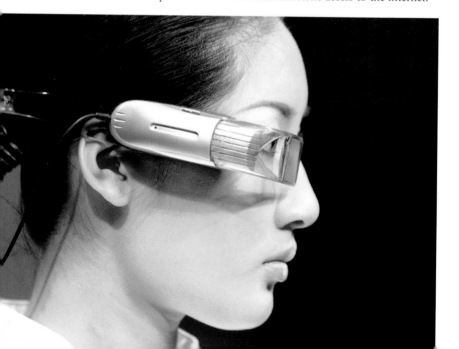

Increasingly, the internet came to be viewed as an information repository. However, it was difficult to access this information. In 1989, Tim Berners-Lee, a young British-born researcher at Cern, a particle physics research facility at Geneva, invented a method of organizing information. He called it the world wide web.

To access information on the web, Berners-Lee used a browser, which incorporated the use of hypertext, a system that links documents from different sources, forming an electronic path that guides users to related information.

The world wide web was initially used only by Cern scientists and was released to the public in the summer of 1991. The information itself was effectively disembodied in cyberspace – existing on computers here, there and everywhere. The world wide web liberated the internet, allowing transfer of text, sound, images, and even video in a straightforward manner. In 1993, the primary users of the internet were academics and scientists; five years later, there were 130 million users all over the world and the web had become a social phenomenon.

A major commercial success was Netscape Corporation, whose Netscape Navigator browser, introduced in December 1994, did much to popularize the internet. Other corporations such as Yahoo and Lycos were commercial spin-offs of "search engines", originally developed in universities to help locate information on the web.

In 1995, Microsoft introduced its Internet Explorer browser and the Microsoft Network (MSN), seeking to dominate the internet. However, as the riches of the internet as a whole dwarfed the content of any one network, full-service providers such as AOL, MSN and CompuServe quickly changed their business model to become Internet Service Providers and mere "portals" to the world wide web.

By 1996, there were 10 million host computers on the internet, a number that was doubling every 18 months. By January 2000, there were more than 70 million and the internet enabled a new commercial paradigm, based on the reduction of economic "friction" by eliminating middle-men and physical inventories. The best-known example was Amazon.com, an online bookstore established in 1995 by Jim Bezos, a 30-year-old entrepreneur.

The mid-1990s internet was a Klondike for so-called dotcom entrepreneurs, with thousands of businesses being formed, such as travel agencies, "e-tailers", stockbrokers and online auctioneers.

Perhaps the most profound change produced by the internet on our everyday lives will be in the delivery of information goods such as newspapers, books, recorded music and movies.

The new MP3 recorded music format, which enables music to be downloaded straight from the web without the need for delivery on a CD, is perhaps the shape of things to come. Many pundits think that internet web-casting will eventually replace broadcast radio and television.

Charge-coupled device

Willard Boyle and George Smith

1969

The charge-coupled device, or CCD, has revolutionized the way in which we capture images. This electronic "eye" of dozens of appliances, from the Hubble space telescope to digital cameras and camcorders, was the brainchild of Willard Boyle and George Smith, who initially developed the technology as a way of storing data.

According to Smith, he and Boyle were working at Bell Laboratories in America in 1969 when they "started batting ideas around and invented charge-coupled devices in an hour". Their invention came to the attention of Nasa, who saw the enormous potential of CCD imaging systems as an alternative to the bulky, vacuum tube-based cameras that were being used on early interplanetary missions.

Under Nasa's sponsorship, a programme was put in place to increase the image resolution of CCDs. Nowadays, these silicon chips have a grid of thousands, or even millions, of microscopic light-sensitive elements, called pixels. The pixels combine to convert light into electrical signals, but the first imaging CCD, produced in 1974 by Fairchild Electronics, had a format of just 100 x 100 pixels. By the late 1980s a second generation of CCDs had one million pixels.

At first confined to military and astronomical use, digital imaging technology gradually entered the arena of commercial photography. By the mid 1990s, the first digital cameras for consumers appeared. The image quality offered by cameras is now sufficient to produce satisfactory snapshots. The digital images they produce are available instantaneously and can be stored on a computer, or transmitted instantly by e-mail.

LCD and LED

George Gray

T he silently changing characters of a liquid crystal display (LCD) have become such a hallmark of our digital age that it may come as a surprise to learn that liquid crystals were first observed more than 150 years ago.

Victorian scientists reported odd optical effects from water containing myelin, the fatty substance that insulates nerve fibres. Under certain lighting conditions, the liquid was seen to produce spectacular colours – like light shining through a crystal.

1970

1970 Debut of the Boeing 747 jumbo jet.

In the 1880s Friedrich Reinitzer, an Austrian botanist, and Otto Lehmann, a German optics expert, showed that the effect was caused by materials as runny as liquids but whose molecules retained enough order to affect light in the same way as crystals.

In the 1930s, scientists were talking about using these "liquid crystals" to make visual displays, using temperature to alter their appearance. The trouble was that the materials were unstable to changes in heat.

George Gray and his colleagues at the University of Hull finally solved the problem in 1970 with compounds that were stable, cheap, and whose appearance could be controlled by voltage rather than heat.

The following year, Hoffmann – LaRoche introduced the twisted nematic LCD. Altering the voltage applied to the crystals changed the alignment of their molecules, so they either allowed light to pass or not, making that part of the display either clear or dark.

LCDs have since found their way on to everything from digital watches to laptop computers, but their visibility in daylight is far from perfect. That has put them in competition with light-emitting diodes (LEDs), which don't just affect light, but create it, through electrons in special semiconductors.

First invented by Nick Holonyak at General Electric in 1962, LEDs are the best bet for making the dream of truly flat television screens a reality. In 1989, researchers at Cambridge showed that a polymer called PPV could be used to make sheets of flat, light-emitting diodes.

The challenge now is to persuade such polymers to emit bright light in all three key colours needed to make a decent television display: red, green and blue.

With both LCDs and LEDs, display technology has come a long way from early digital watches, which sold in 1972 for the equivalent of £1,000 today. However, analogue watches display spatial position, allowing you to judge time at a glance. No wonder most people prefer them.

Floppy disc
Alan Shugart

1971

C omputers got bigger and smarter in the 1960s, but carrying data between them required huge tape drums. That changed in 1971 with the invention of the floppy disc by IBM hardware expert Alan Shugart.

The memory disc held 100 kilobytes of information. Today's floppy disc carries 1.4 mega-bytes of information, or around three million text characters.

1971 The first microprocessor introduced.

Mars 3, a Russian probe, is first spacecraft to land on the red planet.

Prozac
Bryan Molloy and Klaus Schmiegel

1972

T ens of millions of people worldwide have benefited from the discovery more than 25 years ago of fluoxetine hydrochloride, the active ingredient in the antidepressant Prozac.

1972 First disposable lighter.

The quest for a new antidepressant was begun in the 1970s by Bryan Molloy, Klaus Schmiegel and others at Eli Lilly, a pharmaceutical company in America. Older antidepressants had side effects such as dry mouth and blurred vision. In 1972, Fluoxetine, the active ingredient in Prozac, was patented.

Prozac was the first of a new class of antidepressants, called "selective serotonin reuptake inhibitors", approved by the US Food and Drug Administration. It was given the green light for treating depression in 1987 and is now approved in more than 100 countries. It is also approved as a treatment for obsessive compulsive disorder and bulimia nervosa.

Little capsules that have cheered up tens of millions worldwide.

The development of Prozac can be traced to the mid-1960s when it was found that a tranquilizer called reserpine sometimes made people depressed. Studies showed it depleted the supply of chemicals called monoamines.

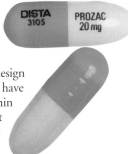

The approach by Lilly and other companies was to design drugs which act more selectively and so, in theory, should have fewer side effects. Prozac inhibits the removal of serotonin – sometimes known as the "pleasure chemical" – without interfering with other brain chemicals. The higher the levels of serotonin, the happier you generally feel.

Genetic engineering

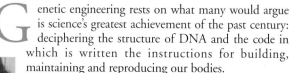

Stanley Cohen and Herbert Boyer

Genetic engineering rests on what many would argue is science's greatest achievement of the past century: deciphering the structure of DNA and the code in which is written the instructions for building, maintaining and reproducing our bodies.

The ability to tinker with DNA has led to the creation of bacteria that can make useful drugs such as insulin and human growth hormone. Plants have been genetically modified to manufacture vitamin A, and even plastic. In humans, diseases such as cancer or hereditary conditions can possibly be corrected by gene therapy, replacing "faulty" genes with "normal" ones.

Before 1860, even the idea of the gene was vague. Charles Darwin, for example, thought that a baby somehow inherited a blended mixture of vital fluid

Herbert Boyer, with fellow American chemist Stanley Cohen, was among the first to alter the recipe of life.

from her parents. Then Gregor Mendel, a Bohemian monk, crossed pea plants and showed that the offspring were not blends but had inherited discrete "factors" (now called genes) from mother and father and passed those factors on unchanged to their offspring in predictable ratios. He published his work in the proceedings of the local society of naturalists in what is now Brno, Czech Republic, but the significance of this work was not appreciated until 1900, 16 years after his death.

The chemical deoxyribonucleic acid (DNA) was found in 1869. It, too, had to be rescued from obscurity by experiments in America in the 1940s, when a Canadian bacteriologist, Oswald Avery, proved that genes are made of DNA and not, as was expected, of protein. The discovery was called "Avery's bombshell".

In the early 1950s, Alfred Hershey and Martha Chase at Cold Spring Harbor in America found (in experiments with viruses that infect bacteria) evidence that DNA was the key to heredity. Yet it was a mystery how DNA could carry the instructions for building and running bodies. The answer came in 1953, from an American, James Watson, and a Briton, Francis Crick, at Cambridge, who discovered DNA's double helix, a chain-like molecule made up of a series of four "bases"

1973 General Motors invents the car airbag.

(adenine, guanine, cytosine and thymine or A, G, C and T). The order of the bases provides a blueprint (a gene) for the proteins that are the building blocks of life. Thus a sequence of bases can spell out a code and that code can be faithfully passed on to offspring by untwining the two helices and copying them.

A key step in genetic engineering was the discovery in the late 1960s of restriction enzymes that enabled scientists to cleave a specific site within DNA. Genetic engineering itself was pioneered in 1973 by the American biochemists Stanley Cohen and Herbert Boyer, who were among the first to cut DNA into fragments, rejoin different fragments, and insert the new genes (in this case into E. coli bacteria), which then reproduced.

James Watson (left) and Francis Crick with a model of the DNA double helix which they discovered in 1953.

Tinkering with the genetic code of a creature can produce animals and plants with desirable properties, for example that grow more quickly or that are disease resistant. Bacteria genes are put in plants to make them insect-resistant; genes are introduced to make human antibodies, vaccines or even plastics.

Gene therapy can be used to implant a gene that is lacking or damaged, correcting hereditary disease or cancer, when the genetic instructions controlling cell division and death go awry. To do this, scientists package the gene in a virus, a fat particle called a liposome, or even inject "naked DNA".

The first human trial took place on September 14 1990, when Ashanti DeSilva received a gene transplant to make up for his lack of a crucial immune system enzyme. The results were equivocal. Despite many trials, human gene therapy has not been successful, though some progress has been made in the treatment of haemophilia and immune deficiency.

Barcode

1973

'Generic Code' by the artist Jana Sterbak.

Bernard Silver and Norman Woodland

I n 1952, Bernard Silver and Norman Woodland, two students at the Drexel Institute of Technology, Philadelphia, came up with the idea of identifying products by marking them with a form of Morse code made from thick and thin stripes that could be read by a beam of light.

Technical difficulties involved in reading the stripes led to luke-warm interest until 1973, when Woodland, then at IBM, devised a sophisticated error-correcting barcode system: the Universal Product Code. On 26 June 1974, a store in Troy, Ohio, became the first to use a scanner to price a product: a pack of chewing gum.

Post-it note
Art Fry

The Post-it note owes its existence to the frustration of an American churchgoer and a "useless" glue. During the early 1970s, Art Fry was a product designer with the huge 3M corporation. Each Sunday, as a member of his local church choir, he would check which hymns he was to sing during the service, and mark them with bits of paper in his hymnbook. Inevitably, just as a hymn was about to start, the paper would drop out and Fry would have a frantic search for the right page.

During one service Fry's mind started to wander, and he suddenly remembered some work done a few years earlier by a 3M colleague, Dr Spencer Silver. While studying adhesives, Silver had made a glue which he discarded when it proved to have poor sticking power. Fry realized in 1973 that this "failure" had precisely the powers he needed to stick the bits of paper in his hymn book: a glue that was sticky but not so sticky that it could not be removed.

The following day, Fry obtained some of the glue and started making his bookmarks with it. It took another 18 months to turn the idea into a commercial product. One key problem was solving the paradox of making this not very sticky glue stick permanently to the message paper, but not to anything else.

Trial marketing began in 1977. People soon found they couldn't live without them.

Monoclonal antibodies
Georges Köhler and César Milstein

One of the most fundamental advances in biotechnology won its inventors plaudits from their peers but little in the way of financial reward, even though it has opened up vast new fields in medicine. The discovery by a Cambridge team was of a way to make proteins called monoclonal antibodies, which can recognize almost any other molecule, ranging from drugs and hormones to microbes and other invaders. But the invention was never

1975 The laser printer invented, a year before the inkjet printer.

patented, leading to speculation that this oversight cost the Government billions of pounds. At the time, the then National Research Development Corporation wrote of the invention: "It is certainly difficult for us to identify any immediate practical application which could be pursued as a commercial venture." Today, it is difficult to know where to start in terms of uses.

Antibodies can be used to detect sites on molecules with exquisite sensitivity, allowing scientists to map variations in the surface components of the influenza virus, to purify substances, to detect hormones in a pregnancy test or to distinguish between cells in the body to reveal tumours.

They provide ways to halt transplant rejection or to prevent diseases such as arthritis, when the body turns on itself. Moreover, poisons can be added to antibodies designed to attach to cancer tissue, highlighting a tumour or killing it.

Antibodies feature in the defences of the body and are secreted by white blood cells, called lymphocytes, to bind to invaders. The blood of a person exposed to a virus contains a mixture of antibodies, all capable of combining with the virus.

Hybridoma cells invented to produce monoclonal antibodies.

However, each lymphocyte can only produce antibodies with a certain specificity. In 1975, at the Medical Research Council Laboratory of Molecular Biology, Cambridge, Georges Köhler, a German and César Milstein, an Argentinian, discovered how to create hybridomas – cells that secrete a single antibody and perpetuate themselves indefinitely. It won them the 1984 Nobel Prize for Medicine. But monoclonal antibodies did not live up to their initial promise in treating disease, when they were hailed as "magic bullets". Early examples were made in mice and, if used in patients, were rejected as foreign. Scientists overcame this problem by perfecting a laboratory method to make the human variety.

Personal computer

Microsoft and Apple

1975

I n 1955, there were 250 computers in use around the world, all using vacuum tubes (or valves). They burned out frequently and attracted moths, which short-circuited the computer and led to the coining of the term for fixing computer problems: debugging.

In the late 1950s the transistor replaced the valve, leading to the "second generation" of computers. Transistors made computers cheaper, smaller and more reliable, and the improved technology enabled IBM and other mainframe computer manufacturers to sell reliable business machines for as little as $100,000 (£482,000 today).

By 1965, there were 20,000 computers in use around the world and the transistor was giving way to the integrated circuit that contained several transistors on one microchip. This development led to the third generation of computers, the most successful of which was the IBM System/360. Its design was widely copied, particularly in Japan, and it remains the dominant design for today's corporate mainframe computers.

The microchip also created the minicomputer, a small computer that could sit on a desk top and cost as little as $20,000 (£67,000 today). As semiconductor manufacture improved, the number of transistors on a microchip doubled each year, so that by 1970 more than a thousand were squeezed on to a chip.

In 1969, Intel, a leading maker of microchips, was asked by a Japanese calculator manufacturer to produce a new calculator chip. The designer, Ted Hoff, made a general-purpose chip that could be programmed with specific calculator functions. It was an astounding invention: the microprocessor. Dubbed the Intel 4004 and advertized as the "computer on a chip", the microprocessor went on sale in November 1971 at $1,000 (£3,150 today).

Within a year, Intel followed up with the more powerful 8008 chip and prices dropped to $100. These first microprocessors were used in smart devices such as cash registers and automatic teller machines in banks.

Microprocessors could also be used as the heart of low-cost "microcomputers". The first enthusiasts for microcomputers were hobbyists: young, technically-savvy males who lived in a region of California, now known as Silicon Valley, where organizations such as the Homebrew Computer Club thrived. The first successful hobby computer was the Altair 8800, which sold in kit form from early 1975. It cost $397 (£890 today), contained an Intel 8008 microprocessor and came with 256 bytes of random access memory (RAM). It had the power of a 1950s mainframe costing $1 million.

The Microsoft team in 1978, with the founders, Bill Gates (bottom left) and Paul Allen (bottom right).

The first hobby computers were fun for geeks, but could not do anything useful, and a sub-industry soon blossomed providing add-ons such as memory boards, disk drives and printers. In 1975 two young entrepreneurs, Bill Gates, aged 20, and Paul Allen, a couple of years older, created an interpreter for the Altair so that people could write programs in the basic programming language. Gates and Allen named their partnership Micro-Soft, which later changed to Microsoft.

The transforming event in the microcomputer's development was the launch of the Apple II in April 1977. Homebrew Computer Club members Steve Jobs, a natural entrepreneur, and Steve Wozniac, a gifted engineer, founded Apple Computer at Cupertino, California, in 1975. Jobs re-invented the microcomputer as the "home/personal computer", an electronic appliance that could be used by anyone. All the electronics were neatly packaged in a plastic case, there was permanent storage using removable floppy discs, a keyboard for input, and a video display. Launched in 1977 at $1,300 (£2,400 today), the Apple II was an instant hit. It made Apple Computer one of the fastest-growing firms of the 20th-century.

During the next couple of years hundreds of personal computers came on to the market. In Britain, the most successful microcomputer was Sir Clive Sinclair's ZX80. It was the first computer to cost less than £100. By 1980, more than a million personal computers had been sold around the world.

The most popular programs at this time were simple word processors and computer games. It was not until full-scale business applications became available that the personal computer moved into the corporate world. The breakthrough was the spreadsheet program, which was invented by Dan Bricklin, a 28-year-old masters student at the Harvard Business School. Bricklin and his colleague Bob Frankston wrote the program for the Apple II computer, calling it VisiCalc. It went on sale in October 1979, and over the next four years it sold 700,000 copies at $250 a copy.

IBM, anxious not to miss out on these developments, launched the IBM PC in 1981. Its entry legitimated and standardized the personal computer, and businesses would eventually buy millions of them. Scores of manufacturers produced IBM "clones", and eventually 85 per cent of the personal computers manufactured were IBM compatible. Fortunes were made and lost but the biggest fortune was made by the small firm of Microsoft, which secured the contract to write the IBM PC's operating system (a small piece of software that co-ordinated the different hardware elements and enabled the computer to run application programs). Microsoft called its operating system MS-DOS and earned about $10 for each copy sold. During the 1980s, it was installed on about 30 million personal computers, powering Microsoft's domination of the software industry.

In 1979, Apple Computer's Steve Jobs visited the Xerox Corporation's research centre in Palo Alto, California (PARC), whose computer design team was led by the visionary Alan Kay, an unsung hero of the computer revolution. There they were inventing a new way of communicating with a computer, called a "graphical user interface", or GUI, pronounced "gooey". With this, the user could point-and-click at objects on the computer screen with a mouse, instead

Apple's co-founder Steve Jobs with the latest iMac.

of having to type the arcane commands used on ordinary computers. Xerox bungled its own attempts to market a computer based on Kay's vision in the early 1980s. In January 1984, Apple launched the Macintosh, at a price of $2,500 (£4,000 today). The Mac's GUI was an instant classic with its user-friendly graphic features. Sadly, Apple failed to capture the market. Not for the first time has the best technology failed to win out, and Microsoft's approach has come to dictate the way most people interact with their computer.

However, Apple is enjoying a renaissance at the moment, and perhaps the breakup of Microsoft, ordered by a US judge, will help this trend.

In vitro fertilization

Patrick Steptoe

1977

When Louise Brown was conceived in a laboratory by in vitro fertilization, she launched a revolution in reproductive medicine that is still a matter of considerable debate.

IVF began with Robert Edwards, of Cambridge University, who was keen to extend work on animals in order to treat women with blocked Fallopian tubes which prevent eggs from travelling from the ovaries to the womb, where they can be fertilized by sperm.

Edwards had worked with several animal species and by 1965, thanks to research with two Baltimore gynaecologists, Howard and Georgeanna Jones, he had adapted these techniques to humans.

Early attempts to fertilize eggs from ovarian tissue, using his own sperm, proved fruitless. Several years later he tested a culture medium supplied by a young student, Barry Bavister, which was successful. Then he discovered that immature eggs, extracted from ovaries, would not develop for long. Here the other IVF pioneer entered the story. Patrick Steptoe, a gynaecologist in Oldham, was routinely encountering ripe eggs in his work on laparoscopy (keyhole surgery of the abdominal cavity).

Louise Brown made history when she was born in 1978, with the aid of the Oldham gynaecologist Patrick Steptoe.

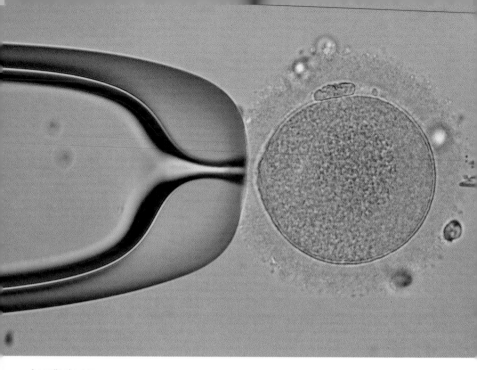

A needle about to inject the DNA of a sperm into a human egg for in vitro fertilization.

The pair met at the Royal Society of Medicine, where Edwards was lecturing, and decided to collaborate. Working together, they gave patients small doses of hormones to produce more than one ripe egg and worked out the best time to harvest eggs for fertilization.

The first pregnancy achieved by the test-tube baby pioneers took place in the summer of 1975 but Marlene Platt suffered an ectopic pregnancy, which had to be terminated.

By 1977, Edwards and Steptoe had endured five years of failure. They decided not to use drugs to stimulate egg production and put their faith in the less productive natural cycle. At that time, Bristol couple Lesley and John Brown were desperate for a child and, inspired by the example of the first heart transplant, felt sure there must be an operation to fix a blocked Fallopian tube.

Lesley Brown underwent an IVF attempt in November 1977 and Steptoe transferred Louise, consisting of just eight cells, to Lesley Brown. The following July, Mrs Brown's blood pressure spiralled because of toxaemia, and Steptoe became anxious. On July 25, he decided to wait no longer and delivered the child by Caesarean section. "Within five seconds of birth she let out the biggest yell you've heard a baby make," said Edwards.

New techniques such as ICSI (intracytoplasmic sperm injection, in which a sperm is injected into an egg) and the ability to grow human sperm in animals, give hope to men with a low sperm count. Pre-implantation genetic diagnosis and methods to sort sperm give hope to carriers of genetic diseases.

MRI Scanner

Raymond Damadian

1977

Unlike X-rays, magnetic resonance scanning can reveal soft tissue detail, allowing doctors to look inside bodies and record on a computer every corporeal detail.

MRI dates back to a Nobel-prize-winning technique called "nuclear magnetic resonance", demonstrated in 1945 by two American groups. When matter is placed in a magnetic field, some atomic nuclei behave like compass needles that can point in only a few directions, each characterized by different energy levels, or "spins".

Nuclear spins can be forced to jump between energy levels when bombarded by radio waves of certain frequencies. Around 1950 it was discovered that these resonance frequencies depended not only on the atomic nuclei, but also on their environment, leading to magnetic resonance's initial use as a tool for chemical analysis.

A coloured MRI scan showing a cross-section through the human head. Unlike X-rays, MRI can show soft tissue, such as the nasal cavity and the tongue. The cerebrum, which controls thought processes, is in the yellow area at the top of the head, while the cerebellum, controlling balance and co-ordination, is the red area by the ear.

The next stage in the development came when the NMR pioneer Felix Bloch stuck his finger into his apparatus and noticed a strong signal created by his finger's high water content. If resonating hydrogen nuclei from the water were giving off radio

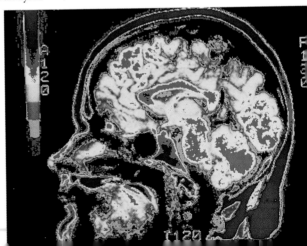

1979 Akio Morita, the head of Sony, invents the personal stereo against the advice of colleagues who do not believe anyone would want to buy a tape player that does not record.

waves, Block reasoned that magnetic resonance imaging could reveal people's insides without opening them up.

In 1971, Raymond Damadian, an American scientist, showed that MRI could be used to detect tumours. In 1977, his team made the first image of an entire human by beaming high-frequency radio waves into a patient in the strong magnetic field of a whole-body scanner.

The technique is now widely used to make detailed pictures of tissue structure. It can also reveal metabolic processes at work within the body.

In 1993, a murderer executed in Texas became the first "virtual man", when a composite three-dimensional "fly-through" computer image of his body was created with the help of MRI.

Abortion pill

Étienne-Emile Baulieu

1980

Few drugs have generated as much controversy as the abortion drug mifepristone, better known by its commercial code-name of RU-486. It became the focus of bitter conflict between the opposing pro-choice and anti-abortion lobbies, the latter describing RU-486 as a "human pesticide".

The controversial RU-486, which terminates a pregnancy within two days.

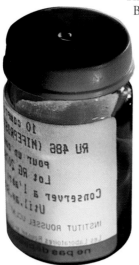

The French pharmacologist Professor Étienne-Emile Baulieu, first synthesized mifepristone in 1980 in tests that revealed that it interfered with progesterone, a hormone essential in the early stages of pregnancy.

A team led by Dr Edouard Sakiz of the French pharmaceutical company Roussel Uclaf had commissioned trials of RU-486 in 15 countries. It was approved for use in abortion by the French government in 1988.

Within weeks anti-abortion activists had threatened boy- cotts of the drug's manufacturer and RU-486 was withdrawn. Two days later, Claude Evin, the French Health Minister, declared the drug to be "the moral property of women, not just of the drug company", and RU-486 was put back on sale.

Taken by mouth under medical supervision and able to induce abortion up to 20 weeks after conception, the drug was soon in widespread use in France, and within a few years more than a third of all abortions were induced by RU-486.

It was subsequently approved in a number of other countries, including Britain, but threats from activists kept it out of America. In 1997, Roussel Uclaf finally yielded to pressure and abandoned production, handing over all patent rights to a new company set up by Dr Sakiz to continue production and research.

Activists continue to protest against the use of the drug, which they say makes abortion too easy. Its defenders say that RU-486 is safer than surgical termination, and has not made abortions more common. Whatever the arguments, the vociferous attacks on RU-486 have left its long-term future in doubt.

1980 Hepatitis B vaccine invented.

Scanning tunnelling microscope

Gerd Binnig and Heinrich Rohrer

The scanning tunnelling microscope was invented by Gerd Binnig and Heinrich Rohrer of IBM's Zurich laboratory in 1981, earning them a Nobel prize.

The probe of the STM is a tiny needle with a tungsten tip so fine that it may consist of only a single atom. It moves over a surface using piezoelectric crystals, ceramics that can be made to contract and expand by subatomic amounts. Electrons tunnel between the sample and tip to produce a contour map of the surface.

In 1990, IBM researchers Don Eigler and Erhard Schweizer used an STM to cajole 35 xenon atoms into forming an IBM trademark, one that was dwarfed by the proteins carrying oxygen around their own bloodstreams. Magnetic discs pack in about 1,000 million bits of information per square inch. Atomic writing has demonstrated that information can be stored at two million times that amount.

1981

1981 The first space shuttles take off.

1982 Genetically engineered insulin, the first product made using human gene technology is introduced.

DNA fingerprinting

Alec Jeffreys

1984

A scientist studies a series of genetic finger-prints through a magni-fying glass to deter-mine the sequence of base pairs (adenine, guanine, cytosine and thymine) that form a code for a section of DNA. The sample has been broken into fragments by an enzyme, then separat-ed in a gel to give a banding pattern that has been preserved in this autoradiogram.

D NA fingerprinting, the biggest advance in crime detection since the Galton–Henry fingerprint system was adopted in 1901, exploits how DNA is found in all cells. Scientists had known since the 1940s that deoxyribonucleic acid (DNA) carries the coded instructions, or genes, that lay down the genetic blueprint of an individual. In 1984, Alec Jeffreys of Leicester University noticed that certain sequences of DNA, called minisatellites, occur in unique patterns in each individual, the exception being identical twins.

Jeffreys found a way of seeking out those parts of the genetic blueprint that vary greatly among individuals in order to make a bar code pattern that he called a "genetic fingerprint".

Samples from a suspect would typically be gathered using a swab to take cells from inside the mouth. The sample would be compared with blood, semen, tissue or hair taken from a crime scene. Tiny traces of degraded DNA can be boosted by a technique called the "polymerase chain reaction".

If a suspect's profile is different from that of the crime sample, then that suspect cannot be the source. In probably the best-known example, DNA fingerprinting provided the physical evidence that derailed Bill Clinton's denials when it proved that the chance of a semen stain on Monica Lewinsky's blue dress not belonging to the President was 7.87 trillion to one.

The technique has found many other uses: paternity testing; identifying body remains; checking whether sashimi originated from an endangered whale species; and investigating family relationships among animals.

High temperature superconductor

Georg Bednorz and Alex Müller

The miraculous ability of certain metals to become "superconductors" of electricity when cooled to ultra-low temperatures was discovered in 1911 by the Dutch physicist Heike Kamerlingh-Onnes. It promised to revolutionize electrical technology, but in practice, the metals lost their superconducting abilities at temperatures higher than just a few degrees above Absolute Zero (−273°C) and when exposed to strong magnetic fields. As a result, superconductivity remained chiefly of academic interest.

In 1986, Georg Bednorz and Alex Müller at the IBM research laboratory in Zurich, stunned the scientific world by revealing a material that became superconducting at 30°C above Absolute Zero. The excitement centred on the fact that the material was a strange, new ceramic material unlike any existing superconductor.

Made from a mixture of copper oxide and a "rare earth element", the material exhibited a new type of superconductivity that other researchers showed persisted above minus 196°C, the temperature of cheap, widely-used liquid nitrogen. But these high-temperature superconductors still suffered from brittleness, and so have yet to have a major technological impact.

1986

1987 Disposable contact lenses invented.

1990 Start of Human Genome Project. The Hubble Space Telescope is launched.

V-Chip
Tim Collings

1991

1991 Digital
answering
machine invented.

S hocked by the massacre of 14 female college students in Montreal in 1989 linked to violence on TV, Tim Collings, an engineer at Simon Fraser University in British Columbia, came up with a system that would enable parents to control the content and quality of TV programmes in their homes.

The prototype developed in 1991 is known as the "V-Chip". Since January 1 2000 it must be built into all televisions sold in America with screen sizes larger than 12 inches (30 cm).

It enables parents to block programmes they regard as too violent, sexual or profane by setting the V-Chip to block or allow viewing according to a six-step scale, based largely on age. All American programmes are required to carry V-Chip codes.

Cockwork radio
Trevor Baylis

1991

1992 The smart
pill invented.

The wind-up radio, bringing news to those without electricity.

I n 1991 the inventor Trevor Baylis was appalled by a BBC documentary about the spread of Aids through Africa. Doctors said that safe-sex education was not getting through: in many parts of Africa there are no phones and no televisions.

The former stuntman, swimming champion and serially unsuccessful inventor dived into his workshop at his home on Eel Pie Island on the River Thames. It took months of experimentation to assemble a hand-cranked, long-life, clockwork escapement that drove a tiny generator, which powered a radio.

The next problem was getting it made. Baylis ran into a brick wall with City financiers. In 1994, he revealed his wind-up wonder on BBC's *Tomorrow's World*. Inundated with offers of help, he could afford to patent the device and set up BayGen

Power Manufacturing with two partners. Nowadays, 120,000 Freeplays a month are made, bringing radio broadcasts to villages without electricity.

Viagra

Nicholas Terrett and Peter Ellis

1994

The small, blue, diamond-shaped pill, popularly thought of as the first pharmaceutical aphrodisiac, was developed to treat heart disease but became a bestseller when it was found it could treat impotence.

Viagra was developed by British scientists at Pfizer's and covered by three patents. The first, attributed to Andrew Bell, David Brown and Nicholas Terrett, was filed in 1991 after they created a compound to treat heart conditions.

The compound, then known as sildenafil or UK-92480, gave insufficient relief to patients suffering from conditions such as angina. But in trials on angina sufferers, men who had difficulty maintaining an erection noticed a beneficial side effect.

Then Terrett and another colleague, Peter Ellis, a biologist, learned of new research into the biological process of erections and realized why their invention could find other uses. In 1994, the pair filed the second patent: as a treatment for impotence. A third patent was filed in 1997 after two Pfizer chemical engineers, Peter Dunn and Albert Wood, discovered a process of mass-producing the compound.

The drug boosts blood-flow to the penis, causing more effective erection. This happens because the drug blocks an enzyme called phosphodiesterase that destroys cyclic GMP, a molecule that makes blood vessels relax. Pfizer had hoped that this relaxing of the cardiac vessels would boost blood flow to the heart, which is restricted in angina. Instead, it boosted blood flow to the penis.

A tablet is taken about an hour before sex and is active for up to five hours. Because it works only on blood vessels that are relaxed by the nervous system, Viagra acts only following sexual stimulation and does not produce erections at inappropriate times.

Cloning

Roslin Institute

It is not every day that the birth of a lamb stuns the President of the United States. In 1997, Bill Clinton asked an advisory commission to review the implications for human beings of the "startling" news that a team in Scotland had successfully cloned Dolly the sheep.

Born on July 5 the previous year, Dolly was not the product of sex. She was cloned from tissue taken from the udder (her name was ungraciously taken from Dolly Parton) of a six-year-old ewe. The ewe was long since dead: four years later, the implications are very much alive.

Clones are highly useful in biology because of their genetic uniformity. Animal breeders welcome the ability to clone top-quality livestock but cloning also makes it much easier to alter animals genetically.

Before the birth of lamb number 6LL3, alias Dolly, the Roslin Institute in Edinburgh was already home to Dolly's forerunners, Morag and Megan. These Welsh mountain ewes were cloned by taking a cell from an early embryo, mass-producing it in the laboratory and using these cells to "reprogram" two eggs emptied of genetic material.

Their birth in August 1995, in turn, had its roots in earlier work: scientists had for some time realized that cloning is fundamental to most living things, since every cell of a plant and animal is a clone ultimately derived from the division of a fertilized seed or egg.

But until Dolly, it had been thought impossible to perform the same feat using cells from an adult animal. Unlike embryo cells, which have the potential to develop into a vast range of cell types, adult cells are differentiated – that is, they have turned into a liver, brain or, in Dolly's case, a cell from the udder.

The simplest way to clone an animal is to split an embryo, the same process that leads to identical twins. The number of clones that can be produced this way is limited, though it has even been tried in humans.

In 1993 Jerry Hall and Robert Stillman from George Washington University split 17 human embryos, the most successful clones reaching the 32-cell stage before dying. The scientists' goal was to help women with low fertility to bear children and, to avoid raising ethical issues, they used faulty human embryos from *in vitro* fertilization treatment. But the work triggered a furore.

Lily, Daffodil, Crocus, Forsythia and Rose, genetically identical quintuplets produced in Iowa by cloning but born to different surrogate mothers.

As head of the team at the Roslin Institute in Scotland, Ian Wilmut was father of Dolly, the first cloned sheep.

Using the "nuclear transfer" technique, Megan and Morag became the first mammals to be grown from cultured, differ-entiated cells. But a century of dogma was overturned when Ian Wilmut, Keith Campbell and Jim McWhir at Roslin and Angelica Schnieke and Alex Kind at PPL Therapeutics cloned Dolly from an adult mammal.

The key to the new procedure was to synchronize the cell cycle (the ordered sequence of events that occur in a cell in preparation for division) of the mammary cell from Dolly with that of the recipient egg. Recognition of the role of the cell cycle proved an important insight in modern biotechnology.

Technically, Dolly was not a true clone: a study showed that the genes in her mitochondria, the powerhouses of her cells, came from the sheep that donated the egg. Still, the Roslin Institute advance and other cloning methods offer the possibility of genetically altering animals much more easily than before. Previously, scientists have used a hit-and-miss affair: injecting DNA into an embryo and hoping that it is taken up, which only happens in 10 per cent of cases or fewer.

Instead, genes can be introduced into large numbers of cloned cells. Then the cells where genetic engineering has been successful are selected and mass-produced, producing a herd of identical "transgenic" animals, for example pigs that are "humanized" to provide organs for human transplantation, or cows that produce drugs in their milk. Indeed, another achievement of the Roslin team was the birth a year after Dolly of Polly, which was both cloned and genetically transformed.

When it comes to humans, cloning has proved controversial. The debate has focused on two applications: the production of transplant organs and tissue from an early cloned embryo, dubbed "therapeutic cloning"; and the cloning of humans, called "reproductive cloning". There is unanimous agreement that the latter should be banned.

1998 Personal video recorder invented by Anthony Wood to record up to 30 hours of programmes on a hard disc drive.

2000 Astronauts take up permanent residency in space.

Critics say that any research on therapeutic cloning will, either directly or indirectly, generate a mass of information that could aid scientists who want to go further and clone a human being. However, proponents point out that the potential benefits of therapeutic cloning are vast. Such work will provide insights into embryo development with applications in the treatment of infertility and miscarriage, and the diagnosis and prevention of birth defects.

Cells made by therapeutic cloning can be used to screen drugs, model human diseases or to test chemicals for toxicity. The most exciting application of therapeutic cloning is to allow scientists to mint a patient's cells that could be used for repair and transplant without running any risk of rejection. The technique could mass produce any cell type: insulin-producing cells to treat diabetes; nerve cells for brain disorders such as stroke, Parkinson's disease, and Alzheimer's disease; muscle cells to treat heart attacks; liver cells for the treatment of hepatitis; and so on. In the medium term, such work could show how to reprogramme adult cells so scientists can make a patient's tissue without the need for cloning.

Picture Credits

Nicholson/ - p227 top; Marc Granger/ - p171; W.A. Raymond/ - p176; Philip de Bay/ - p199 bottom; The Mariner's Museum/ - p205; p254; Paul Almasy/ - p231; p232; Michael Freeman/ - p231; Bill Ross/ - p239; The Burnstein Collection/ - p316; Robert Landau/ - p352; Baldwin H. Ward & Katherine C. Ward/ - p377 main picture; Schenectady Museum/ - p370; Roger Ressmeyer/ - p405; p447 top; main picture; Adrian Arbib/ - p477; Reuters New Media Inc./ - p476; *The Daily Telegraph:* p367; p397; p429; p450; p474; *Dorling Kindersley:* p56/7 bottom; p116; p139; p136 bottom left; David Ashby/ - p200; p297; Jason Lewis/ - p285; p385; p457; *Drake Well Museum/Pennsylvania Historical and Museum Commission:* p286; *Dyson:* p331 far right; *Eve Arnold/Magnum:* p243; *Henry Ford Museum and Greenfield Village:* p311; *Trevor Baylis/Freeplay Energy Europe Limited:* p484; *John Gleave:* p84; *Phil Green:* p19 bottom; p78/9 bottom; p295; *Susanna Hickling: Hulton Getty:* p190 bottom; p214; p219; p226; p227; p246, p255, p256; p310 and 321; p249; p275; p298; p307; p386; p344; p351; p365; p369 main picture; p399; p404; *The Hulton Deutsch Collection:* p209; p253; p358; /Corbis - p323; p278; p340; p350; p389; p398; p406; p409; p438; p439; *The Image Bank:* p309; *Donald McGill:* p19 bottom; *Magnum:* p355; Robert Cape/ - p47; *Mary Evans Picture Library:* p10; p18; p153 top; p162; p93; p178; p289; p388; *Michael Fisher/Custom Medical Stock Photo/Science:* p479; *Rick Maybury:* p444; *NASA:* p379; AFP/ - p1; *National Gallery, London:* p40; *National Maritime Museum:* p42; p100; p196; p198 bottom; /The Art Archive - p195; *National Museum of Photography, Film & Television:* p338; p381; *Northern Territory Tourist Commission:* - p4; *Tim O'Sullivan:* p143; *Dr Rodolphe Gombergh & Dr Albert Castro/Paris Match:* p324; *Petrie Museum of Egyptian Archaeology, University College London:* p7 right; *Popperfoto:* p322; p393; p414; *Private Collection:* p137; *'Products of Our Time; (August Media):* p161; *Reuters:* p407; *Rex Features:* p456; *Rob White:* p199 top; *Robert Doisneau/Rapho/Network:* p290; *Robert Opie Collection:* p212; p237; p238; p266; p287; p294; p288; p300; p303; p308; p320; p326; p325; /Dorling Kindersley - p317; p330 1st, 2nd & 3rd left; p330 right; p342 bottom middle & right; p343; p347; p357; p360; p372; p376; p434 bottom; p435; *Royal Armouries Museum:* p126/7 top 3 guns; *Rural University of Reading:* p177; *Science and Society Picture Library:* Science Museum/ - p51; p52; p65; p147; p154; p157 left; p159; p158; p168/9; p169 right; p170/1 3 pictures in box; far right; p182; p184; p191 bottom; p218; p98 far left; p250; p257; p258; p265; p268; p277; p280/1 & p282 ; p333; p348; p387; National Railway Museum/ - p221; p361; National Museum of Photography, Film and Television/ - p302; *Science Museum:* p229; *Science Photo Library:* Peter Menzel/ - p363; p396; Peter Menzel/ - p420; Mehau Kulyk/ - p418; Sam Ogden/ - p421; Syred/ - p447 bottom; Salisbury District Hospital/ - p452; John Greim/ - 446; A. Barrington Brown/ - p469; Geoff Tompkinson/ p482; Dr Jeremy Burgess/ - p472; CC Studio/ - p478; Philippe Plailly/ - p480; *Sportsphoto Agency:* p382; *Stanley Gibbons:* p262; *Jana Sterbak:* p470; *Sygma:* p61 bottom; p207; *Tim Berners-Lee website:* p461; *Tony Stone Images:* Andrew Hall/ - p223; p233; *Topham Photo Library:* p428; *Graham Trott:* p108; *Bob Venables:* p69; *Vitra Design Museum:* p342 bottom left; *Vitruvius's De Architectura:* p76; *Werner Forman Archive:* /The Egyptian Museum, Cairo: - p5 top; p29; p50; /Musée Royal de l'Afrique Centrale, Tervuren - p23 left; /Musée de Rennes - p67

Index